SEVERAL COMPLEX VARIABLES

PROCEEDINGS OF SYMPOSIA
IN PURE MATHEMATICS
Volume XXX, Part 2

SEVERAL COMPLEX VARIABLES

AMERICAN MATHEMATICAL SOCIETY
PROVIDENCE, RHODE ISLAND
1977

PROCEEDINGS OF THE SYMPOSIUM IN PURE MATHEMATICS
OF THE AMERICAN MATHEMATICAL SOCIETY

HELD AT WILLIAMS COLLEGE
WILLIAMSTOWN, MASSACHUSETTS
JULY 28 – AUGUST 15, 1975

EDITED BY
R. O. WELLS, JR.

Prepared by the American Mathematical Society
with partial support from National Science Foundation grant MPS 75–01436

Library of Congress Cataloging in Publication Data

Symposium in Pure Mathematics, Williams College, 1975.
Several complex variables.

(Proceedings of symposia in pure mathematics ; v. 30, pt. 1-2)
"Twenty-third Summer Research Institute."
Includes bibliographies and index.
1. Functions of several complex variables--Congresses.
I. Wells, Raymond O'Neil, 1940- II. American Mathematical Society. III. Title. IV. Series.
QA331.S937 1975 515'.94 77-23168
ISBN 0-8218-0249-6 (v. 1)
ISBN 0-8218-0250-X (v. 2)

AMS (MOS) subject classifications (1970). Primary 32–XX.

CONTENTS

Value Distribution Theory

Group Representation and Harmonic Analysis

These proceedings are dedicated to

Lillian Casey

who was the guiding force
behind six years of AMS meetings

PREFACE

The American Mathematical Society held its twenty-third Summer Research Institute at Williams College, Williamstown, Massachusetts, from July 28 to August 15, 1975. Several Complex Variables was selected as the topic for the institute. Members of the Committee on Summer Institutes at the time were Louis Auslander (chairman), Richard E. Bellman, S. S. Chern, Richard K. Lashof, Walter Rudin, and John T. Tate. The institute was supported by a grant from the National Science Foundation.

ORGANIZATION. The Organizing Committee for the institute consisted of Ian Craw, Hans Grauert, Robert C. Gunning (cochairman), David Lieberman, James Morrow, R. Narasimhan, Hugo Rossi (cochairman), Yum-Tong Siu, and R. O. Wells, Jr. (Editor of these PROCEEDINGS).

PROGRAM. The topic of the 1975 summer institute was the theory of functions of several complex variables. The emphasis in arranging the program was on the more analytical aspects of that subject, with particular attention to the relations between complex analysis and partial differential equations, to the properties of pseudoconvexity and of Stein manifolds, and the relations between currents and analytic varieties. However, there were also lectures and seminars on other aspects of that broad and active field of investigation, such as deformation theory, singularities of analytic spaces, value distribution theory, compact complex manifolds, and approximation theory.

There were six series of invited expository lectures, as well as twenty-two hour lectures of a general or survey nature; there were also eight series of seminars on current developments in the subject, six of which were planned and partially arranged in advance.

PRINCIPAL LECTURE SERIES.
Chern-Moser invariants by Daniel M. Burns, Jr. and Steven Shnider (2 lectures);
Power series methods in deformation theorems by Otto Forster (5 lectures);
Holomorphic chains and their boundaries by Reese Harvey (4 lectures);

Methods of PDE in complex analysis by Joseph J. Kohn (5 lectures);
Tangential Cauchy-Riemann equations by Masatake Kuranishi (3 lectures);
Analysis on noncompact Kähler manifolds by Hung-Hsi Wu and Robert E. Greene (4 lectures).

HOUR LECTURES.

Theta functions with characteristic and distinguished subspaces of the Heisenberg manifold by Louis Auslander;
Hardy spaces and local estimates on boundaries of strongly pseudo-convex domains. I by Ronald Coifman;
Some recent developments in the theory of the Bergman kernel function. A survey by Klas Diederich;
Fibering of residual currents by Miguel Herrera;
The Monge-Ampere equation and complex analysis by Norbert Kerzman;
Stable homology and positive currents on abelian varieties by H. Blaine Lawson, Jr.;
A method of inverse functions for plurisubharmonic functions; applications to positive and closed currents by Pierre Lelong;
Holomorphic vector fields on projective varieties by David Lieberman;
Diffusion estimates in complex analysis by Paul Malliavin;
Rational singularities, Moisheson spaces, and compactifications of $\mathbf{C}^3$ by James A. Morrow;
Rankin-Selberg method in the theory of automorphic forms by Mark Novodvorsky;
Hardy spaces and local estimates on boundaries of strongly pseudo-convex domains. II by Richard Rochberg;
Analysis of the boundary Laplacian and related hypoelliptic differential operators by Linda Rothschild;
Classification of quasi-symmetric domains by Ichiro Satake;
The Levi problem by Yum-Tong Siu;
Boundary values of the solutions of the $\bar{\partial}$ *equation and characterization of zeros of functions in the Nevanlinna class* by Henri Skoda;
Value distribution on parabolic spaces by Wilhelm Stoll;
Extending functions from submanifolds of the boundary by Edgar L. Stout;
Equisingularity and local analytic geometry by J. Stutz;
Constructability by Jean-Louis Verdier;
Boundary values of holomorphic functions on a Siegel domain and Cauchy-Riemann tangential equations by Michelle Vergne;
Deformations of strongly pseudoconvex domains in $\mathbf{C}^2$ by R. O. Wells, Jr.

SEMINAR SERIES.

(1) Singularities of analytic spaces (Egbert Brieskorn, chairman);
(2) Function theory and real analysis (Stephen Greenfield, chairman);
(3) q-convexity and noncompact manifolds (Yum-Tong Siu, chairman);
(4) Value distribution theory in several variables (Wilhelm Stoll, chairman);
(5) Compact complex manifolds (A. Van de Ven, chairman);
(6) Problems in approximation (John Wermer, chairman);
(7) Group representations and harmonic analysis (Kenneth I. Gross, chairman);
(8) Differential geometry and complex variables (Peter Gilkey, chairman).

SUMMARY. The broad areas emphasized by this summer institute can be summarized as follows:

The $\bar{\partial}$-equation. The relation between complex analysis and partial differential equations centers about the Cauchy-Riemann or $\bar{\partial}$-equations. Kohn gave a series of lectures which provided a survey of the role of the $\bar{\partial}$-equations. Several lectures and seminar talks were devoted to the regularity up to the boundary of solutions of the $\bar{\partial}$-equation, with applications to complex analysis. The structure imposed by the $\bar{\partial}$-equation on the smooth boundary of a region in C^n was discussed in several seminars as well. Most lectures on these topics were in the seminars led by S. Greenfield and Y.-T. Siu.

Holomorphic chains. Reese Harvey gave a survey of the characterization of holomorphic chains and their boundaries among other currents, with a number of applications to classical problems in complex analysis; other lectures discussed recent research on holomorphic chains, which includes, for instance, value distribution theory.

Differential geometry. R. E. Greene and H. Wu gave a survey of the characterization of Stein manifolds by curvature conditions on an underlying Kähler metric. A seminar was organized by P. Gilkey on differential geometry and complex analysis in an extension of this. The survey lectures by D. M. Burns, and S. Shnider discussed recent developments in the theory of higher order invariants associated to real hypersurfaces in C^n.

Singularities. A seminar led by E. Brieskorn was devoted to the singularities of analytic spaces, with an emphasis on the complex analytic properties rather than the algebraic properties treated in the preceding summer institute on algebraic geometry.

Value distribution theory. Wilhelm Stoll lectured on general value distribution theory and led a seminar in which a number of recent developments in this area were discussed.

Compact complex manifolds. A number of analytic results on compact manifolds were discussed in a seminar led by A. Van de Ven.

Approximation theory. While no attempt was made to cover recent developments in complex analysis and function algebras, there was some discussion of approximation theory in a seminar led by J. Wermer.

Harmonic analysis. Symmetric spaces and boundaries of homogeneous domains have been much investigated in complex analysis and lead quite naturally to recent work on group representation and harmonic analysis, some of which was discussed in a seminar led by K. Gross.

PROCEEDINGS. The proceedings of the 1975 summer institute are published here in two volumes. The hour lectures and seminar papers accepted for publication appear in the seminar series most appropriate to the subject matter of the given paper. These are principally research reports describing current research of the authors, while some are of a general expository nature in a given area. The principal lecture series are represented by six survey articles which have been interlaced in these volumes with the seminar series, with an attempt being made for some relationship between the seminar series and the survey articles they juxtapose.

Volume 1: *Seminar Series*: Singularities of Analytic Spaces
Principal Lecture 1: M. Kuranishi

Seminar Series: Function Theory and Real Analysis
Principal Lecture 2: J. J. Kohn
Seminar Series: Compact Complex Manifolds
Principal Lecture 3: Reese Harvey

Volume 2: *Seminar Series*: Noncompact Complex Manifolds
Principal Lecture 4: R. Greene and H. Wu
Seminar Series: Differential Geometry and Complex Analysis
Principal Lecture 5: D. Burns, Jr., and S. Shnider
Seminar Series: Problems in Approximation
Principal Lecture 6: O. Forster
Seminar Series: Value Distribution Theory
Seminar Series: Group Representation and Harmonic Analysis

The detailed list of authors and titles of papers is given in the table of contents for each volume. At the conclusion of each volume is an author index for authors of articles, as well as authors of papers cited in the bibliographies for each particular part.

R. O. WELLS, JR.
Houston, Texas
May 1976

SEMINAR SERIES

NONCOMPACT COMPLEX MANIFOLDS

Proceedings of Symposia in Pure Mathematics
Volume 30, 1977

NONLINEAR CAUCHY-RIEMANN EQUATIONS AND q-CONVEXITY*

RICHARD F. BASENER

Let D be a smoothly bounded domain in C^n. The geometric property of pseudoconvexity has a fundamental role in complex analysis in several variables. For example, one knows that D is holomorphically convex if and only if D is pseudoconvex. In [2] Andreotti and Grauert introduced a generalization of pseudoconvexity. Following the terminology of [1], one says that D is q-convex if at each x in the boundary of D the Levi form (in the complex tangent space to ∂D at x) of a defining function for D has at most q negative eigenvalues.

In view of the importance of pseudoconvexity in complex function theory, it seems reasonable to ask whether there is a natural class F of functions on D for which "F-convexity" of D is equivalent to q-convexity of D. (D is F-convex if for each compact subset K of D the set $\{x \in D\colon \text{for all } f \in F, |f(x)| \leqq \max_K |f|\}$ is again compact.) In order to obtain a result of this type we introduce a generalization of the holomorphic functions on D. If $f \in C^\infty(D)$, let $M(f)$ denote the $n \times (n+1)$ matrix obtained by adjoining to the complex Hessian of f the antiholomorphic gradient $f_{\bar z} = (f_{\bar z_1}, \cdots, f_{\bar z_n})$; so $M(f) = ((f_{\bar z_i z_j}) f_{\bar z})$. Then the "$q$-holomorphic" functions on D, $\mathcal{O}_q(D)$, are defined by

$$\mathcal{O}_q(D) = \{f \in C^\infty(D)\colon \operatorname{rank}(M(f)) \leqq q \text{ on } D\}.$$

(Obviously the condition "f is q-holomorphic" means that f satisfies a certain system of partial differential equations on D. When $q = 0$ the system is equivalent to the Cauchy-Riemann equations, and for $0 < q < n$ it is nonlinear.) By developing some properties of q-holomorphic functions, most importantly a maximum principle (Theorem 2), we obtain some analogues of the relationship between pseudo-

AMS (*MOS*) *subject classifications* (1970). Primary 32A30, 32F10; Secondary 32E99, 35G20.
Key words and phrases. Cauchy-Riemann equations, several complex variables, q-convexity.
*Research supported in part by NSF grants GP–30671 and MPS 75–07922.

convexity and holomorphic convexity (Theorem 3). We now outline these results, full details of which will appear elsewhere [**5**].

The class $\mathcal{O}_q(D)$ is preserved under holomorphic maps: If $\phi: D \to \Omega$ is holomorphic and if $f \in \mathcal{O}_q(\Omega)$, then $f \circ \phi \in \mathcal{O}_q(D)$. It follows then that any function $f \in C^\infty(D)$ for which locally one can find holomorphic coordinates $w_1, \cdots, w_n$ such that f depends holomorphically on $w_1, \cdots, w_{n-q}$ is q-holomorphic. Thus, for example, the function $f(z, w) = (\bar{z} + \bar{w})/(|z|^2 + |w|^2)$ is 1-holomorphic on $C^2 \backslash \{(0, 0)\}$. It is a generalization of the one-variable function $1/z$ in the sense that it has an isolated nonremovable singularity at the origin.

It is to some extent true that a q-holomorphic function is actually locally a holomorphic function of $(n - q)$ variables.

THEOREM 1. *Let D be a domain in $\mathbf{C}^2$, let f be 1-holomorphic on D, and suppose that $f_{\bar{z}}(0) \neq 0$. Then near 0 there are local holomorphic coordinates ζ, η so that $f(\zeta, \eta)$ is holomorphic in η.*

This result requires only the most naive methods; to obtain a similar foliation by $(n - q)$-dimensional varieties along which f is holomorphic seems to require more sophisticated techniques when $q > 1$. (Compare the complex analogue of a theorem of Hartman and Nirenberg in [**7**] which is given in [**8**] by Kalka as Theorem 7.1; the simplest real case is that of a developable surface, a readable account of which can be found in [**10**].)

One does not in general expect $\mathcal{O}_q(D)$ to be a linear space since the equations which define $\mathcal{O}_q(D)$ are not in general linear. It is true, however, that $f \in \mathcal{O}_p(D)$, $g \in \mathcal{O}_q(D)$ imply $f + g$, $fg \in \mathcal{O}_{p+q}(D)$.

The most important result thus far obtained for q-holomorphic functions is the maximum principle.

THEOREM 2. *Let $f \in C^2(D)$ and suppose that $\operatorname{rank}((f_{\bar{z}_i z_j}))$ is less than n on all of D (which is, of course, true if $f \in \mathcal{O}_q(D)$ for some $q < n$). Then for any compact subset K of D, the maximum of $|f|$ over K is achieved on the boundary of K.*

A connection between q-holomorphic convexity and q-convexity is now easily established. In addition to the above results about q-holomorphic functions, all that is needed is the existence of "nice" local holomorphic coordinates at a strictly q-convex boundary point x of D: coordinates $w_1, \cdots, w_n$ such that $\bar{D} \cap \{w_1 = \cdots = w_{q+1} = 0\} = \{x\}$. (This result is mentioned in [**1**], and is a straightforward generalization of corresponding results of Levi in the pseudoconvex case (see [**6**, Chapter IX]).)

THEOREM 3. *Let D be a smoothly bounded domain in $\mathbf{C}^n$. If D is q-holomorphically convex, then D is q-convex. If D is strictly q-convex, then each x in the boundary of D has a neighborhood U in $\mathbf{C}^n$ so that $D \cap U$ is q-holomorphically convex.*

Comments. The condition rank $M(f) \leqq q$ is equivalent to $\bar{\partial} f \wedge (\partial \bar{\partial} f)^q = 0$. Thus the nonlinear Cauchy-Riemann equations are closely related to the equation $(\partial \bar{\partial} u)^{q+1} = 0$ which has been studied by some of the other authors who have contributed to this volume (see, e.g., [**8**] and [**11**]).

An early notion of q-convexity in terms of $(q + 1)$-tuples of holomorphic functions was developed by Rothstein in [**9**]. It was in part this definition which suggested that the q-convexity of Andreotti and Grauert might have some relevancy to problems in the theory of uniform algebras which the author had worked on. (The generalized Šilov boundary in **[3]** is defined in terms of $(q + 1)$-tuples of functions.) Some tentative suggestions about the relationships among q-convexity, q-holomorphic functions, and uniform algebras were made in **[4]**, but these connections need further elaboration.

BIBLIOGRAPHY

1. A. Andreotti, *Whittemore lectures*, Yale University, 1974.
2. A. Andreotti and H. Grauert, *Théorèmes de finitude pour la cohomologie des espaces complexes*, Bull. Soc. Math. France **90** (1962), 193–259. MR **27** #343.
3. R. Basener, *A generalized Shilov boundary and analytic structure*, Proc. Amer. Math. Soc. **47** (1975), 98–104.
4. ———, *Nonlinear Cauchy-Riemann equations and notions of convexity*, Seminar in Functional Analysis and Function Theory, Kristiansand S., Norway, June 9–14, 1975.
5. ———, *Nonlinear Cauchy-Riemann equations and q-pseudoconvexity* (preprint).
6. R. Gunning and H. Rossi, *Analytic functions of several complex variables*, Prentice-Hall, Englewood Cliffs, N. J., 1965. MR **31** #4927.
7. P. Hartman and L. Nirenberg, *On spherical image maps whose Jacobians do not change sign*, Amer. J. Math. **81** (1959), 901–920. MR **23** #A4106.
8. M. Kalka, *Some nonlinear problems concerning intrinsic norms on complex manifolds*, Ph.D. Dissertation, New York University, 1975.
9. W. Rothstein, *Zur Theorie der analytischen Mannigfaltigkeiten in Raume von n komplexen Veränderlichen*, Math. Ann. **129** (1955), 96–138. MR **17**, 84.
10. J. J. Stoker, *Differential geometry*, Pure and Appl. Math., vol. 20, Interscience, New York, 1969. MR **39** #2072.
11. **Eric** Bedford and B. A. Taylor, *The Dirichlet problem for a complex Monge-Ampère equation*, Proc. Sympos. Pure Math., vol. 30, Part 1, Amer. Math. Soc., Providence, R. I., 1977, pp. 109–113.

LEHIGH UNIVERSITY

Proceedings of Symposia in Pure Mathematics
Volume 30, 1977

HOMOMORPHISMS INTO ANALYTIC RINGS

JOSEPH A. BECKER AND WILLIAM R. ZAME*

A celebrated theorem of Bers [1] asserts that if X, Y are plane domains and $\mathcal{O}(X)$, $\mathcal{O}(Y)$ are the rings of analytic functions on X, Y, respectively, then any isomorphism of $\mathcal{O}(X)$ with $\mathcal{O}(Y)$ is induced by a conformal (or anticonformal) equivalence of Y with X. This result has been generalized in a number of ways. Nakai [9] and Rudin [11] have shown that it holds for open Riemann surfaces, Iss'sa [7] has shown that it holds for reduced Stein spaces, and Forster [3], [4] has shown that it holds for unreduced Stein spaces (and linear isomorphisms). In a slightly different direction, Kra [8] showed that if X and Y are Riemann surfaces, any homomorphism of $\mathcal{O}(X)$ into $\mathcal{O}(Y)$ which induces an automorphism of the complex numbers and whose range contains a nonconstant function is induced by an analytic (or antianalytic) mapping of Y into X. (Somewhat different results have been obtained by Edwards [2] and Royden [10].)

An important feature of these results is that the homomorphism (or isomorphism) is not assumed to be linear or continuous. Indeed, it is relatively easy to show, using Cartan's Theorems A and B, that a continuous linear homomorphism of $\mathcal{O}(X)$ into $\mathcal{O}(Y)$ (where X and Y are Stein spaces) is induced by an analytic mapping of Y into X. Thus the essential feature of the results cited above is that certain ring homomorphisms are necessarily linear (or conjugate-linear) and continuous.

In this note we announce some very general results on the linearity and continuity of homomorphisms into rings of analytic functions (and rings of germs of analytic functions). Our results easily imply those already mentioned, but also apply in many other situations, since we obtain our results for homomorphisms whose domains are F-algebras which may not consist of analytic functions at all. In this more general context, it is of course meaningless to speak of homomorphisms being induced by analytic mappings, and our techniques are quite different from those in the papers cited above.

AMS (*MOS*) *subject classifications* (1970). Primary 32B05, 32A05, 46H99; Secondary 32E25.
*Supported in part by NSF grant PO 37961–001.

By an F-algebra we shall mean a complete, metrizable, commutative topological algebra with unit (over the complex numbers); we make no local convexity assumptions. For each n, we let ${}_0\mathcal{O}_n$ be the algebra of germs of analytic functions at the origin of $\boldsymbol{C}^n$, endowed with its natural direct limit topology. We will identify ${}_0\mathcal{O}_n$ with the algebra of convergent power series in n variables. Recall that every ideal J of ${}_0\mathcal{O}_n$ is finitely generated and closed; thus, the quotient space ${}_0\mathcal{O}_n/J$ is again a Hausdorff topological algebra. Our main result is as follows.

THEOREM 1. *Let A be an F-algebra, J an ideal in ${}_0\mathcal{O}_n$ and $\phi: A \to {}_0\mathcal{O}_n/J$ a ring homomorphism. Assume that the range of ϕ contains a nonunit, nonzero divisor. Then ϕ is linear (or conjugate-linear) and continuous.*

Now suppose that $(X, \mathcal{O}_X)$ is an unreduced analytic space in the sense of Grauert [5] (i.e., the structure sheaf may contain nilpotent elements). For each $x \in X$, there is an integer n and an ideal $J \subset {}_0\mathcal{O}_n$ so that ${}_x\mathcal{O}_X \simeq {}_0\mathcal{O}_n/J$. Theorem 1 thus asserts the linearity (or conjugate-linearity) and continuity of certain homomorphisms into the local rings ${}_x\mathcal{O}_X$. There are obvious parallel results for homomorphisms into the ring $\Gamma(X, \mathcal{O}_X)$ of global sections, but we can in fact obtain better results. Recall that if $\mathcal{N}$ denotes the sheaf of nilpotents in $\mathcal{O}_X$, then $(X, \mathcal{O}_X/\mathcal{N})$ is a reduced analytic space (i.e., an analytic space in the sense of Gunning and Rossi [6]). The quotient homomorphism $\mathcal{O}_X \to \mathcal{O}_X/\mathcal{N}$ induces a homomorphism $\Gamma(X, \mathcal{O}_X) \to \Gamma(X, \mathcal{O}_X/\mathcal{N})$. The image of a section $f \in \Gamma(X, \mathcal{O}_X)$ is called the reduction of f and may be viewed as an analytic function on X.

THEOREM 2. *Let A be an F-algebra, $(X, \mathcal{O}_X)$ a connected, reduced analytic space and $\phi: A \to \Gamma(X, \mathcal{O}_X)$ a ring homomorphism. Assume that the range of ϕ contains at least one nonconstant function. Then ϕ is linear (or conjugate-linear) and continuous.*

THEOREM 3. *Let A be an F-algebra with the property that for each continuous, linear homomorphism $\alpha: A \to \boldsymbol{C}$ and each integer k, the ideal* $(\ker \alpha)^k$ *is closed and of finite codimension in A. Let $(X, \mathcal{O}_X)$ be a connected, unreduced analytic space and let $\phi: A \to \Gamma(X, \mathcal{O}_X)$ be a linear ring homomorphism. Assume that the range of ϕ contains at least one section whose reduction is nonconstant. Then ϕ is continuous.*

Note that the analytic spaces in the above theorems are not assumed to be Stein, irreducible, or of bounded regular dimension.

Theorem 2 seems like a very satisfying result, which makes the additional hypotheses in Theorem 3 all the more annoying. They are unavoidable, however, since the conclusion of Theorem 3 does not obtain under weaker hypotheses. More precisely, we have constructed examples of the following phenomena:

(a) there is an irreducible, unreduced Stein space $(X, \mathcal{O}_X)$ and a ring homomorphism $\phi: \Gamma(\boldsymbol{C}, \mathcal{O}_{\boldsymbol{C}}) \to \Gamma(X, \mathcal{O}_X)$ which is neither linear nor conjugate-linear but whose range contains a section with nonconstant reduction;

(b) there is an irreducible, unreduced Stein space $(X, \mathcal{O}_X)$, a semisimple Banach algebra B and a discontinuous linear homomorphism $\phi: B \to \Gamma(X, \mathcal{O}_X)$ whose range contains a section with nonconstant reduction.

We also have examples which show that the hypotheses of Theorem 1 cannot be weakened.

Examples of algebras A satisfying the hypotheses of Theorem 3 abound; for example, the algebra $\Gamma(Y, \mathcal{O}_Y)$, where $(Y, \mathcal{O}_Y)$ is an unreduced Stein space. This leads to the following striking application.

THEOREM 4. *Let $(X, \mathcal{O}_X)$ be an unreduced Stein space. Then there is a unique topology on $\Gamma(X, \mathcal{O}_X)$ in which it is an F-algebra.*

Theorems 2, 3 and 4 are proved by applications of Theorem 1 and some functional-analytic arguments. The proof of Theorem 1 is quite involved and we merely indicate the general outline.

We deal first with the case of a ring homomorphism $\phi: A \to {}_0\mathcal{O}_1$; there is clearly no loss in assuming that $\phi(\sqrt{-1}) = \sqrt{-1}$. The hypothesis assures us that there is an element $a \in A$ such that the power series $\phi(a) = \sum \alpha_j z^j$ has zero constant term. We show first that, if $b_0, b_1, \cdots$ is a sequence in A for which the series $\sum b_i a^i$ converges, then

$$\phi(\sum b_i a^i) = \sum \phi(b_i)\phi(a)^i,$$

where this equality is interpreted merely in the sense of formal power series. We next show that ϕ is linear. To accomplish this, we consider the linear functionals $\gamma_k: {}_0\mathcal{O}_1 \to \boldsymbol{C}$ that assign to each convergent power series its kth coefficient. If $\gamma_k \circ \phi$ is continuous (for each $k = 0, 1, \cdots$) when restricted to the scalars in A, we can conclude that ϕ is continuous when restricted to the scalars and hence linear. But if $\gamma_k \circ \phi$ is not continuous when restricted to the scalars (for some k) then the additivity of $\gamma_k \circ \phi$ guarantees the existence of a sequence $\lambda_0, \lambda_1, \cdots$ of scalars tending to zero for which the sequence $|\gamma_k \circ \phi(\lambda_i)|$ tends to infinity. A careful inductive choice of a subsequence, together with the formal power series equality already established, then leads to an element of ${}_0\mathcal{O}_1$ whose coefficients grow so fast that its radius of convergence is necessarily zero. This contradiction establishes the linearity of ϕ. We then utilize a variant of this power series argument to show that the mapping ϕ has closed graph, and now the closed graph theorem (which is valid for ${}_0\mathcal{O}_1$ since it is the direct limit of a sequence of Fréchet spaces) yields the continuity of ϕ.

We can now establish the theorem for homomorphisms $\phi: A \to {}_0\mathcal{O}_n/J$, where J is a radical ideal, by considering the restriction to a sufficient number of curves and appealing to the results of the paragraph above.

To obtain the general result, we first show that by judicious choices of slices and projections, it suffices to prove the theorem for a homomorphism $\phi: A \to {}_0\mathcal{O}_2/J$, where the (geometric) locus of J is one-dimensional and irreducible. If $\pi: {}_0\mathcal{O}_2/J \to {}_0\mathcal{O}_2/\text{radical}\ (J)$ is the natural homomorphism, our results above assert the linearity and continuity of $\pi \circ \phi$. A rather complicated argument involving general functional analysis and the equivalence of free sheaves with vector bundles reduces the situation above to one in which the locus of J is nonsingular and ${}_0\mathcal{O}_2/J$ is a cyclic module over ${}_0\mathcal{O}_1$. We now use still another variant of our original power series argument to obtain the desired condition.

Complete details will appear elsewhere.

REFERENCES

1. L. Bers, *On rings of analytic functions*, Bull. Amer. Math. Soc. **54** (1948), 311–315. MR **9**, 575.

2. R. E. Edwards, *Algebras of holomorphic functions*, Proc. London Math. Soc. (3) **7** (1957), 510–517. MR **19**, 1194.

3. O. Forster, *Primärzerlegung in Steineschen Algebren*, Math. Ann. **154** (1964), 307–329. MR **29** #2671.

4. ———, *Uniqueness of topology in Stein algebras*, Function Algebras (Proc. Internat. Sympos., Tulane Univ., 1965), Scott-Foresman, Chicago, Ill., 1966, pp. 157–163. MR **33** #5935.

5. H. Grauert, *Ein Theorem der analytischen Garbentheorie und die Modulräume komplexer Strukturen*, Inst. Hautes Études Sci. Publ. Math. No. 5 (1960). MR **22** #12544.

6. R. C. Gunning and H. Rossi, *Analytic functions of several complex variables*, Prentice-Hall, Englewood Cliffs, N. J., 1965. MR **31** #4927.

7. H. Iss'sa, *On the meromorphic function field of a Stein variety*, Ann. of Math. (2) **83** (1966), 34–46. MR **32** #2613.

8. I. Kra, *On the ring of holomorphic functions on an open Riemann surface*, Trans. Amer. Math. Soc. **132** (1968), 231–244. MR **37** #1633.

9. M. Nakai, *On rings of analytic functions on Riemann surfaces*, Proc. Japan Acad. **39** (1963), 79–84. MR **27** #295.

10. H. L. Royden, *Rings of analytic and meromorphic functions*, Trans. Amer. Math. Soc. **83** (1956), 269–276. MR **19**, 737.

11. W. Rudin, *An algebraic characterization of conformal equivalence*, Bull. Amer. Math. Soc. **61** (1955), 543.

PURDUE UNIVERSITY

STATE UNIVERSITY OF NEW YORK AT BUFFALO
TULANE UNIVERSITY

Proceedings of Symposia in Pure Mathematics
Volume 30, 1977

CONTINUATION OF SMOOTH HOLOMORPHIC FUNCTIONS OVER ANALYTIC HYPERSURFACES

ERIC BEDFORD*

A smooth hypersurface in C^n is pseudoconvex from both sides if and only if it is Levi flat, i.e., an analytic hypersurface (see Vladimirov [8]). We shall study analytic hypersurfaces S with singularities contained in a variety of codimension 2. That is, the surface S is a closed C^2 surface in $\Omega \backslash E$, where $\Omega \subset C^n$ is open, and $E \subset \Omega$ is a variety of codimension 2. The point of this work is to show that the geometric nature of the singularity of S determines whether $\bar{S}$ bounds a domain of holomorphy of type A^∞.

An analytic hypersurface is locally foliated by complex manifolds of codimension 1, but we shall assume that each leaf M, when continued indefinitely, forms a closed submanifold of S. Since E has codimension 2, it is a removable singularity for M, and $\bar{M} \cap \Omega$ is a variety in Ω. Thus we assume:

$$(*) \qquad \begin{array}{l} \text{there exists } f: \bar{\Omega} \times R \to C \text{ such that } f(z, \theta) \in C^2, \\ f \text{ is periodic in } \theta, \text{ and } f \text{ defines } S \text{ in the sense} \\ S = \{z \in \Omega \backslash E : f(z, \theta) = 0 \text{ for some } \theta\}. \end{array}$$

If we define $V(\theta) = \{z \in \Omega \backslash E: f(z, \theta) = 0\}$ and $\bar{V}(\theta) \in \{z \in \Omega: f(z, \theta) = 0\}$, then $S = \bigcup_\theta V(\theta)$. There are two kinds of behavior at a singularity.

THEOREM 1. *Let S satisfy $(*)$ with E irreducible. Then either*
(1) *for $z \in E$ there exists at most one θ such that $z \in \bar{V}(\theta)$, or*
(2) *$E \subset \bar{V}(\theta)$ for all θ.*

In the first case E will be said to be an *inessential* singularity of S; otherwise E is *essential.* Note that although a singularity may be inessential, the variety $\bar{V}(\theta_0)$

AMS (MOS) subject classifications (1970). Primary 32D05, 32D20; Secondary 32C40.

*This research was supported in part by a Sloan Foundation Grant to the Courant Institute of Mathematical Sciences, New York University, and Army Research Office contract DAHC04–74–G–0159.

passing through E may be singular, and $\bar{S} \cap \Omega$ may not even be a manifold. For instance, the analytic hypersurface $S = \{z \in C^2 : 0 < |z| < 1, z_1 z_2 = t \text{ for some } t \in (-1, 1)\}$ fails to be a manifold at $(0, 0)$ because $S \cap \{|z| = \varepsilon\} = \{(re^{i\theta}, \pm\sqrt{\varepsilon^2 - r^2}\, e^{-i\theta}), |r| \leqq \varepsilon, 0 \leqq \theta \leqq 2\pi\}$ is a torus.

The main result of this paper is the following theorem which allows us to view an analytic hypersurface as the restriction of a larger foliation.

THEOREM 2. *Let S satisfy $(*)$ and suppose that $d_\theta f$ and $d_z f \neq 0$ on S. Then there is a C^1 foliation $\{V(\theta, \eta): |\eta| < \delta\}$ of a neighborhood of S in $\Omega \backslash E$ by varieties of codimension 1 that extends the foliation $\{V(\theta)\}$ of S.*

One way to define the extended foliation $\{V(\theta, \eta)\}$ is to consider a complex variable $\zeta = \theta + i\eta$ and define

$$f(z, \zeta) = f(z, \theta) + i\eta d_\theta f(z, \theta).$$

If we set $V(\zeta) = \{z \in \Omega \backslash E : f(z, \zeta) = 0\}$, then we must show that $V(\zeta_1) \cap V(\zeta_2) = \emptyset$ for $\zeta_1 \neq \zeta_2$, $|\mathrm{Im}\, \zeta_1|$, $|\mathrm{Im}\, \zeta_2|$ sufficiently small. If E is essential, an equivalent statement is that

$$\bar{V}(\zeta_1) \cap \bar{V}(\zeta_2) = E.$$

It is immediate that the set of (ζ_1, ζ_2) for which this holds is closed and contains the set $\mathrm{Im}\, \zeta_1 = \mathrm{Im}\, \zeta_2 = 0$. Using the techniques of degree theory of analytic mappings (e.g., see Rabinowitz [5]) one can show that the set of ζ is also open (if $|\mathrm{Im}\, \zeta|$ is small), and this proves the theorem.

For an open set $D \subseteq C^n$, let $\mathcal{O}(D)$ denote the holomorphic functions and $A^j(D) = \mathcal{O}(D) \cap C^j(\bar{D})$ denote the holomorphic functions that have j continuous derivatives up to the boundary. A domain of holomorphy $D \subseteq C^n$ will be called a *domain of holomorphy of type A^j* if there exists a function $f \in A^j(D)$ that cannot be continued analytically to a larger domain. An immediate sufficient condition for a domain to be A^∞ is that it have a fundamental neighborhood basis of Stein open sets (see Diederich and Fornaess, [1] and [2]).

A domain of interest in this context is $\Omega^+ = \{z \in C^2, |z_1|^{p/q} < |z_2| < 1\}$. This domain of holomorphy is bounded by the analytic hypersurface

$$S = \{z \in C^2, 0 < |z| < 1, |z_1|^p = |z_2|^q\} = \bigcup_{t \in [0,1]} V(t)$$

where $V(t) = \{0 < |z| < 1,\ z_1^p = e^{2\pi i t} z_2^q\}$. Observe that in general the domain $\Omega^+(\alpha) = \{z \in C^2 : |z_1|^\alpha < |z_2| < 1\}$ is bounded by an analytic hypersurface. But if α is irrational, the leaves $V(\theta) = \{e^{i\theta} z_1^\alpha = z_2\}$ are dense in the set $\{|z_1|^\alpha = |z_2|\}$ and cannot be given as varieties in the set $\{0 < |z| < 1\}$.

It was shown by N. Sibony [6], using a Laurent expansion in z_2, that Ω^+ is not a domain of holomorphy of type A^∞. The following theorem tells more generally when analytic hypersurfaces bound domains of holomorphy of type A^∞.

THEOREM 3. *Let S satisfy the hypotheses of Theorem 2, let Ω be of type A^∞, and let $\Omega^\pm$ be the connected components of $\Omega \backslash \bar{S}$. Then $\Omega^\pm$ are of type A^∞ if and only if E is an inessential singularity.*

The main technical tool used to prove Theorem 3 is reminiscent of the sliding "disk" lemma; see Vladimirov [8, p. 150]. In the present context, however, the disk is not slid but rather wiggled, using the foliation of Theorem 2.

Whether or not E is an essential singularity, the sets $\Omega^{\pm}$ are easily seen to be domains of holomorphy. In fact more is true.

THEOREM 4. *If S and $\Omega^{\pm}$ are as in Theorem 3, then $\Omega^{\pm}$ are domains of holomorphy of type A^j for all $j < \infty$.*

Consider the domain

$$\Omega = \{z \in \Omega' : |f_1(z)| < \cdots < |f_j(z)|\}$$

where Ω' is a domain of type A^∞, $f_1, \cdots, f_j$ are in $A^\infty(\Omega')$, $f_1 \not\equiv 0$ on any component of Ω', and $f_2 \neq 0$ on $\bar{\Omega}'$. Then Ω can be shown to be of type A^∞. On the other hand Theorems 3 and 4 give the following corollary which handles the case where f_2 is allowed to vanish.

COROLLARY. *Let $\Omega' \subseteq C^n$ be an open set of type A^∞, let $f_1(z), \cdots, f_j(z)$ be in $A^\infty(\Omega')$ and suppose that*

$$\Omega = \{z \in \Omega' : |f_1(z)| < \cdots < |f_j(z)|\}$$

where $f_1 \not\equiv 0$ on any component of Ω'. If the set

$$T = \{z \in \Omega' : f_1(z) = f_2(z) = 0 < |f_3(z)| < \cdots < |f_j(z)|\}$$

has codimension 2, then every $f \in A^\infty(\Omega)$ may be continued to a neighborhood of T. On the other hand, if $m < \infty$ then Ω is a domain of holomorphy of type A^m.

It would be of interest to know what happens to Theorems 3 and 4 if the assumption (*) is dropped or if the set E is allowed to be larger. Observe that the singularity E is the smallest possible, for if two varieties $V_1, V_2 \subseteq C^n$ of complex dimension $n - 1$ intersect, then the intersection $V_1 \cap V_2$ must have dimension $\geqq n - 2$.

REFERENCES

1. K. Diederich and J. E. Fornaess, *Exhaustion functions and Stein neighborhoods for smooth pseudoconvex domains*, Proc. Nat. Acad. Sci. U.S.A. **72** (1975), 3279–3280.

2. ———, *A strange bounded smooth domain of holomorphy*, Bull. Amer. Math. Soc. **82** (1976), 74–76.

3. J. J. Kohn, *Global regularity for $\bar{\partial}$ on weakly pseudo-convex manifolds*, Trans. Amer. Math. Soc. **181** (1973), 273–292. MR **49** #9442.

4. P. Lelong, *Fonctionnelles analytiques et fonctions entières* (*n variables*) , Presses Univ. Montréal, Montréal, Qué., 1968, p. 31.

5. P. Rabinowitz, *A note on topological degree theory for holomorphic maps*, Israel J. Math. **16** (1973), 46–52. MR **49** #10909.

6. N. Sibony, *Prolongement analytique des fonctions holomorphes bornées*, C. R. Acad. Sci. Paris Sér. A-B **275** (1972), A973–A976. MR **47** #7062.

7. K. Spallek, *Differenzierbare und holomorphe Funktionen auf analytischen Mengen*, Math. Ann. **161** (1965), 143–162. MR **33** #4328.

8. V. S. Vladimirov, *Methods for the theory of functions of many complex variables*, "Nauka", Moscow, 1964; English transl., M.I.T. Press, Cambridge, Mass., 1966. MR **30** #2163; **34** #1551.

9. H. Whitney, *Complex analytic varieties*, Addison-Wesley, Reading, Mass., 1972.

COURANT INSTITUTE OF MATHEMATICAL SCIENCES

Proceedings of Symposia in Pure Mathematics
Volume 30, 1977

THE LEVI PROBLEM IN HOLOMORPHIC BUNDLES WITH COMPACT FIBRE

JÉRÔME BRUN

A central conjecture in the Levi problem is the following: If $f: A \to S$ is a Stein morphism and if S is Stein, then A is Stein (recall that a holomorphic mapping of complex spaces $f: A \to S$ is a Stein morphism if, for every $s \in S$, there exists a neighborhood U of s such that $f^{-1}(U)$ is Stein). If f is a locally trivial fibration, this is the Serre conjecture. We are interested in another particular case that we shall describe now.

Let $\pi: X \to S$ be a locally trivial fibration with base a Stein manifold S and with fibre a compact manifold F. Let A be an open set of X and $\pi_A: A \to S$ the restriction of π. We would like to deduce from the assumption: π_A is Stein, that A is Stein, thus verifying the conjecture in this particular case. In fact, this leads to a classical Levi problem since, if π_A is Stein, then A is locally pseudoconvex (locally Stein), and moreover $A \cap \pi^{-1}(s)$ is Stein for every $s \in S$. With some assumption of F, the next two theorems give the solution to this Levi problem.

THEOREM 1. *Let $\pi: X \to S$ be a locally trivial fibration with base a Stein manifold S and with fibre a compact homogeneous manifold F. Let A be a locally pseudoconvex open set of X such that $A \cap \pi^{-1}(s)$ is Stein for every $s \in S$. Then A is Stein.*

THEOREM 2. *Let $\pi: X \to S$ be a locally trivial fibration with base a Stein manifold S and with fibre a compact Riemann surface F. Let A be a locally pseudoconvex open set of X such that $\pi^{-1}(s)$ is not contained in A for any $s \in S$. Then A is Stein.*

I would like to thank André Hirschowitz who introduced me to this matter.

Matsugu **[8]** gave a proof of these theorems in the particular case $X = S \times T$, where T is a 1-dimensional torus.

REMARKS ON THEOREM 1. (The details are in **[1]**.) In the proof, we use results of

---

AMS (MOS) subject classifications (1970). Primary 32E10, 32F15; Secondary 32M10.

Hirschowitz [5] on infinitesimally homogeneous manifolds. (Recall that a manifold is infinitesimally homogeneous if its tangent space is generated in every point by global holomorphic vector fields; a locally trivial bundle with Stein base and compact homogeneous fibre is infinitesimally homogeneous.)

Note that this theorem raises the Levi problem in the homogeneous space F. The natural question is: Let U be a locally pseudoconvex open set of F, different from F. Is U Stein?

The answer is yes if F is a complex projective space $\boldsymbol{P}_n(\boldsymbol{C})$ (Fujita [4], Takeuchi [9], Kiselman [7]), if F is an irreducible rational homogeneous manifold, for instance a complex grassmannian (Hirschowitz [6]). Grauert has given a counterexample in a p-dimensional torus (cf. [10]).

REMARK ON THEOREM 2. (The details are in [2].) In the proof, we construct plurisubharmonic functions, removing points from the fibres and using results of Elencwajg [3].

BIBLIOGRAPHY

1. J. Brun, *Sur le problème de Levi dans certains fibrés*, Manuscripta Math. **14** (1974), 217–222.
2. ———, *Le problème de Levi dans les fibrés à base de Stein et à fibre une courbe compacte*, Ann. Inst. Fourier (Grenoble) (to appear).
3. G. Elencwajg, *Pseudoconvexité locale dans les variétés Kälhériennes*, Ann. Inst. Fourier (Grenoble) **15** (1975), 295–314.
4. R. Fujita, *Domaines sans point critique intérieur sur l'espace projectif complexe*, J. Math. Soc. Japan **15** (1963), 443–473. MR **28** #2252.
5. A. Hirschowitz, *Pseudoconvexité au dessus d'espaces plus ou moins homogènes*, Invent. Math. **26** (1974), 303–322.
6. ———, *Le probléme de Levi pour les espaces homogènes*, Bull. Soc. Math. France (to appear).
7. C. O. Kiselman, *On entire functions of exponential type and indicators of analytic functionals*, Acta Math. **117** (1967), 1–35. MR **35** #1825.
8. Y. Matsugu, *The Levi problem for a product manifold*, Pacific J. Math. **46** (1973), 231–233. MR **48** #569.
9. A. Takeuchi, *Domaines pseudoconvexes infinis et la mètrique riemannienne dans un espace projectif*, J. Math. Soc. Japan **16** (1964), 159–181. MR **30** #3997.
10. R. Narasimhan, *The Levi problem in the theory of functions of several complex variables*, Proc. Internat. Congr. Math. (Stockholm, 1962), Almqvist & Wiksells, Uppsala, 1963, pp. 385–388.

UNIVERSITÉ DE NICE

Proceedings of Symposia in Pure Mathematics
Volume 30, 1977

PROPERTIES OF PSEUDOCONVEX DOMAINS WITH SMOOTH BOUNDARIES

KLAS DIEDERICH AND JOHN ERIK FORNAESS

1. Exhaustion functions. Let U be any Hausdorff space. A function $\rho: U \to (a, b)$ of U into some open interval $(a, b) \subseteqq \boldsymbol{R}$ is called an exhaustion function if it is a proper mapping.

For the purposes of complex analysis, plurisubharmonic exhaustion functions of domains in $\boldsymbol{C}^n$ are of particular interest, and among them especially those which are bounded from above. It is well known that a bounded domain $D \subset \boldsymbol{C}^n$ is a domain of holomorphy if and only if the exhaustion function $-\log \delta(z)$ is plurisubharmonic in D, where $\delta(z)$ denotes the euclidean distance between z and ∂D. On the other hand, a simple argument shows that the bounded pseudoconvex domain

$$D = \{(z, w) \in \boldsymbol{C}^2 \mid 1 > |w| > |z|\}$$

has no plurisubharmonic exhaustion function which is bounded from above. This situation changes if one supposes that ∂D is smooth, because one has:

THEOREM 1 (SEE ALSO [3]). *Let $D \subset\subset \boldsymbol{C}^n$ be a pseudoconvex domain with smooth boundary. Then there exists a smooth strictly plurisubharmonic exhaustion function $\rho: D \to (-\infty, 0)$ on D. The function ρ can be chosen to be of the form $\rho = -h\,\delta^{1/m}$ with a strictly positive function $h \in C^\infty(\bar{D})$ and a (large) positive integer m.*

In the proof of this theorem at first for each $z_0 \in \partial D$ an exhaustion function on D with values in $(-\infty, 0)$ is constructed, which is strictly plurisubharmonic in D only near z_0, but which is nevertheless not too bad far away from z_0. Then, a suitable finite sum of such functions has the desired properties.

As an application of Theorem 1 one can prove a Docquier-Grauert lemma for

AMS (MOS) subject classifications (1970). Primary 32F15, 32F05, 32L05; Secondary 32E25, 46J15.

relatively compact pseudoconvex domains D with smooth boundary in a Stein manifold X (see also [5]). Furthermore, together with the criterion of J.-L. Stehlé from [6], one gets at once:

COROLLARY. *Let $B \to^{\pi} X$ be a locally trivial holomorphic fibre bundle over a Stein space X. If the typical fibre is a relatively compact pseudoconvex smooth domain in a Stein manifold, then B is again Stein.*

2. Stein neighborhoods. We now deal with the question: Which bounded pseudoconvex domains $D \subset C^n$ have the property that $\bar{D}$ has a neighborhood basis consisting of pseudoconvex domains? That this is not always true can again be seen from the example $D = \{(z, w) \mid 1 > |w| > |z|\}$, because any Stein domain $\Omega \subset C^2$ containing $\bar{D}$ must also contain the dicylinder $\{(z, w) \mid |z| < 1, |w| < 1\}$. Behnke and Thullen have used this example in [1] to characterize a whole class of Stein domains whose closure is not the intersection of Stein neighborhoods. The boundaries of all these examples are not smooth, and whether the additional hypothesis of smoothness of the boundary guarantees the existence of a Stein neighborhood basis has been an open question. We want to show at first that this supposition is not enough. For this purpose, we fix a smooth function λ: $R \to R^+ - \{0\}$ with the properties $\lambda(r) = 0$ for $r \leqq 0$, $\lambda''(r) > 0$ for $r > 0$ and such that λ is sufficiently convex. We define for $r > 1$ the smooth function

$$\rho_r: (C - \{0\}) \times C \to R$$

by putting

$$\rho_r(z, w) = |w + e^{i \cdot \ln z\bar{z}}|^2 - 1 + \lambda(1/|z|^2 - 1) + \lambda(|z|^2 - r^2).$$

Finally, we define

$$D_r = \{(z, w) \in (C - \{0\}) \times C \mid \rho_r(z, w) < 0\}.$$

Then, we have

THEOREM 2 (SEE ALSO [4]). *For each $r > 1$ the domain D_r is bounded and pseudoconvex with smooth boundary. The set of points, where ∂D_r is not strictly pseudoconvex, is*

$$M_r = \{(z, w) \mid 1 \leqq |z| \leqq r, w = 0\}.$$

If $r > 1$ is large enough, then any holomorphic function on $\overline{D_r}$ can be holomorphically continued to the set

$$\Omega_r = \overline{D_r} \cup \{(z, w) \mid e^{\pi} < |z| < e^{2\pi}, |w| < 2\}.$$

In particular, $\overline{D_r}$ is not an intersection of Stein domains.

The domains D_r also have some other strange properties (for more details see [4]).

After exhibiting this counterexample, we want to ask for sufficient conditions on bounded pseudoconvex domains $D \subset C^n$ with smooth boundary that allow us to construct a Stein neighborhood basis of $\bar{D}$. For this purpose we specialize the question in the following way:

Let $n(p)$ denote the exterior (real) normal on ∂D at $p \in \partial D$ and let $f > 0$ be any smooth function on ∂D. Then

$$\partial D_\varepsilon = \{p + \varepsilon f(p)n(p) \mid p \in \partial D\} \tag{1}$$

is for all small $\varepsilon > 0$ the smooth boundary of a domain D_ε. Under which conditions is it possible to choose f in such a way that D_ε is strictly pseudoconvex for all small $\varepsilon > 0$?

DEFINITION. The bounded pseudoconvex domain $D \subset \boldsymbol{C}^n$ with smooth boundary is called regular if the set

$$M = \{p \in \partial D \mid \partial D \text{ is not strictly pseudoconvex at } p\}$$

can be written as $M = M_0 \cup \cdots \cup M_k$ with $M_j \subset \partial D$ closed, and such that for each $j = 0, \cdots, k$ there exists a real closed CR submanifold X_j of $\partial D - M_{j-1}$ with $M_j - M_{j-1} \subset X_j$ and $L_{\partial D} \mid T^{1,0} X_j$ positive definite ($L_{\partial D}$ denotes the Levi form of ∂D with respect to any defining function).

THEOREM 3 [3]. *For a regular domain D there exists a smooth function $f > 0$ on ∂D such that ∂D_ε as defined in* (1) *is strictly pseudoconvex for all small $\varepsilon > 0$.*

A careful investigation shows that, in the case of a real analytic boundary of a pseudoconvex domain $D \subset\subset \boldsymbol{C}^2$, the set M of points in ∂D where ∂D is not strictly pseudoconvex cannot be too bad. Therefore one can show:

THEOREM 4 [3]. *For any pseudoconvex domain $D \subset\subset \boldsymbol{C}^2$ with real analytic boundary a smooth function $f > 0$ can be constructed on ∂D with the same properties as in Theorem* 3.

Whether this theorem remains true in $\boldsymbol{C}^n$ with $n \geqq 3$ is unknown.

Immediate consequences of the construction of the domains D_ε in Theorems 3 and 4 are:

(a) For all $0 \leqq \varepsilon < \eta$ small enough, D_ε is relatively Runge in D_η.
(b) $\bar{D}$ is holomorphically convex in D_ε for all small $\varepsilon > 0$.
(c) $\bar{D}$ is uniformly H-convex in the sense of Čirka [2].

REFERENCES

1. H. Behnke and P. Thullen, *Zur Theorie der Singularitäten der Funktionen mehrerer komplexer Veränderlicher. Das Konvergenzproblem der Regularitätshüllen*, Math. Ann. **108** (1933), 91–104.
2. E. M. Čirka, *Approximation by holomorphic functions on smooth manifolds in $\boldsymbol{C}^n$*, Mat. Sb. **78 (120)** (1969), 101–123 = Math. USSR Sb. **7** (1969), 95–113. MR **39** #480.
3. K. Diederich and J. E. Fornaess, *Exhaustion functions and Stein neighborhoods for smooth pseudoconvex domains*, Proc. Nat. Acad. Sci. U.S.A. **72** (1975), 3279–3280.
4. ———, *A strange bounded smooth domain of holomorphy*, Bull. Amer. Math. Soc. **82** (1976), 74–76.
5. H. Rossi, *A Docquier-Grauert lemma for strongly pseudoconvex domains in complex manifolds*, Rocky Mountain J. Math. **6** (1976), 171–176.
6. J.-L. Stehlé, *Fonctions plurisousharmoniques et convexité holomorphe de certains fibrés analytiques*, C. R. Acad. Sci. Paris Sér A **279** (1974), 235–238.

WESTFÄLISCHE WILHELMS-UNIVERSITÄT MÜNSTER
PRINCETON UNIVERSITY

Proceedings of Symposia in Pure Mathematics
Volume 30, 1977

KÄHLER MANIFOLDS AND THE LEVI PROBLEM

G. ELENCWAJG

1. Statement of the problem. Let X be a complex manifold and $D \subset\subset X$ an open subset of X. We shall say that D is locally Stein in X if for every $x \in \partial D$ (= boundary of D) there is an open subset $U \subset X$ such that the open subset $U \cap D$ of X is a Stein manifold. To solve a Levi-type problem is to give sufficient conditions on X which will imply that D is Stein.

2. Some known results. If $X = \boldsymbol{C}^n$ and D is a locally Stein open subset of $\boldsymbol{C}^n$, then D is Stein. This is the solution to the original Levi problem and is due to K. Oka in the case $n = 2$ and to K. Oka, H. Bremermann and F. Norguet for arbitrary n.

F. Docquier and H. Grauert **[2]** generalized this result to the case where X is an arbitrary Stein manifold (actually they solved a more general problem, with D spread over X).

However Grauert showed that for every $n \geqq 2$ there exists an n-dimensional torus T_n and an open subset U (with nonempty boundary) of T_n which is locally Stein but such that every holomorphic function on U is constant, so that U is not even holomorphically convex. (An account of Grauert's example can be found in Andreotti **[1]**.)

Fujita **[4]** proved that a locally Stein open subset (with nonempty boundary) of a projective space $\boldsymbol{P}_n(\boldsymbol{C})$ is Stein (cf. also Kiselman **[7]** and Takeuchi **[8]**). A. Hirschowitz **[6]** recently generalized this result: He proved that a locally Stein open subset $U \subset X$ ($\partial U \neq \emptyset$) is Stein if X is a compact homogeneous irreducible rational manifold. The hyperquadrics and the grassmannians are examples of such admissible X.

3. The use of Kähler structures in the Levi problem. Let (X, g) be a Kähler manifold. Its tangent bundle is said to have positive curvature or simply to be positive

AMS (MOS) subject classifications (1970). Primary 32E05, 32F15; Secondary 32E10, 32F05.

if, for every $x \in X$, every chart at x and every couple of nonzero tangent vectors $u = \sum u^i(\partial/\partial z^i)$, $v = \sum v^i(\partial/\partial z^i)$ we have the following inequality involving the curvature tensor $R = (R_{ijkl})$ of g:

$$\sum_{i,j,k,l} R_{ijkl} u^i \bar{u}^j v^k \bar{v}^l > 0.$$

(Recall that the components R_{ijkl} of R in the chart z are given by

$$R_{ijkl} = -\frac{\partial^2 g_{ij}}{\partial z^k \partial \bar{z}^l} + \sum g^{\alpha\beta} \frac{\partial g_{i\beta}}{\partial z^k} \frac{\partial g_{\alpha j}}{\partial \bar{z}^l}.)$$

This notion of positivity (there are many!) is the one given by P. Griffiths in **[5]**. For example $P_n(C)$ with its standard Fubini-Study Kähler structure is well known to have positive tangent bundle.

Recall that a manifold D is said to be 0-convex if it admits a C^∞ function f which is strictly plurisubharmonic off a compact subset of D and such that the sets $\{z \in D \mid f(z) < d\}$ are compact for all $d \in R$.

Using techniques initiated by A. Takeuchi in **[9]** we can then prove

THEOREM 1. *Let D be a relatively compact locally Stein open subset of the Kähler manifold X. If the tangent bundle of X has positive curvature, then D is* 0-*convex.*

The theorem is proved by showing that the function $-\log(\text{distance to } \partial D)$ is strictly plurisubharmonic on the intersection of D and a neighbourhood of ∂D.

The proof of this assertion rests on a sharp estimate which also yields the following extension of a theorem of A. Takeuchi **[9]**.

THEOREM 2. *Let $D \subset\subset X$ be a locally Stein open subset of a manifold X on which there exists a strictly plurisubharmonic function f. Then D is Stein.*

Let us emphasize that X is not necessarily Stein, since f was not assumed to be an exhaustion function.

COROLLARY. *Let $\pi : X \to Y$ be a holomorphic map between complex manifolds such that*:

(i) *There is a strictly plurisubharmonic function on Y.*

(ii) *There is a covering (Y_α) of Y such that there exists a strictly plurisubharmonic function on each $\pi^{-1}(Y_\alpha)$.*

Then any locally Stein open subset $D \subset\subset X$ is Stein.

The best known example of a map $\pi : X \to Y$ satisfying (i) and (ii) is a locally trivial fibration with Stein basis and Stein fibre; in this example the corollary would of course also be a consequence of the Serre conjecture. Proofs of the theorems and corollary will appear in **[3]**.

BIBLIOGRAPHY

1. A. Andreotti, *Nine lectures in complex analysis*, (C.I.M.E. I Ciclo 1973), Complex Analysis, Edizioni Cremonese, Roma, 1974.

2. F. Docquier and H. Grauert, *Levisches Problem und Rungescher Satz für Teilgebiete Steinscher Mannigfaltigkeiten*, Math. Ann. **140** (1960), 94–123. MR **26** #6435.

3. G. Elencwajg, *Pseudo convexité locale dans les variétés kählériennes*, Ann. Inst. Fourier (Grenoble) **25** (1975), fasc. 2.

4. R. Fujita, *Domaines sans point critique intérieur sur l'espace projectif complexe*, J. Math. Soc. Japan **15** (1963), 443–473. MR **28** #2252.

5. P. A. Griffiths, *Hermitian differential geometry, Chern classes and positive vector bundles*, Global Analysis (Papers in Honor of K. Kodaira), Univ. of Tokyo Press, Tokyo, 1969, pp. 185–251. MR **41** #2717.

6. A. Hirschowitz, *Le problème de Levi pour les espaces homogènes*, Bull. Soc. Math. France **103** (1975), 191–201.

7. C. O. Kiselman, *On entire functions of exponential type and indicators of analytic functionals*, Acta Math. **117** (1967), 1–35. MR **35** #1825.

8. A. Takeuchi, *Domaines pseudoconvexes infinis et la métrique riemannienne dans un espace projectif*, J. Math. Soc. Japan **16** (1964), 159–181. MR **30** #3997.

9. ———, *Domaines pseudoconvexes sur les variétés kählériennes*, Math. Kyoto Univ. **6** (1967), 323–357. MR **36** #426.

Université de Nice

Proceedings of Symposia in Pure Mathematics
Volume 30, 1977

THE HOLOMORPHIC CONVEXITY OF PSEUDOCONVEX COMPLEX MANIFOLDS

ALAN T. HUCKLEBERRY*

0. Introduction. If a complex manifold X can be exhausted by relatively compact open sets $\{\varphi < c\}$, where φ is strictly plurisubharmonic, then it is holomorphically convex [2]. However, if the word "strictly" is dropped, then X is in general not holomorphically convex. Several examples of this are outlined below. If the rank of the Levi form is allowed to vary then it is not yet clear how such examples fit into a general setting. In the constant rank case (i.e., $dd^{\perp}\varphi \geqq 0$ when restricted to the complex tangent space CT_p, $p \in \{\varphi = c\}$, and the rank of this form is independent of p), there are rather natural necessary and sufficient conditions for the holomorphic convexity of X. For example, in the case where $dd^{\perp}\varphi \equiv 0$, X is holomorphically convex if and only if it possesses a nonconstant holomorphic function. The purpose of this note is to describe these conditions and to give some idea of the techniques which are used in the proofs.

1. Basic terminology and the main result. Let X be a complex manifold. A smooth map φ of X into the nonnegative real numbers is called an exhaustion of X if the sets $X_c = \{\varphi < c\}$ are relatively compact. If $d = \partial + \bar\partial$ then $d^{\perp} := (\partial - \bar\partial)/2i$ and $dd^{\perp}\varphi$ is the full complex Hessian of φ. For a point p in the boundary manifold $B_c = \partial X_c = \{\varphi = c\}$, it is useful to consider the restriction of the Hermitian form $dd^{\perp}\varphi(p)$ to the complex tangent space, CT_p, of B_c at p. For simplicity this form, $L(\varphi)_{(p)}$, is called the Levi form of φ at p. If X is equipped with an exhaustion φ such that, for all $c \gg 0$, $L(\varphi)_{(p)} \geqq 0$ for all $p \in B_c$, then X is called pseudoconvex. If, in addition, rank $L(\varphi)_{(p)} \equiv k - 1$ for some fixed k, with $1 \leqq k \leqq n = \dim X$, for all $p \in B_c$ with $c \gg 0$ then X is called k-Levi-flat. The 1-Levi-flat manifolds are called Levi-flat and the $(n-1)$-Levi-flat manifolds are called, as is customary, strongly pseudoconvex.

AMS (MOS) subject classifications (1970). Primary 32C10; Secondary 32E05, 32F99.
*Partially supported by NSF grant MPS 75–07086.

For an arbitrary n-dimensional complex manifold X, there exists a holomorphic map $F: X \to \mathbf{C}^{2n+1}$ so that $F(p) = F(q)$ if and only if $f(p) = f(q)$ for all $f \in \mathcal{O}(X)$. Given $p \in X$, $\operatorname{rank}_p \mathcal{O}(X) := \operatorname{rank}_p F$ and $\operatorname{rank} \mathcal{O}(X) := \max_{p \in X} \operatorname{rank}_p \mathcal{O}(X)$. It is now possible to state the main result:

THEOREM 1. *Let X be a k-Levi-flat complex manifold. Then X is holomorphically convex if and only if, for all $p \in X$ sufficiently near infinity, $\operatorname{rank}_p \mathcal{O}(X) = k$.*

In the sections that follow, examples of such manifolds are discussed and the idea of the proof of this theorem is described. Details will appear in **[5]**.

2. Examples. It was pointed out in **[6]** that every complex Lie group is k-Levi-flat. The number k can be explicitly determined from the maximal complex Lie algebra in the real Lie algebra of the maximal compact *real* subgroups of the given group. All possible k's are realized even for abelian groups. Since the proofs of these facts rely on some prior knowledge of Lie theory, they are not given here. Instead, some constructions of Grauert **[1]** will be discussed.

(a) *A slit torus.* Consider the "standard" torus in $\mathbf{R}^{2n}$ (see the figure) as the unit "square" with sides appropriately identified. Remove the strip $\{|x_1 - \frac{1}{2}| < \varepsilon\}$ from the square. No matter how $\mathbf{R}^{2n}$ is made into $\mathbf{C}^n$ (using linear structure), the boundary of this strip is Levi-flat, because the defining function is linear. Then the complex manifold X, which corresponds to the shaded area in the figure, is seen to be a Levi-flat manifold by using the natural exhaustion which depends only on x_1. In this case the boundary manifolds B_c are just $(2n - 1)$-dimensional *real* tori.

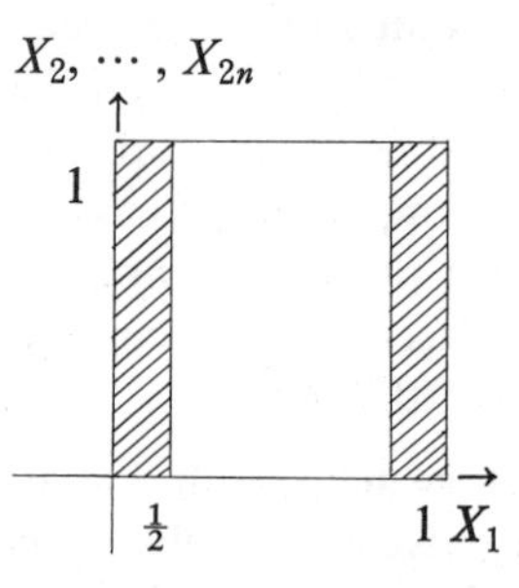

If the complex structure is chosen appropriately then each such B_c will be foliated by *everywhere dense* real $(2n - 2)$-planes, each of which is in fact a complex $(n - 1)$-plane. Now let $f \in \mathcal{O}(X)$, where the complex structure is chosen in this way. Then $\max_{p \in B_c} |f(p)| = |f(p_0)|$. But $p_0 \in H$, where $H \cong \mathbf{C}^{n-1}$ is dense in B_c. By the maximum principle, f must be constant on H and, by the density of H, f must therefore be constant on B_c. Since B_c is a real hypersurface in X, f must be constant on X. Hence $\mathcal{O}(X) \cong \mathbf{C}$. The point of this example is that there are pseudoconvex manifolds ($dd^{\perp}\varphi \equiv 0$ in this case) with no nonconstant holomorphic functions.

(b) *Topologically trivial line bundles over Kähler manifolds.* Let M be a compact Kähler manifold with $\dim H^1(M, \mathbf{Z}) > 0$. Thus M has a nontrivial Picard variety $\mathcal{P} := H^1(M, \mathcal{O})/j_* H^1(M, \mathbf{Z})$, where j is the natural injection of $\mathbf{Z}$ into $\mathcal{O}$. Distinct points of $\mathcal{P}$ correspond to distinct topologically trivial (i.e., the Chern form is cohomologous to zero) line bundles over Y. It turns out that, in a natural way, each such bundle space X_a, $a \in \mathcal{P}$, is a Levi-flat manifold. If a is a torsion element with respect to the group structure on $\mathscr{P}$ (i.e., if some power of the bundle X_a is analytically trivial) then there exist nonconstant analytic functions on X_a. In fact X_a is holomorphically convex. If, on the other hand, a is not a torsion element then $\mathcal{O}(X_a) \cong \boldsymbol{C}$. Since $\mathscr{P}$ is really a moduli space for the bundles X_a, this shows that rank $\mathcal{O}(X)$ is not stable under small deformations of a Levi-flat X.

The proofs of the above facts will now be sketched. First, one obtains an exhaustion of the total space of any line bundle over a compact complex manifold by choosing a hermitian metric on the fibers. If $\{f_{ij}\}$ is the set of transition functions with respect to some trivialization of the bundle on a cover $U = \{U_i\}$ then such a metric is given by $h_i \langle\ ,\ \rangle$ on U_i, where h_i is a smooth function satisfying $h_i = |f_{ij}|^{-2}h_j$. The exhaustion of the bundle space is then given by $\varphi = \log\|\cdot\|^2$ and a tubular neighborhood of radius r about the zero section is just the set $\{\varphi < r\}$. An easy calculation shows that the Levi form, $L(\varphi)_{(p)}$, is just $\partial\bar{\partial} \log h_i(\pi(p))$, where π is the projection map for the bundle. Thus the Chern form of the bundle with respect to the same metric is just $L(\varphi)$.

If M is a compact Kähler manifold and X is a topologically trivial line bundle over M then, using classical Hodge theory techniques, one can choose *unitary* transition functions (i.e., $|f_{ij}| = 1$). In particular this means that we can choose $h_i \equiv 1$ and $L(\varphi) \equiv 0$, where $\varphi = \log\|\cdot\|^2$. Thus each X_a, $a \in \mathscr{P}$, is Levi-flat.

Suppose X_a^n is analytically trivial for some $n > 0$. Then there exists a nowhere vanishing section $s \in \Gamma(M, X_a^{-n})$ and, if z_i is a coordinate in the fiber of X_a over U_i, $(s_i \circ \pi)\cdot z_i^n$ is a well-defined nonconstant holomorphic function on X_a. On the other hand, if a is arbitrary and $\mathcal{O}(X_a) \ncong \boldsymbol{C}$ then the Taylor series of a nonconstant $f \in \mathcal{O}(X_a)$ is

$$f(p) = \sum_{n=0}^{\infty} s_i(\pi(p)) \cdot z_i(p)^n,$$

where s_i and z_i are as above. In particular, for some $n > 0$, X_a^{-n} has a nontrivial section. But, since M is a Kähler manifold, the Chern form of X_a^{-n} is dual to the divisor defined by s. Since X_a^{-n} is topologically trivial, s is thus nowhere vanishing, X_a^n is analytically trivial, and a is a torsion element. In other words, for nontorsion elements a, $\mathcal{O}(X_a) \cong \boldsymbol{C}$.

(c) *Bundles on bundles.* By combining the topologically trivial bundles described above with appropriately chosen negative bundles, Grauert [**1**] constructed complex manifolds which are strongly pseudoconvex except along certain divisors where they are flat. It turns out that, for such a space X, $\mathcal{O}(X)$ separates points and gives local coordinates on $X - D$, where D is the flat divisor. Every $f \in \mathcal{O}(X)$ is constant on D. Thus, since D is noncompact, X is *not* holomorphically convex. This example is mentioned here in order for the reader to gain a better feeling for the conditions of Theorem 1.

Due to lack of space, it is impossible to give the details of Grauert's construction. The idea is, however, not too complicated. One starts with some (say) algebraic manifold M. Then one takes some negative line bundle over M (e.g., the dual of the hyperplane section bundle), makes it into a $\boldsymbol{P}_1$-bundle, and calls it G. Since the line bundle was negative, the zero section of G can be blown down to a compact, analytic space with one normal isolated singularity. One can visualize this as the "cone" over M coming from an embedding of M in $\boldsymbol{P}_n$ by taking the inverse image of M in C^{n+1} by the usual defining projection for P_n. Let F_1' be any negative line bundle over G' and let F_1 be its lift to G. Using the norm-exhaustion of F_1 described in (b) above, we see that F_1 has an exhaustion φ_1 so that $L(\varphi_1) > 0$ everywhere except over the zero section of G, where F_1 is the trivial bundle.

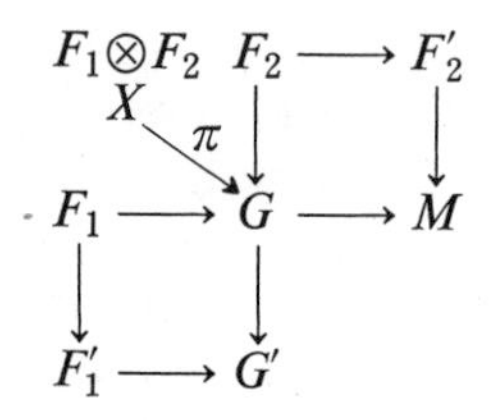

Let F_2' be a topologically trivial line bundle over M and assume that no power of F_2' is analytically trivial. For this one needs $\dim H^1(M, Z) > 0$. Let F_2 be its lift to G. Define $X := F_1 \otimes F_2$. Use the tensor product norm for X, where the norm on F_1 is as above and the norm on F_2 is the lift of the norm on F_2' which gives a Levi-flat exhaustion of F_2'. Since $X \cong F_2'$ over the zero section of G, it is clear that every $f \in \mathcal{O}(X)$ is constant on D, where D is defined as the zero section of X along with π^{-1} of the zero section of G. However, the strong pseudoconvexity of F_1 at all other points allows one to construct functions in $\mathcal{O}(X)$ which separate points and give local coordinates on $X - D$. For details of this, see Grauert's paper [**1**].

3. The proof of the main result. A detailed proof of Theorem 1 will not be given. However, assuming a rather fundamental lemma which has a rather technical proof and assuming further that the exhaustion function is plurisubharmonic, the idea of the proof is simple. A few remarks must be made before the fundamental lemma can be stated.

Let X be an irreducible complex space and K any subset of X. Then the analytic dimension of K is defined as $\min\{k: K \subset \bigcup_{r=1}^{\infty} A_r, \dim A_r = k\}$, where the A_r's are *local* analytic sets in X. For K compact and A an algebra of continuous functions on K which contains the constants and separates points, the Šilov boundary of A is defined as the smallest closed subset of K on which all of the functions in A achieve their maximum modulus. This set is denoted by $S(A)$. A point $p \in K$ is said to be a peak point for A if some $f \in A$ has an *absolute* maximum modulus at p. The set of peak points is denoted by $\mathcal{P}$. In this setting $S(A)$ is the topological closure of $\mathcal{P}$.

LEMMA [**4**]. *Let X be an irreducible complex space, K a compact subset, and A an algebra of continuous functions which contains the constants. Assume that the holomorphic (in a neighborhood of K) elements of A separate points on K. Then* anal dim K = anal dim $S(A)$.

It is now possible to sketch the proof of Theorem 1. If X is holomorphically convex then the fibers of the separation map $F: X \to C^{2n+1}$ (see §2) are compact. Let $\mathcal{F}$ be a connected component of such a fiber. The restriction of φ to $\mathcal{F}$ must therefore have a maximum at some point $p_0 \in \mathcal{F}$. If p_0 is near enough to infinity for $L(\varphi)_{p_0}$ to have rank $k - 1$ then $\dim \mathcal{F} \leq n - k$. Thus, for p near infinity $\operatorname{rank}_p \mathcal{O}(X) \geqq k$. However, since X is k-pseudoflat [**3**], rank $\mathcal{O}(X) \leq k$. Thus holomorphic convexity implies that, for p near infinity, $\operatorname{rank}_p \mathcal{O}(X) = k$.

The other direction of the proof is not quite so simple. First, one considers only the points $p \in X$ so that the boundary manifolds B_c, $\varphi(p) = c$, are k-Levi-flat. Such real hypersurfaces are foliated by k-codimensional complex manifolds which are tangent to the zero eigenspace of the Levi form. The first step of the proof is to show that, under the assumption that $\operatorname{rank}_p \mathcal{O}(X) = k$ for all p as above, each leaf of the foliation of B_c is compact. For this it is enough to show that the separation map $F: X \to \mathbf{C}^{2n+1}$ is constant on each leaf. The reader should think of the "slit-torus" as a paradigm.

Let $K = F(B_c)$ and let A be any algebra satisfying the hypotheses of the above lemma. Let $p \in B_c$ such that $F(p) \in \mathscr{P}$. Then, by the maximum principle, F is constant on the leaf of the foliation of B_c which contains p. Let $\mathscr{I}$ be the analytic subvariety of X where the differential geometric rank of F is less than k. By Sard's theorem and the lemma above, there is a peak point in $K - F(\mathscr{I})$. Thus there is a leaf of the foliation of B_c, L, such that F is constant on L and $F(L) \notin F(\mathscr{I})$. Since $\operatorname{rank} F = k$ and $\dim L = n - k$, L must be compact. Since F is not singular in a neighborhood of L, L and the fiber of F at any point of L must agree.

Since L is compact, there is a neighborhood W of L in X on which F is proper. But φ restricted to any compact fiber of F in W is plurisubharmonic and thus is constant. Hence all of the fibers of F which intersect B_c near L must be in B_c and agree with the appropriate leaf of B_c's foliation. Hence the set C of compact leaves of the foliation of B_c each of which does not intersect $\mathscr{I}$ is a nonempty open set. The complement of C is nowhere dense, as otherwise we could apply the above argument to it. Hence C is open and dense. The density of C and the continuity of F imply that F is constant on all leaves of B_c's foliation. Thus they are compact.

To complete the proof one need only to notice that the compactness of the leaves implies that the connected components of the fibers of F are compact. Applying Stein factorization to F, pushing down the exhaustion of X to the "Zerlegung" space, and using the classical solution of the Levi problem, one shows that X is properly fibered over a Stein space and thus is holomorphically convex.

References

1. H. Grauert, *Bemerkenswerte pseudokonvexe Mannigfaltigkeiten*, Math. Z. **81** (1963), 377–391. MR **29** #6054.

2. ———, *On Levi's problem and the imbedding of real-analytic manifolds*, Ann. of Math. (2) **68** (1958), 460–472. MR **20** #5299.

3. A. Huckleberry and R. Nirenberg, *On k-pseudoflat complex spaces*, Math. Ann. **200** (1973), 1–10. MR **48** #2417.

4. A. Huckleberry and W. Stoll, *On the thickness of the Shilov boundary*, Math. Ann. **207** (1974), 213–231. MR **49** #3210.

5. A. T. Huckleberry, *The Levi problem on pseudoconvex manifolds which are not strongly pseudoconvex*, Math. Ann. **219** (1976), 127–137.

6. ———, *Über Funktionenkörper auf komplexen Mannigfaltigkeiten*, Schriftenreihe Math. Inst. Univ. Munster **9** (1975), 1–43.

University of Notre Dame

Proceedings of Symposia in Pure Mathematics
Volume 30, 1977

HÖLDER ESTIMATES FOR $\bar{\partial}$ ON CONVEX DOMAINS IN C^2 WITH REAL ANALYTIC BOUNDARY

R. MICHAEL RANGE*

1. In 1969 Grauert and Lieb [1] and Henkin [3] independently proved the following result.

THEOREM. *Let D be a bounded domain in* $\mathbf{C}^n$ *with smooth, strictly pseudoconvex boundary. Then for every bounded* $\bar{\partial}$*-closed form* $f \in C^1_{0,1}(D)$ *there is a solution* $u \in C^1(D)$ *of* $\bar{\partial}u = f$ *such that*

$$\|u\|_{L^\infty(D)} \leqq K \cdot \|f\|_{L^\infty(D)},$$

where $K < \infty$ *is independent of* f.

Here

$$\|f\| = \sum \|f_j\|, \quad \text{if } f = \sum f_j d\bar{z}_j.$$

The method of proof involves the representation of a solution u by means of integral kernels constructed by Ramirez [7] and Henkin [2], which generalize the Cauchy kernel from the plane to higher dimensions. This result has been generalized in many ways, always assuming strict pseudoconvexity; in particular, there are solutions u which are Hölder continuous of order α for any $\alpha < \frac{1}{2}$ (Kerzman [5]) and of order $\alpha = \frac{1}{2}$ (Henkin and Romanov [4]). In general, it is not possible to obtain Hölder continuity of order $\alpha > \frac{1}{2}$, as an example in [5], due to E. M. Stein, shows.

2. One would like to prove corresponding results on pseudoconvex domains which are not necessarily strictly pseudoconvex. The following examples indicate how the order of convexity in the direction of the holomorphic tangent vectors affects the regularity of solutions of $\bar{\partial}u = f$ near the boundary.

AMS (*MOS*) *subject classifications* (1970). Primary 32F15, 35N15; Secondary 32A25.

*Author partially supported by NSF grant MPS 75–07062.

EXAMPLE 1. For a positive integer q let $B^q = \{(z_1, z_2) \in C^2 : |z_1|^2 + |z_2|^{2q} < 1\}$. After choosing a fixed branch for the logarithm, the function $v = \bar{z}_2/\log(z_1 - 1)$ is defined on B^q, and $\bar{\partial} v$ is a bounded, $\bar{\partial}$-closed $C^{\infty}_{0,1}$ form on B^q. There is no solution u on B^q of $\bar{\partial} u = \bar{\partial} v$ which is Hölder continuous of order $\alpha > 1/2q$.

For $q = 1$ (the case of the unit ball), this example appears in **[5]**. The general case is obtained by a straightforward modification. Similarly one obtains:

EXAMPLE 2. Let $B^{\infty} = \{(z_1, z_2) : |z_1|^2 + e \cdot \exp - 1/|z_2|^2 < 1\}$, and let v be as in Example 1. There is no solution u on B^{∞} of $\bar{\partial} u = \bar{\partial} v$ which is Hölder continuous of any positive order.

Thus, if one looks for Hölder estimates, a smooth, pseudoconvex, nowhere Levi-flat boundary is not sufficient. This situation should be compared with the problem of finding necessary and sufficient conditions for the subellipticity of the $\bar{\partial}$-Neumann problem (cf. Kohn **[6]**).

3. We have obtained the following positive results.

THEOREM 1. *For every bounded, $\bar{\partial}$-closed form $f \in C^1_{0,1}(B^q)$ there is $u \in C^1(B^q)$ such that $\bar{\partial} u = f$ and*

$$|u(z) - u(z')| \leqq K_q \|f\|_{L^{\infty}} |z - z'|^{1/2q}$$

for $z, z' \in B^q$.

Example 1 shows that the Hölder exponent in Theorem 1 cannot be improved.

We do not know whether on B^{∞} there are supremum estimates for solutions of $\bar{\partial} u = f$.

THEOREM 2. *Let $D \subset C^2$ be a bounded convex domain with real analytic boundary. Then there are $\alpha > 0$ and $K < \infty$ such that for every bounded $\bar{\partial}$-closed $f \in C^1_{0,1}(D)$ there is $u \in C^1(D)$ such that $\bar{\partial} u = f$ and*

$$|u(z) - u(z')| \leqq K \|f\|_{L^{\infty}(D)} |z - z'|^{\alpha}, \qquad z, z' \in D.$$

In particular, u extends continuously to the boundary.

4. The proofs are based on constructing explicit solutions u by means of integral kernels. Crucial for the estimation of u is a precise estimate for a defining function ρ for the domain D, i.e., a smooth function ρ in a neighborhood W of ∂D, such that $W \cap D = \{z \in W : \rho(z) < 0\}$ and $d\rho \neq 0$ on ∂D. For $\zeta \in \partial D$, the space of holomorphic tangent vectors to ∂D at ζ is denoted by $H_{\zeta}(\partial D)$.

PROPOSITION 1. *Let $D \subset\subset C^n$ be convex with real analytic boundary. Then there are a defining function ρ, a constant $c > 0$ and a positive integer m such that*

$$(*) \qquad \rho(\zeta + v) \geqq c|v|^m \quad \text{for } \zeta \in \partial D \text{ and } v \in H_{\zeta}(\partial D), |v| \leqq 1.$$

For $\rho_q(z) = |z_1|^2 + |z_2|^{2q} - 1$, one can choose $m = 2q$.

The proof of the general case makes essential use of the real analyticity of the boundary; the case of ρ_q involves only an elementary, though nontrivial, argument.

From $(*)$ one obtains an estimate for an analytic support function Φ as follows.

PROPOSITION 2. *Suppose $D \subset C^n$ is defined by a smooth function ρ which satisfies*

(*) *in Proposition* 1. *Then there are positive constants* γ, δ, ε *such that the function* $\Phi(\zeta, z) = \sum_{j=1}^{n}(\partial\rho/\partial z_j)(\zeta)(\zeta_j - z_j)$ *satisfies*

$$|\Phi(\zeta, z)| \geqq \gamma\,[|\mathrm{Im}\,\Phi(\zeta, z)| + \mathrm{dist}\,(z, \partial D) + \delta|\zeta - z|^m]$$

for $\zeta \in \partial D$ *and* $z \in D$ *with* $|\zeta - z| < \varepsilon$.

The function Φ can be used to construct Cauchy-Fantappié kernels on convex domains and an explicit solution u of $\bar{\partial}u = f$. In dimension 2 we are able to estimate the kernels and prove Hölder continuity of u of order $\alpha = 1/m$ by a method similar to the one used in **[8]**.

In higher dimensions our method of estimation fails. However, Theorem 2 should remain true; there is hope that a more careful estimation of the kernels will settle this case as well.

To treat more general domains, one can introduce a local condition for the boundary of the domain, closely related to the estimate (*), which allows us to prove a local result on Hölder estimates for solutions of $\bar{\partial}u = f$, at least in dimension two. These extensions, as well as detailed proofs of the above results, will appear elsewhere.

REFERENCES

1. H. Grauert and I. Lieb, *Das Ramirezsche Integral und die Lösung der Gleichung* $\bar{\partial}f = \alpha$ *im Bereich der beschränkten Formen*, Rice Univ. Studies **56** (1970), no. 2, 29–50 (1971). MR **42** #7938.

2. G. M. Henkin, *Integral representations of functions holomorphic in strictly pseudoconvex domains and some applications*, Mat. Sb. **78** **(120)** (1969), 611–632 = Math. USSR Sb. **7** (1969), 597–616. MR **40** #2902.

3. ———, *Integral representations of functions in strictly pseudoconvex domains and applications to the* $\bar{\partial}$ *problem*, Mat. Sb. **82** **(124)** (1970), 300–308 = Math. USSR Sb. **11** (1970), 273–281. MR **42** #534.

4. A. V. Romanov and G. M. Henkin, *Exact Hölder estimates for the solutions of the* $\bar{\partial}$*-equation*, Izv. Akad. Nauk SSSR Ser. Mat. **35** (1971), 1171–1183 = Math. USSR Izv. **5** (1971), 1180–1192. MR **45** #2200.

5. N. Kerzman, *Hölder and* L^p *estimates for solutions of* $\bar{\partial}u = f$ *in strongly pseudoconvex domains*, Comm. Pure Appl. Math. **24** (1971), 301–379. MR **43** #7658.

6. J. J. Kohn, *Subellipticity on pseudoconvex domains with isolated degeneracies*, Proc. Nat. Acad. Sci. U.S.A. **71** (1974), 2912–2914. MR **50** #7840.

7. E. Ramírez de Arellano, *Ein Divisionsproblem und Randintegraldarstellungen in der komplexen Analysis*, Math. Ann. **184** (1969/70), 172–187. MR **42** #4767.

8. R. M. Range and Y.-T. Siu, *Uniform estimates for the* $\bar{\partial}$*-equation on domains with piecewise smooth strictly pseudoconvex boundaries*, Math. Ann. **206** (1973), 325–354. MR **49** #3214.

STATE UNIVERSITY OF NEW YORK AT ALBANY

Proceedings of Symposia in Pure Mathematics
Volume 30, 1977

LEFSCHETZ THEOREMS AND A VANISHING THEOREM OF GRAUERT-RIEMENSCHNEIDER

MICHAEL SCHNEIDER

Introduction. If X is a projective algebraic complex manifold and if $Y \subset X$ is a nonsingular hyperplane section then the cohomology groups $H^i(X, Y; \boldsymbol{Z})$ vanish for $i < \dim X$. This is the (weak) Lefschetz theorem for hyperplane sections. A modern proof using Morse theory was given by Andreotti and Frankel in **[1]**. Barth extended this theorem to submanifolds Y of complex projective space of higher codimension **[5]**:

$$H^i(\boldsymbol{P}_n, Y; \boldsymbol{C}) = 0 \quad \text{for } i \leqq 2 \dim Y - n + 1.$$

Barth and Larsen made a first step towards a homotopy Lefschetz theorem in higher codimension by proving $\pi_1(Y) = 0$ for $2 \dim Y \geqq n + 1$ **[6]**. Using this, Barth's techniques and Morse theory, Larsen finally established

$$\pi_i(\boldsymbol{P}_n, Y) = 0 \quad \text{for } i \leqq 2 \dim Y - n + 1$$

and therefore also

$$H^i(\boldsymbol{P}_n, Y; \boldsymbol{Z}) = 0 \quad \text{for } i \leqq 2 \dim Y - n + 1.$$

In their investigation on vector bundles of rank two on Grassmannians, Barth and Van de Ven deduced from Larsen's theorem a Lefschetz theorem for 2-codimensional submanifolds of a Grassmann manifold **[7]**. Recently Hartshorne used the strong Lefschetz theorem for hyperplane sections to give a very elegant proof of Barth's theorem.

It is the aim of this note to show that there is a connection between Lefschetz theorems and a vanishing theorem of Grauert-Riemenschneider for hyperconvex manifolds **[9]**:

Let X be a compact Kähler manifold, and let $Y \subset X$ be a closed analytic sub-

AMS (MOS) subject classifications (1970). Primary 32M15, 32F10, 32L10; Secondary 32M10.

manifold. Assume that $X\backslash Y$ is hyper-q-convex and that the holomorphic tangent bundle T_X of X is Nakano semipositive. Then $H^i(X, Y; \boldsymbol{C}) = 0$ for $i \leqq \dim X - q$. For $X = \boldsymbol{P}_n$ and k-codimensional $Y \subset \boldsymbol{P}_n$ the complement $\boldsymbol{P}_n\backslash Y$ is hyper-$(2k - 1)$-convex and one obtains the Lefschetz theorem of Barth.

Notation. For a complex manifold X the sheaf of holomorphic p-forms will be denoted by Ω_X^p. No distinction will be made between holomorphic vector bundles and locally free sheaves over the holomorphic functions. The canonical bundle K_X corresponds to Ω_X^n, $n = \dim X$. For a holomorphic vector bundle E let $\boldsymbol{P}(E)$ denote the associated projective fibre bundle. For $x \in X$ the fibre $\boldsymbol{P}(E)_x$ consists of all complex lines through the origin of E_x. There is a holomorphic line bundle $L(E)$ on $\boldsymbol{P}(E)$ whose fibre at a point $l \in \boldsymbol{P}(E)$ is given by the line l. All manifolds in this paper are supposed to be of pure dimension.

1. Lefschetz theorems and hyperconvexity. We begin by recalling two notions [9], [13]. A holomorphic vector bundle E on a complex manifold X is *Nakano semipositive* if there exists a hermitian metric h on E such that for each $x_0 \in X$, with respect to normal trivialization, the hermitian form

$$\sum_{i,j,\alpha,\beta} \frac{\partial^2 h_{\alpha\beta}}{\partial z_i \partial \bar{z}_j} \xi^{(\beta,i)} \overline{\xi^{(\alpha,j)}}$$

is negative semidefinite. The range of summation is $1 \leqq i, j \leqq \dim X, 1 \leqq \alpha, \beta \leqq \operatorname{rank} E$. The hermitian metric h is represented via a trivialization of E by a differentiable hermitian matrix $(h_{\alpha\beta})$ in a neighborhood of x_0. Such a trivialization is called normal in x_0 if $(h_{\alpha\beta}(x_0))$ is the unit matrix and if $dh_{\alpha\beta}(x_0) = 0$ for all α, β. The matrix $-((\partial^2 h_{\alpha\beta}/\partial z_i \partial \bar{z}_j)(x_0))$ is the curvature matrix $\Theta^\alpha_{\beta i j}$ in x_0.

Let X now be an n-dimensional Kähler manifold and let $G \subset X$ be an open relatively compact set. Let $q \geqq 1$ be an integer. G is called *hyper-q-convex* if for each $x_0 \in \partial G$ there exists an open neighborhood U of x_0 in X, a smooth function φ in U with $d\varphi(x) \neq 0$ for $x \in U$ and $U \cap G = \{x \in U: \varphi(x) < 0\}$ and normal coordinates (with respect to the Kähler metric) $z_1, \cdots, z_n$ at x_0 such that:

(i) $\{z_n = 0\}$ is the complex tangent plane to ∂G at x_0.

(ii) The Levi form of φ on this tangent plane has diagonal form

$$\frac{\partial^2 \varphi}{\partial z_i \partial \bar{z}_j}(x_0) = \delta^i_j \lambda_i, \qquad 1 \leqq i, j \leqq n - 1.$$

(iii) $\lambda_{i_1} + \cdots + \lambda_{i_q} > 0$ for $1 \leqq i_1 < \cdots < i_q \leqq n - 1$.

In particular at least $n - q$ of the eigenvalues $\lambda_1, \cdots, \lambda_{n-1}$ are positive. Hence a hyper-q-convex domain G is q-convex in the sense of Andreotti and Grauert [2]. For $q = 1$ both notions coincide. The vanishing theorem of Grauert and Riemenschneider in [9] is as follows:

Let $G \subset X$ be hyper-q-convex and let E be a Nakano semipositive holomorphic vector bundle on X. Then $H^i(G, E \otimes K_X) = 0$ for $i \geqq q$. If X is a compact Kähler manifold and if $Y \subset X$ is a closed submanifold then $X\backslash Y$ will be called hyper-q-convex if there is a fundamental system of open neighborhoods V_n of Y in X such that $X\backslash \bar{V}_n$ is hyper-q-convex in X.

THEOREM 1. *Let X be an n-dimensional compact Kähler manifold and let $Y \subset X$ be a closed analytic submanifold. If $X\backslash Y$ is hyper-q-convex and if the holomorphic tangent bundle T_X of X is Nakano semipositive then*

$$H^i(X, Y; \boldsymbol{C}) = 0 \quad \text{for } i \leqq n - q.$$

PROOF. The Nakano semipositivity of T_X implies that all exterior powers $\bigwedge^p T_X$ are Nakano semipositive. The vanishing theorem of Grauert-Riemenschneider therefore implies

$$\begin{aligned} H^i(X\backslash Y, \Omega_X^p) &= H^i(X\backslash Y, (\Omega_X^p \otimes K_X^{-1}) \otimes K_X) \\ &= H^i\Big(X\backslash Y, \bigwedge^{n-p} T_X \otimes K_X\Big) = 0 \quad \text{for } i \geqq q. \end{aligned}$$

By considering the exact sequence

$$0 \to \boldsymbol{C} \to \mathcal{O} \overset{d}{\to} \Omega^1 \overset{d}{\to} \Omega^2 \to \cdots \to \Omega^n \to 0$$

one obtains $H^i(X\backslash Y, \boldsymbol{C}) = 0$ for $i \geqq q + n$ and hence $H_i(X\backslash Y; \boldsymbol{C}) = 0$ for $i \geqq q + n$. Poincaré duality implies $H_c^i(X\backslash Y, \boldsymbol{C}) = 0$ for $i \leqq n - q$. The isomorphism $H_c^i(X\backslash Y, \boldsymbol{C}) = H^i(X, Y; \boldsymbol{C})$ concludes the proof of the theorem.

REMARK. $H^i(X, Y; \boldsymbol{C}) = 0$ for $i \leqq n - q$ is equivalent to $H^i(X, \boldsymbol{C}) \to H^i(Y, \boldsymbol{C})$ is an isomorphism for $i < n - q$ and a monomorphism for $i = n - q$.

2. Submanifolds of projective space. In this section it will be shown that the tangent bundle of the complex projective space is Nakano semipositive and that the complement of a k-codimensional submanifold Y of $\boldsymbol{P}_n$ is hyper-$(2k - 1)$-convex.

THEOREM 2. *The holomorphic tangent bundle of the complex projective space $\boldsymbol{P}_n$ is Nakano semipositive.*

PROOF. The curvature matrix of $T_{\boldsymbol{P}_n}$ in normal coordinates is given by [8] $\Theta^\sigma_{\rho i\bar{j}} = \delta^\sigma_j \delta^\rho_i + \delta^\rho_\sigma \delta^i_j$. We have to check that

$$\sum_{\rho,\sigma,i,j} \Theta^\rho_{\sigma i\bar{j}}\, \xi^{(\sigma,i)}\, \overline{\xi^{(\rho,j)}} \geqq 0$$

for all $\xi = (\xi^{(\sigma,i)}) \in \boldsymbol{C}^{n^2}$.

$$\begin{aligned} \sum \Theta^\rho_{\sigma i\bar{j}}\, \xi^{(\sigma,i)}\, \overline{\xi^{(\rho,j)}} &= \sum_{i,j} \xi^{(i,j)}\, \overline{\xi^{(j,i)}} + \sum_{i,j} \xi^{(i,j)}\, \overline{\xi^{(i,j)}} \\ &= 2 \sum_i |\xi^{(i,i)}|^2 + \sum_{i \neq j} \xi^{(i,j)}\, \overline{(\xi^{(i,j)} + \xi^{(j,i)})} \\ &= 2 \sum_i |\xi^{(i,i)}|^2 + \sum_{i<j} |\xi^{(i,j)} + \xi^{(j,i)}|^2. \end{aligned}$$

THEOREM 3. *Let $Y \subset \boldsymbol{P}_n$ be a closed analytic submanifold of codimension k. Then $\boldsymbol{P}_n\backslash Y$ is hyper-$(2k - 1)$-convex.*

PROOF. The proof follows the lines of [14]. Blow up $\boldsymbol{P}_n$ along Y to obtain a hypersurface

$$\begin{array}{ccc} \tilde{Y} & \hookrightarrow & \tilde{\boldsymbol{P}}_n \\ \big\downarrow & & \big\downarrow \pi \\ Y & \hookrightarrow & \boldsymbol{P}_n \end{array}$$

If N is the normal bundle of Y in $\boldsymbol{P}_n$ and $\tilde{N}$ that of $\tilde{Y}$ in $\tilde{\boldsymbol{P}}_n$, one can identify $\tilde{Y}$ with $\boldsymbol{P}(N)$ and $\tilde{N}$ with $L(N)$.

On $\boldsymbol{P}_n$ we have the exact sequence $0 \to \mathcal{O} \to \bigoplus_{n+1} \mathcal{O}(1) \to T_{\boldsymbol{P}_n} \to 0$. On Y there is the exact sequence $0 \to T_Y \to T_{\boldsymbol{P}_n} | Y \to N \to 0$. Hence N appears as a quotient of $\bigoplus_{n+1} \mathcal{O}_Y(1)$ and therefore inherits a hermitian metric h making N into a positive vector bundle (in the sense of Griffiths [**10**]). The metric h induces a metric $\tilde{h}$ on $L(N)$. It is first shown that $L(N)$ becomes a "hyper-$(2k-1)$-concave" line bundle, i.e., the eigenvalues $\mu_1, \cdots, \mu_{n-1}$ of the Levi form of $\tilde{h}$ with respect to a normal trivialization around a point $x_0 \in \tilde{Y}$ satisfy

$$\mu_{i_1} + \cdots + \mu_{i_{2k-1}} < 0 \quad \text{for } 1 \leqq i_1 < \cdots < i_{2k-1} \leqq n-1.$$

This will be done as in the lemma of [**14**].

Let $(h_{\alpha\beta})$ be the hermitian matrix representing h in a normal trivialization of N around $y_0 := \pi(x_0)$. Choose local coordinates $z_1, \cdots, z_d$ of Y in a neighborhood U of y_0 and fibre coordinates $\xi_1, \cdots, \xi_k$ of N. The metric $\tilde{h}$ will then be represented in $U \times \{\xi_i \neq 0\}$ by the function $|\xi_i|^{-2} \sum h_{\alpha\beta}(x)\, \bar{\xi}_\alpha \xi_\beta$. We may assume that x_0 is given by $(y_0, (0 : \cdots : 0 : 1)) \in Y \times \boldsymbol{P}_{k-1}$. An explicit calculation shows

$$\partial\bar{\partial} \log \tilde{h}(x_0) = \begin{pmatrix} \dfrac{\partial^2 h_{kk}}{\partial z_i \partial \bar{z}_j}(y_0) & 0 \\ 0 & \delta^\alpha_\beta \end{pmatrix}$$

where $1 \leqq i, j \leqq d$, $1 \leqq \alpha, \beta \leqq k - 1$.

On $\bigoplus_{n+1} \mathcal{O}_{\boldsymbol{P}_n}(1)$ we have the usual metric. In normal trivialization the curvature matrix of this metric is the identity matrix. Since N is a quotient of $\bigoplus \mathcal{O}_Y(1)$, the eigenvalues of $(\partial^2 h_{kk}(x_0)/\partial z_i \partial \bar{z}_j)$ are not bigger than -1 (see [**10**]). This implies that a sum of at least $2k - 1$ different eigenvalues of $\partial\bar{\partial} \log \tilde{h}(x_0)$ has to be negative. Therefore $L(N)$ is hyper-$(2k - 1)$-concave. Now one can proceed exactly as in [**14**, Satz 1]. One obtains a smooth function $\varphi : \boldsymbol{P}_n \backslash Y \to \boldsymbol{R}$ such that the sets $\{x \in \boldsymbol{P}_n \backslash Y : \varphi(x) < c\}$ are hyper-$(2k - 1)$-convex for all $c \in \boldsymbol{R}$. This implies the hyper-$(2k - 1)$-convexity of $\boldsymbol{P}_n \backslash Y \simeq \tilde{\boldsymbol{P}}_n \backslash \tilde{Y}$.

If one takes Theorems 1, 2 and 3 together one gets the Lefschetz theorem of Barth.

Corollary (Barth [**5**]). *Let $Y \subset \boldsymbol{P}_n$ be a closed analytic submanifold of pure dimension d. Then $H^i(\boldsymbol{P}_n, Y; \boldsymbol{C}) = 0$ for $i \leqq 2d - n + 1$.*

As noticed earlier, any hyper-q-convex manifold is also q-convex. For each $q \geqq 2$ we will give an example of a q-convex manifold which is not hyper-q-convex.

Let $j : \boldsymbol{P}_1 \times \boldsymbol{P}_q \to \boldsymbol{P}_{2q+1}$ be the Segre embedding,

$$\begin{aligned} &j((x_0 : x_1), (y_0 : \cdots : y_q)) \\ &\quad = (x_0 y_0 : \cdots : x_0 y_q : x_1 y_0 : \cdots : x_1 y_q). \end{aligned}$$

The image of j is nonsingular and of codimension q. A theorem of Barth [**4**] shows that $\boldsymbol{P}_{2q+1} \backslash (\boldsymbol{P}_1 \times \boldsymbol{P}_q)$ is q-convex. By Theorem 3, $\boldsymbol{P}_{2q+1} \backslash (\boldsymbol{P}_1 \times \boldsymbol{P}_q)$ is hyper-$(2q - 1)$-convex. But $\boldsymbol{P}_{2q+1} \backslash (\boldsymbol{P}_1 \times \boldsymbol{P}_q)$ is not hyper-r-convex for any $r < 2q - 1$! Assume that $\boldsymbol{P}_{2q+1} \backslash (\boldsymbol{P}_1 \times \boldsymbol{P}_q)$ is hyper-r-convex for some $r < 2q - 1$. Theorem 1 would imply $H^i(\boldsymbol{P}_{2q+1}, \boldsymbol{P}_1 \times \boldsymbol{P}_q) = 0$ for $i \leqq 3$. In particular we would have $H^2(\boldsymbol{P}_{2q+1}) \simeq H^2(\boldsymbol{P}_1 \times \boldsymbol{P}_q)$, which is impossible.

The manifolds $\boldsymbol{P}_{2q+1} \backslash (\boldsymbol{P}_1 \times \boldsymbol{P}_q)$, $q \geqq 2$, give also examples of q-convex manifolds which are for no $r < 2q - 1$ (cohomologically) r-complete despite the fact that they do not contain compact analytic sets of dimension greater or equal to q. A first but more complicated example of this type has been noticed by Vo Van Tan.

REMARK. If X is a compact Kähler manifold and if $Y \subset X$ is a closed submanifold, the proof of Theorem 3 shows that the knowledge of the eigenvalues of the curvature matrix of the normal bundle of Y in X gives information about the hyperconvexity of $X \backslash Y$. This can be worked out in particular for the compact irreducible hermitian symmetric manifolds [**15**]. Unfortunately the tangent bundles of these manifolds are in general not Nakano semipositive (but one can show that the tangent bundle of a homogeneous manifold is semipositive in the sense of Griffiths). It is an open problem if our method applies to submanifolds of irreducible tori. The results of Barth [**3**] seem to indicate that one can expect the same Lefschetz theorem as in projective space.

BIBLIOGRAPHY

1. A. Andreotti and T. Frankel, *The Lefschetz theorem on hyperplane sections*, Ann. of Math. (2) **69** (1959), 713–717.

2. A. Andreotti and H. Grauert, *Théorèmes de finitude pour la coholomogie des espaces complexes*, Bull. Soc. Math. France **90** (1962), 193–259. MR **27** #343.

3. W. Barth, *Verallgemeinerung des Bertinischen Theorems in abelschen Mannigfaltigkeiten*, Ann. Scuola Norm. Sup. Pisa (3) **23** (1969), 317–330. MR **40** #7476.

4. ———, *Der Abstand von einer algebraischen Mannigfaltigkeit im komplex-projektiven Räum*, Math. Ann. **187** (1970), 150–162. MR **42** #3080.

5. ———, *Transplanting cohomology classes in complex-projective space*, Amer. J. Math. **92** (1970), 951–967. MR **44** #4239.

6. W. Barth and M. E. Larsen, *On the homotopy groups of complex projective algebraic manifolds*, Math. Scand. **30** (1972), 88–94. MR **49** #5395.

7. W. Barth and A. Van de Ven, *On the geometry in codimension* 2 *of Grassmann manifolds*, Classification of Algebraic Varieties and Compact Complex Manifolds, Lecture Notes in Math., vol. 412, Springer-Verlag, Berlin, 1974, pp. 1–35. MR **50** #7139.

8. E. Calabi and E. Vesentini, *On compact, locally symmetric Kähler manifolds*, Ann. of Math. (2) **71** (1960), 472–507. MR **32** #1922b.

9. H. Grauert and O. Riemenschneider, *Kählersche Mannigfaltigkeiten mit hyper-q-konvexem Rand*, Proc. in Analysis (Lectures Sympos. in Honor of S. Bochner, 1969), Princeton Univ. Press, Princeton, N. J., 1970, pp. 61–79. MR **50** #7584.

10. P. A. Griffiths, *Hermitian differential geometry, Chern classes, and positive vector bundles*, Global Analysis (Papers in Honor of K. Kodaira), Univ. of Tokyo Press, Tokyo, 1969, pp. 185–251. MR **41** #2717.

11. R. Hartshorne, *Varieties of small codimension in projective space*, Bull. Amer. Math. Soc. **80** (1974), 1017–1032.

12. M. E. Larsen, *On the topology of complex projective manifolds*, Invent. Math. **19** (1973), 251–260. MR **47** #7058.

13. S. Nakano, *On complex analytic vector bundles*, J. Math. Soc. Japan **7** (1955), 1–12. MR **17**, 409.

14. M. Schneider, *Über eine Vermutung von Hartshorne*, Math. Ann. **201** (1973), 221–229.

15. ———, *Lefschetzsätze und Hyperkonvexität*, Invent. Math. **31** (1975), 183–192.

MATHEMATISCHES INSTITUT DER UNIVERSITÄT GÖTTINGEN

Proceedings of Symposia in Pure Mathematics
Volume 30, 1977

EMBEDDING STRONGLY (1,1)-CONVEX-CONCAVE SPACES IN $C^N \times P_M$

ALESSANDRO SILVA

We wish to present here an embedding theorem for two classes of complex analytic spaces into $C^N \times P_M$. If X is a compact analytic space, Grauert has proved in [6], extending Kodaira's results, that it can be embedded in some P_M if and only if X carries a line bundle satisfying a positivity condition. On the other hand, Narasimhan has proved in [8] that every Stein space with bounded embedding dimension can be embedded in some C^N. Very little is known for the intermediate case between Stein and compact, for instance about the embeddibility properties of (p,q)-convex-concave spaces. Namely, Andreotti-Tomassini and Andreotti-Siu proved in [3] and [2], respectively, that a strongly 1-pseudoconcave manifold and a strongly 1-pseudoconcave normal space with noncompact irreducible components of dimension $\geqq 3$ admit a projective embedding if they carry a line bundle satisfying a kind of positivity condition. The following results provide a first step towards embeddibility properties in the strongly pure convex case and in the mixed convex-concave case, both in degree 1:

(3.2) *Let X be a strongly 1-pseudoconvex complex-analytic space and $\boldsymbol{L} = \{L, X, \pi\}$ a holomorphic line bundle such that $\boldsymbol{L}$ restricted to the exceptional set E of X is positive. Then X is biholomorphic to an analytic subset of $C^N \times P_M$.*

(4.3) *Let X be a normal (1,1)-convex-concave complex-analytic space and $\boldsymbol{L} = \{L, X, \pi\}$ a (1,1)-positive holomorphic line bundle. Suppose that the noncompact irreducible components of X have dimension $\geqq 3$. Then X can be immersed into an analytic subset of $C^N \times P_M$.*

1. Preliminaries. In the following X will be a complex analytic space with bounded complex dimension n and countable topology. A function φ on X is said to be strongly p-pseudoconvex if for every $x \in X$ there is an analytic isomorphism

AMS (MOS) subject classifications (1970). Primary 32F10, 32L10.

τ of an open neighborhood V of x onto an analytic subset of an open set G of $\boldsymbol{C}^n$ (with coordinates $(z_1, \cdots, z_n)$) and there is a real-valued C^2 function ψ on G such that $\varphi = \psi \circ \tau$ and the hermitian matrix $(\partial^2\psi/\partial z_i \partial \bar{z}_j)$ has at least $n - p + 1$ positive eigenvalues at every point of G. X is called strongly (p, q)-convex-concave with respect to the pair $(\varphi, (a', b'))$ if there is a proper C^2 function $\varphi: X \to (a, b)$, with $a \in \{-\infty\} \cup \boldsymbol{R}$ and $b \in \boldsymbol{R} \cup \{+\infty\}$, such that there are $a', b' \in \boldsymbol{R}$ with $a < a' < b' < b$ for which φ is strongly p-pseudoconvex on $B^{b'} = \{x \in X : \varphi(x) > b'\}$ and strongly q-pseudoconvex on $B_{a'} = \{x \in X : \varphi(x) < a'\}$, and $\{x \in X: d \leqq \varphi(x) \leqq c\} = \bar{B}^d_c$ for $b' < d < b$ and $a < c < a'$. If $B_{a'} = \emptyset$, X is called strongly p-pseudoconvex; if $B^{b'} = \emptyset$, X is called strongly q-pseudoconcave. We will omit in the following discussion the prefix "pseudo."

The following basic results hold:

(1.1) (ANDREOTTI-GRAUERT [1], CF. ALSO [9]). *If X is (p, q)-convex-concave and $\mathscr{F}$ is a coherent analytic sheaf on X, the complex vector spaces $H^r(X, \mathscr{F})$ have finite dimension for $p \leqq r \leqq \operatorname{prof}(\mathscr{F}) - q - 1$.*

(1.2). (GRAUERT [6]). *Let X be a strongly 1-convex space. Then there is a Stein space S, a finite set of points of S, $\{s_1, \cdots, s_k\}$, and a proper surjective holomorphic map $f : X \to S$ such that $E = f^{-1}(\{s_1, \cdots, s_k\})$ is a maximal compact analytic subset of X and $f: X - E \to S - \{s_1, \cdots, s_k\}$ is an analytic isomorphism.*

E is called the exceptional set of X.

2. Vanishing theorems and other ingredients. Let $\boldsymbol{L} = \{L, X, \pi\}$ be a holomorphic line bundle. We will call $\boldsymbol{L}$ negative if there is a hermitian metric h along the fibers of $\boldsymbol{L}$ such that the real-valued function on L given by the square length of a vector in that metric is differentiable and strongly 1-convex outside the zero section of $\boldsymbol{L}$. We will say that $\boldsymbol{L}$ is positive if its dual bundle $\boldsymbol{L}^*$ is negative.

(2.1) *If X is a strongly 1-convex analytic space and $\boldsymbol{L} = \{L, X, \pi\}$ is a line bundle such that $\boldsymbol{L}_{|E}$ is positive, then for every coherent analytic sheaf $\mathscr{F}$ on X there is a positive integer k_0 such that, for every $k \geqq k_0$,*

$$H^r(X, \mathscr{F} \otimes_{\mathcal{O}_X} \mathcal{O}(\boldsymbol{L}^k)) = 0, \quad \textit{for } r \geqq 1.$$

Indeed, by [7, Theorem 2.21], L is strongly 1-convex; hence (2.1) follows at once from [4, §2].

Let X be a strongly (p, q)-convex-concave space and let $\boldsymbol{L} = \{L, X, \pi\}$ be a holomorphic line bundle. We will call L (p, q)-positive if the zero section of $\boldsymbol{L}^*$ has a tubular neighborhood T which is a strongly (p,q)-convex-concave space; then:

(2.2) *If X is a strongly (p, q)-convex-concave space and $\boldsymbol{L} = \{L, X, \pi\}$ is a (p, q)-positive line bundle, then for every coherent sheaf $\mathscr{F}$ on X there is a positive integer k_0 such that for every k, $k \geqq k_0$,*

$$H^r(X, \mathscr{F} \otimes_{\mathcal{O}_X} \mathcal{O}(\boldsymbol{L}^k)) = 0$$

for $p \leqq r \leqq \operatorname{prof}(\mathscr{F}) - q - 1$.

Indeed, the natural injective homomorphism

$$\bigoplus_{r=1}^{+\infty} H^r(X, \mathscr{F} \otimes_{\mathcal{O}_X} \mathcal{O}(\boldsymbol{L}^k)) \to H^r(L^*, \pi^*\mathscr{F})$$

factors through $H^r(T, \pi^*\mathscr{F})$; hence the conclusion follows from (1.1).

As a consequence of the vanishing theorems and of the coherence of the direct images under convex-concave mappings, one obtains:

(2.3) *If X and $\mathbf{L}$ are as in* (2.1) (*or X and $\mathbf{L}$ are as in* (2.2) *and* $p = q = 1$), *then for every $x \in X$ and for every sheaf of ideals $\mathscr{I}$ on X (for every sheaf of ideals $\mathscr{I}$ such that for* $\operatorname{prof}(\mathscr{I}) \geqq 3$) *there is an integer $k_0 = k_0(\mathscr{I})$ such that the natural map*

$$\Gamma(X, \mathscr{I} \otimes_{\mathcal{O}_X} \mathcal{O}(\mathbf{L}^k)) \to (\mathscr{I} \otimes_{\mathcal{O}_X} \mathcal{O}(\mathbf{L}^k))_x$$

is surjective.

(2.4) COROLLARY. *Suppose X and $\mathbf{L}$ are as in* (2.1) (*or X and $\mathbf{L}$ are as in* (2.2) *and* $p = q = 1$). *Let $A = \{x \in X : \varphi_1(x) = \cdots = \varphi_s(x) = 0\}$, $\varphi_i \in \Gamma(X, \mathcal{O}(\mathbf{L}^h))$ for every i and some h, and $A \cap B_{b'} = \varnothing$. Then there is a set $\{\psi_0, \cdots, \psi_r\}$ in $\Gamma(X, \mathcal{O}(\mathbf{L}^k))$ for sufficiently large k such that $\psi_i(x) \neq 0$ for every i and for every $x \in A$.*

3. The pure 1-convex case.

(3.1) *Let X be a strongly* 1-*convex space and $\mathbf{L} = \{L, X, \pi\}$ be a holomorphic line bundle such that $\mathbf{L}_{|E}$ is positive. Then there are an integer k_1 and a finite number $\varphi_1, \cdots, \varphi_s$ of sections of $\mathcal{O}(\mathbf{L}^{k_1})$ over X such that $\varphi_1, \cdots, \varphi_s$ restricted to an open neighborhood of $\bar{B}_{b'}$, give a projective embedding of $\bar{B}_{b'}$.*

The positivity of $\mathbf{L}$ together with a repeated use of (2.1) shows that a finitely generated subalgebra of $\bigoplus_{h=0}^{+\infty} \Gamma(X, \mathcal{O}(\mathbf{L}^h))$ gives local coordinates and separates points in a neighborhood of $\bar{B}_{b'}$; then one applies Grauert's theorem [**6**].

We are now ready to prove

(3.2) THEOREM. *Let X be a strongly l-convex space and $\mathbf{L}$ a holomorphic line bundle on X such that $\mathbf{L}_{|E}$ is positive. Then X is biholomorphic to an analytic subset of $\mathbf{C}^N \times \mathbf{P}_M$.*

Let $f : X \to S$ be the blowing down of X along E to a Stein space S, and $g : S \to \mathbf{C}^N$ be the embedding of S. The composition $F = g \circ f$ is then regular in an open neighborhood of $X - B_{b'}$, and it is injective on X. Suppose the contrary; then, for some $x_1 \neq x_2 \in X$, $F(x_1) = F(x_2)$ so that $f(x_1) = f(x_2)$. The only bad case is if $x_1 \in X - E$ and $x_2 \in E$, but $f(x_1) = f(x_2)$ implies $f(x_1) = s_i$ for some i; thus we have a contradiction. Let us consider the map $G : X \to \mathbf{P}_M$ given by

$$G = (\varphi_0^k, \cdots, \varphi_s^k, \psi_1^h, \cdots, \psi_r^h, \chi_1, \cdots, \chi_t)$$

where $\{\varphi_0, \cdots, \varphi_s\}$ in $\Gamma(X, \mathcal{O}(\mathbf{L}^h))$ give the projective embedding of $\bar{B}_{b'}$, according to (3.1); $\{\psi_1, \cdots, \psi_r\}$ in $\Gamma(X, \mathcal{O}(\mathbf{L}^k))$, for k sufficiently large, is a set of sections with no common zeros on the zero set of the φ_i's, according to (2.4); and the set $\{\chi_1, \cdots, \chi_t\}$ in $\Gamma(X, \mathcal{O}(\mathbf{L}^{hk}))$ completes $\{\varphi_0^k, \cdots, \varphi_r^k\}$ to a projective embedding. The set $\{x \in X : \varphi_i(x) = 0, \psi_j(x) = 0, \chi_l(x) = 0\}$ is empty; then G is well defined on X and it is injective and regular on $B_{b'}$. The product map $F \times G : X \to \mathbf{C}^N \times \mathbf{P}_M$ is then holomorphic, regular and injective.

(3.3) Eto, Kazama and Watanabe in [**5**] proved (3.2) for X a manifold and $\mathbf{L}$ positive in the whole X. Their proof that the embedding is also proper seems to be faulty.

ADDED IN PROOF. A proof that the embedding in (3.2) is proper will appear in A. Silva, *Some properties of positive line bundles on noncompact analytic spaces*, Abh. Math. Sem. Hamburg.

4. The mixed (1, 1)-convex-concave case.

(4.1) *Let X be a normal space strongly* (1,1)-*convex-concave with respect to the pair* $(\varphi, (a', b'))$ *and* $\boldsymbol{L} = \{L, X, \pi\}$ *be a* (1,1)-*positive holomorphic line bundle. Suppose that the noncompact irreducible components of* $B_{a'}$ *have dimension* $\geqq 3$. *There are then an integer* k_1 *and a finite number* $\varphi_1, \cdots, \varphi_s$ *of sections of* $\mathcal{O}(\boldsymbol{L}^{k_1})$ *over X such that* $(\varphi_1, \cdots, \varphi_s)$ *restricted to* $B_{b'}$ *gives a projective embedding of* $B_{b'-\varepsilon}$.

By positivity and (2.2) a finitely generated subalgebra of $\bigoplus_{k=0}^{+\infty} \Gamma(X, \mathcal{O}(\boldsymbol{L}^k))$ gives local coordinates and separates points in a neighborhood of $\bar{B}_{b'-\varepsilon}$. By **[2]**, $B_{b'-\varepsilon}$ is then isomorphic to an open subset of a projective variety.

(4.2) Let us denote by $\mathrm{Hol}(X, S)$ the set of all holomorphic maps from an analytic space X to a Stein space S. We need the following:

LEMMA. *Let X be a normal analytic space which is* (1,1)-*convex-concave with respect to the pair* $(\varphi, (a'b'))$. *Then the restriction map* $\mathrm{Hol}(X, S) \to \mathrm{Hol}(B^c, S)$ *is an isomorphism for every* $c \in (a, b)$.

We can now prove, along the same lines as (3.2)

(4.3) PROPOSITION. *Let X be a normal strongly* (1, 1)-*convex-concave space and* $\boldsymbol{L}$ *a* (1, 1)-*positive holomorphic line bundle on X. Suppose that the noncompact irreducible components of X have dimension* $\geqq 3$. *Then X can be immersed into an analytic subset of* $\boldsymbol{C}^N \times \boldsymbol{P}_M$.

Indeed, blow down $B^{b'-\varepsilon}$ to a Stein space S along a compact subvariety in $B_{b'}^{b'-\varepsilon}$. Let $g : S \to \boldsymbol{C}^N$ be the embedding of S. By (4.2) extend f to $\tilde{f} : X \to S$. The composition $F = g \circ \tilde{f}$ is regular on $B^{b'-\varepsilon}$. Using (4.1) and (2.4) construct a map $G : X \to \boldsymbol{P}_M$ in the same way as in (3.2). G is regular in an open neighborhood of $\bar{B}_{b'-\varepsilon}$. Then $F \times G : X \to \boldsymbol{C}^N \times \boldsymbol{P}_M$ is holomorphic and regular.

REFERENCES

1. A. Andreotti and H. Grauert, *Théorèmes de finitude pour la cohomologie des espaces complexes*, Bull. Soc. Math. France **90** (1962), 193–259. MR **27** #343.

2. A. Andreotti and Y.-T. Siu, *Projective embedding of pseudoconcave spaces*, Ann. Scuola Norm. Sup. Pisa (3) **24** (1970), 231–278. MR **42** #542.

3. A. Andreotti and G. Tomassini, *Some remarks on pseudoconcave manifolds*, Essays on Topology and Related Topics (Mémoires dédiées à G. de Rham), Springer, New York, 1970, pp. 85–104. MR **42** #541.

4. ———, *A remark on the vanishing of certain cohomology groups*, Compositio Math. **21** (1969), 417–430. MR **41** #5651.

5. S. Eto, H. Kazama and K. Watanabe, *On strongly q-pseudoconvex spaces with positive vector bundles*, Mem. Fac. Sci. Kyushu Univ. Ser. A **28** (1974), 135–146.

6. H. Grauert, *Über Modifikationen und exzeptionelle analytische Mengen*, Math. Ann. **146** (1962), 331–368. MR **25** #583.

7. D. Lieberman and H. Rossi, *Deformations of strongly pseudoconvex manifolds*, 1975 (preprint).

8. R. Narasimhan, *Imbedding of holomorphically complete complex spaces*, Amer. J. Math. **82** (1960), 917–934. MR **26** #6438.

9. J. P. Ramis, *Théorèmes de séparation et de finitude pour l'homologie et la cohomologie des espaces (p, q)-convexes concaves*, Ann. Scuola Norm. Sup. Pisa **27** (1973), 933–997.

THE INSTITUTE FOR ADVANCED STUDY

Proceedings of Symposia in Pure Mathematics
Volume 30, 1977

THE LEVI PROBLEM

YUM-TONG SIU*

By the Levi problem one usually means the problem of proving the holomorphic convexity or the Steinness of a complex space either under the assumption of strict pseudoconvexity (of an exhaustion function or the boundary) or under the assumption of local Steinness (which is of course a form of weak pseudoconvexity). Solutions to the Levi problem under the first kind of assumption were given by Grauert [**9**] and Narasimhan [**18**]. For the second kind of assumption, the general Levi problem is the following. Suppose $\pi: X \to Y$ is a holomorphic map of complex spaces and Y is Stein. Suppose for every $y \in Y$ there is an open neighborhood U of y in Y such that $\pi^{-1}(U)$ is Stein. Is X Stein? It is conjectured that the answer is in the affirmative. The main difficulty in proving it is to show that X is holomorphically convex. Two extreme cases of the problem have been studied by a number of people.

The first extreme case is the case where X is an open subset of Y. When Y is a manifold, this was solved by Oka [**20**], Norguet [**19**], Bremermann [**3**], and Docquier and Grauert [**5**]. The case where Y has isolated singularities was solved by Andreotti and Narasimhan [**1**]. The case where Y has general singularities remains unsolved.

The second extreme case is the case where $\pi : X \to Y$ is a holomorphic fiber bundle. This case is known as the Serre conjecture, which Serre first posed in 1953 [**24**]. Through the years, various special cases have been proved. Matsushima and Morimoto [**17**] proved the case where the structure group of the fiber bundle is a *connected* complex Lie group. Stein [**28**] proved the case where the fiber is 0-dimensional, i.e., a topological covering space of a Stein space is again Stein. These results are still far from the complete resolution of the Serre conjecture.

AMS (*MOS*) *subject classifications* (1970). Primary 32E10; Secondary 32L05, 32C35, 32F05.

Key words and phrases. Levi problem, Serre conjecture, almost positive line bundle, Moišezon space.

*Partially supported by an NSF grant and a Sloan Fellowship.

For example, they cannot provide any conclusion even for the case where the fiber is a bounded Stein domain in C^n, because the automorphism group of a bounded Stein domain is a real Lie group which cannot be made into a complex Lie group unless it is regarded as a discrete group. With the case of a bounded Stein domain in C^n in mind, Fischer **[6]**, **[7]**, **[8]** introduced the concept of a Banach-Stein space and affirmed the Serre conjecture when the fiber is Banach-Stein. Königsberger **[16]** showed that, if the fiber admits a strictly plurisubharmonic function which transforms in a certain way under the action of the structure group, then the Serre conjecture is true. Hirschowitz **[12]** announced, with a sketch of the proof, the resolution of the Serre conjecture in the case where the fiber is a bounded domain in C^n which is strongly complete with respect to the Carathéodory metric and also in the case where the fiber is an open Riemann surface whose quotient with respect to the action of its automorphism group is again an open Riemann surface. Stehlé **[27]** solved the case where the fiber is hyperconvex in the sense that the fiber admits a plurisubharmonic function which defines a proper map from the fiber into $[-\infty, 0)$. Siu **[25]** obtained the case where the fiber is an open subset of C. Pflug **[21]** proved that, if the fiber is a bounded Stein domain in C^n with certain boundary conditions and with the property that the Jacobian determinant of any automorphism is bounded, then the Serre conjecture is true. Another recent contribution toward the resolution of the Serre conjecture is the paper of Hirschowitz **[13]** in which he proved that the Serre conjecture is true if the fiber is an open subset of C (with a proof different from that of **[25]**) or if the fiber is strongly complete with respect to the Carathéodory metric. Diederich and Fornaess **[4]** announced that every bounded Stein domain in C^n with C^2 boundary is hyperconvex in the sense of Stehlé and thus by Stehlé's result the Serre conjecture is true for the case where the fiber is a bounded Stein domain in C^n with C^2 boundary. All of these results, which may be applicable to the case of a bounded Stein domain as fiber (except the case of the fiber being an open subset of C), assume some special analytic properties of the domain or the smoothness of its boundary.

We consider here the case of the Serre conjecture where the fiber is a bounded Stein open subset of C^n whose first Betti number is zero and obtain the following.

THEOREM 1. *A holomorphic fiber bundle B over a Stein space X is Stein if its fiber F is a bounded Stein open subset of C^n whose first Betti number is zero.*

The idea of its proof is as follows. Since the transition functions of the fiber bundle are locally constant, as in **[28]** we can reduce the general case to the case where X is an open subset of some C^k. Let X^* be the universal covering of X. Then the pull-back B^* of B to X^* is isomorphic to $X^* \times F$ and is therefore Stein. If we can represent B as a Riemann domain over C^{k+n}, then since the topological covering B^* of B is Stein, by using $-\log$ of the Euclidean distance of C^{k+n}, we can conclude by Oka's theorem that B is Stein. However, in general, B cannot be represented as a Riemann domain over C^{k+n}. To overcome this difficulty, we employ a variation of the techniques of Andreotti and Narasimhan **[1]**. This variation requires the existence of certain holomorphic multivector fields on B. To get these multivector fields, we have to introduce certain almost invariant pseudometrics of F and have to take log of the Jacobian determinants, and for the process of taking log we need the vanishing of the first Betti number of F.

The above method of proof yields also the following.

THEOREM 2. *Suppose B is a holomophic fiber bundle over a Stein space whose fiber is a bounded Stein open subset F of C^n. If the structure group of B is the connected component of the identity in the group of all biholomorphisms of F, then B is Stein.*

Theorem 2 can be regarded as the analog of the result of Matsushima and Morimoto for the case of a bounded Stein domain in C^n. In both results, the connectedness of the structure group has to be assumed. (One can easily relax the connectedness condition to the condition that the number of connected components is finite.)

Details of the proofs of Theorems 1 and 2 will appear in [**26**].

We discuss here also the following conjecture of Grauert and Riemenschneider [**10**] which is related to the Levi problem: A compact complex manifold M is Moišezon if it admits a holomorphic line bundle L which is almost positive (in the sense that L has a Hermitian metric whose curvature form is semipositive everywhere on M and strictly positive at some point of M). Riemenschneider has proved the following two special cases:

(i) M is Kähler [**23**];

(ii) the subset A of M where the curvature form of L fails to be strictly positive is finite [**22**].

The general case remains unsolved. Consider the dual bundle L^* of L and the open subset G of L^* consisting of all vectors of length < 1. Then the boundary ∂G of G is (weakly) pseudoconvex everywhere and is strictly pseudoconvex at some point. One way to attack the conjecture is to try to prove that G is holomorphically convex near the points of ∂G where ∂G is strictly pseudoconvex in the sense that for every sequence in G approaching such a boundary point of G one can find a holomorphic function on G whose values on the sequence go to infinity. This is also a kind of Levi problem. So far I have not succeeded in resolving the conjecture by this kind of approach. We give here the following partial result on the conjecture of Grauert and Riemenschneider.

THEOREM 3. *If A is a subvariety of dimension at most one, then M is Moišezon.*

The idea of the proof is as follows. First one uses $\dim A = 1$ to construct a Kähler metric on some open neighborhood of A in M. Then one uses the compact case of the formula in [**11**, p. 429, Theorem 7.2] and some standard functional analysis techniques [**14**, Paragraph 1.1 and the proof of Lemma 3.4.2] and the uniqueness theorem of Aronszajn [**2**] to prove that $H^1(M, L^k \otimes K) = 0$ for k sufficiently large (K being the canonical line bundle of M), from which one produces as in [**15**], [**23**] enough holomorphic cross sections of $L^k \otimes K$ for some k to show that M is Moišezon.

REFERENCES

1. A. Andreotti and R. Narasimhan, *Oka's Heftungslemma and the Levi problem for complex spaces*, Trans. Amer. Math. Soc. **111** (1964), 345–366. MR **28** #3176.

2. N. Aronszajn, *A unique continuation theorem for solutions of elliptic partial differential equations or inequalities of second order*, J. Math. Pures Appl. (9) **36** (1957), 235–249. MR **19**, 1056.

3. H. J. Bremermann, *Über die Äquivalenz der pseudokonvexen Gebiete und der Holomorphiegebiete im Raum von n komplexen Veränderlichen*, Math. Ann. **128** (1954), 63–91. MR **17**, 82.

4. K. Diederich and J. E. Fornaess, *Exhaustion functions and Stein neighborhoods for smooth pseudoconvex domains*, Proc. Nat. Acad. Sci. U.S.A. **72** (1975), 3279–3280.

5. F. Docquier and H. Grauert, *Levisches Problem und Rungescher Satz für Teilgebiete Steinscher Mannigfaltigkeiten*, Math. Ann. **140** (1960), 94–123. MR **26** #6435.

6. G. Fischer, *Holomoph-vollständige Faserbündel*, Math. Ann, **180** (1969), 341–348. MR **39** #3033.

7. ———, *Fibrés holomorphes au-dessus d'un espace de Stein*, Espaces Analytiques (Séminaire, Bucharest, 1969), Editura Acad. R.S.R., Bucharest, 1971, pp. 57–69. MR **44** #2944.

8. ———, *Hilbert spaces of holomorphic functions on bounded domains*, Manuscripta Math. **3** (1970), 305–314. MR **42** #4772.

9. H. Grauert, *On Levi's problem and the imbedding of real-analytic manifolds*, Ann. of Math. (2) **68** (1958), 460–472. MR **20** #5299.

10. H. Grauert and O. Riemenschneider, *Verschwindungssätze für analytische Kohomologiegruppen auf komplexen Räumen*, Invent. Math. **11** (1970), 263–292. MR **46** #2081.

11. P. A. Griffiths, *The extension problem in complex analysis*. II:*Embeddings with positive normal bundle*, Amer. J. Math. **88** (1966), 366–446. MR **34** #6796.

12. A. Hirschowitz, *Corrections: "Sur certains fibrés holomorphes à base et fibre de Stein"*, C.R. Acad. Sci. Paris Sér. A **278** (1974), 89–91. MR **50** #7598.

13. ———, *Domaines de Stein et fonctions holomorphes bornées*, Math. Ann. **213** (1975), 185–193.

14. L. Hörmander, *L^2 estimates and existence theorems for the $\bar\partial$ operator*, Acta Math. **113** (1965), 89–152. MR **31** #3691.

15. K. Kodaira, *On Kähler varieties of restricted type* (*an intrinsic characterization of algebraic varieties*), Ann. of Math. (2) **60** (1954), 28–48. MR **16**, 952.

16. K. Königsberger, *Über die Holomorphie-Vollständigkeit lokal trivialer Faserräume*, Math. Ann. **189** (1970), 178–184. MR **42** #3308.

17. Y. Matsushima and A. Morimoto, *Sur certains espaces fibrés holomorphes sur une variété de Stein*, Bull. Soc. Math. France **88** (1960), 137–155. MR **23** #A1061.

18. R. Narasimhan, *The Levi problem for complex spaces*. I, Math. Ann. **142** (1960/61), 355–365; II, Math. Ann. **146** (1962), 195–216. MR **26** #6439; **32** #229.

19. F. Norguet, *Sur les domaines d'holomorphie des fonctions uniformes de plusieurs variables complexes*. (*Passage du local au global*.), Bull. Soc. Math. France **82** (1954), 137–159. MR **17**, 81.

20. M. Oka, *Sur les fonctions analytiques de plusieurs variables*. VI. *Domaines pseudoconvexes*, Tôhoku Math. J. **49** (1942), 15–52. MR **7**, 290.

21. P. Pflug, *Quadratintegrable holomorphe Funktionen und die Serre-Vermutung*, Math. Ann. **216** (1975), 285–288.

22. O. Riemenschneider, *Characterizing Moišezon spaces by almost positive coherent analytic sheaves*, Math. Z. **123** (1971), 263–284. MR **45** #3782.

23. ———, *A generalization of Kodaira's embedding theorem*, Math. Ann. **200** (1973), 99–102. MR **48** #4355.

24. J.-P. Serre, *Quelques problèms globaux relatifs aux variétés de Stein*, Colloq. sur les fonctions de plusieurs variables (Bruxelles, 1953), Georges Thone, Liège; Masson, Paris, 1953, pp. 57–68. MR **16**, 235.

25. Y.-T. Siu, *All plane domains are Banach-Stein*, Manuscripta Math. **14** (1974), 101–105. MR **50** #13603.

26. ———, *Holomorphic fiber bundles whose fibers are bounded Stein domains with zero first Betti number*, Math. Ann. **219** (1976), 171–192.

27. J.-L. Stehlé, *Fonctions plurisousharmoniques et convexité holomorphe de certains fibrés analytiques*, C. R. Acad. Sci. Paris Sér. A **279** (1974), 235–238.

28. K. Stein, *Überlagerungen holomorph-vollständiger komplexer Räume*, Arch. Math. **7** (1956), 354–361. MR **18**, 933.

YALE UNIVERSITY

Proceedings of Symposia in Pure Mathematics
Volume 30, 1977

ON HOLOMORPHIC JET BUNDLES

ANDREW JOHN SOMMESE

A large part of the study of compact complex manifolds has been concerned with first-order invariants such as holomorphic forms and vector fields. This is natural because of the very tight ways such invariants are linked with a complex manifold's geometry. In this article I will describe my work ([10], for full details) on the following conjecture about higher order structure.

CONJECTURE. *Let L be a holomorphic line bundle on a connected compact complex manifold X. Assume there is an integer $k > 0$ such that $J_k(X, L)$, the kth holomorphic jet bundle of L on X, is the trivial bundle. Then either:*

(A) *X is $\mathbf{C}P^N$ and $L = H^k$ where H is the hyperplane section bundle of $\mathbf{C}P^N$, or*

(B) *X is the cocompact quotient of a complex Lie group by a discrete subgroup and L is the trivial bundle.*

Below I will give the basic definitions, and sketch a proof of the conjecture if either X is homogeneous, or $k = 1$, or X is projective and k is relatively prime to $1 + \dim_C X$, or X has a nontrivial holomorphic one-form, or Hartshorne's conjecture that a projective manifold with ample holomorphic tangent bundle is projective space is true, or if $\dim_C X \leq 2$.

I. Let F be a coherent sheaf on a connected complex manifold X. Let $\mathcal{J}_\Delta$ denote the sheaf of ideals of Δ, the diagonal of $X \times X$. Letting p denote the projection of $X \times X$ onto its first factor, one defines, following Malgrange [8], the kth jet sheaf of F, $\mathcal{J}_k(X, \mathcal{F})$, or $\mathcal{J}_k(\mathcal{F})$ for short, as $p^*\mathcal{F}/p^*\mathcal{F} \otimes \mathcal{J}_\Delta^{k+1}$ where the tensor product is of course with respect to $\mathcal{O}_{X\times X}$, the holomorphic structure sheaf of $X \times X$. If $\mathcal{F}$ is locally free with associated vector bundle F, then $\mathcal{J}_k(\mathcal{F})$ is locally free and the associated vector bundle will be denoted by $J_k(X, F)$, or $J_k(F)$ for short. I will also denote $\mathcal{J}_k(X, \mathcal{O}(F))$ by $\mathcal{J}_k(X, F)$ for simplicity where $\mathcal{O}(F)$ denotes the sheaf of

AMS (MOS) subject classifications (1970). Primary 14M99, 32J25, 32L10.
Key words and phrases. Jet bundles, holomorphic differential equations.

germs of holomorphic sections of F, i.e., $\mathscr{F}$. I refer to [7] for full details and the various connections with partial differential equations.

One can interpret $J_k(F)$ as the bundle of k jets of F, i.e., Taylor expansions of holomorphic sections of F truncated after the kth term. This gives [7, p. 69ff.]:

$$(\mathrm{A}_l) \qquad 0 \to T_X^{*(l)} \otimes F \to J_l(F) \to J_{l-1}(F) \to 0$$

and

$$(\mathrm{B}) \qquad J_k(F) \to J_l(F) \to 0$$

for $l \leqq k$ where T_X^* is the holomorphic cotangent bundle of X and $F^{(l)}$ denotes the lth symmetric power of F. Let T_X denote the holomorphic tangent bundle of X, and in general let F^* denote the dual bundle to a holomorphic vector bundle F.

A line bundle L on X is called very ample if $H^0(X, L)$ gives an embedding of X into projective space; L is called ample if there is some positive integer m such that L^m is very ample. A vector bundle E on X is called ample (very ample) if the dual of the tautological bundle of lines of π^*E^* over $\pi : P(E) \to X$ is ample (very ample) where $P(E)$ is the projectivization of E^*, i.e., the quotient of E^* minus its zero section by the natural $\boldsymbol{C}^*$ action. The above definition of ample is the same as the ample of Hartshorne [4], and is equivalent to Griffiths' coholomogically positive [2]. The above notion of ample is also equivalent to Grauert's [1] notion of weakly positive.

The first step in the study of the conjecture is the following structure theorem.

PROPOSITION I. *Let L be a holomorphic line bundle on a connected compact complex manifold X and assume $J_k(X, L)$ is trivial for a fixed $k > 0$. Then there exists a holomorphic surjection $g : X \to Y$ where*:

(a) *$g : X \to Y$ is a holomorphic fibre bundle over a projective manifold Y, with fibre of the form G/Γ where G is a connected complex Lie group and G is a cocompact discrete subgroup of G,*

(b) *L is the pull-back of an ample line bundle $\tilde{H}$ on Y and $L^{n+1} = K_X^{-k}$ where $\dim_{\boldsymbol{C}} X = n$ and K_X is the canonical bundle of X,*

(c) *the tangent bundle T_Y is ample. Further if T_X^* has a nontrivial section or L is trivial, then Y reduces to a point. If T_X^* has only the trivial section and X is projective, then $g : X \to Y$ is an isomorphism.*

How does one show this? To start one notes one has by A_k and B above $J_k(X, L) \to L \to 0$. Thus the sections of L give a holomorphic map from X into projective space. By using the Remmert-Stein factorization [9], [11] one gets a holomorphic map g from X into a normal analytic space Y with connected fibres where $L = g^*\tilde{H}$ for some ample line bundle $\tilde{H}$ on Y. Now one does a detailed analysis of g using A_l and the dual sequence. Roughly given a point $y \in Y$ one constructs enough holomorphic vector fields on $g^{-1}(U(y))$ for some neighborhood $U(y)$ of y to span $T_{U(y)}$. By integrating these one shows that Y is a manifold and g is a fibre bundle. $L^{n+1} = K_X^{-k}$ follows from a direct calculation of $\det[J_k(X, L)]$ and an argument showing $\mathrm{Pic}(Y) \approx \boldsymbol{Z}^m$ for some m.

Next one shows:

PROPOSITION II. *Let L and $g : X \to Y$ be as in Proposition* I. *X is biholomorphic to* *projective space and g is a biholomorphism if either X is homogeneous, $k = 1$, Y is biholomorphic to projective space, or Hartshorne's conjecture that a compact complex manifold with ample tangent bundle is projective space is true. If $k = 2$, then g is a biholomorphism.*

One first reduces to the case Y is projective space. If $k = 1$ one dualizes A_1 and tensors with L to conclude that X is homogeneous. If X is homogeneous one shows Y is also. But Hartshorne's conjecture is true for homogeneous spaces **[5]**, **[6]**. Now one assumes Y is projective space and reduces to the case Y is CP^1. This is done by a few lemmas.

LEMMA A. *Let $g : X \to Y$ be as in Proposition* II *and assume $Y = CP^m$ for some $m > 0$. Then $L = g^*(H^k)$ where H is the hyperplane section bundle of CP^m, and k is such that $J_k(X, L)$ is trivial. Also $K_X = g^*(H^{-n-1})$ where* $\dim_C X = n$.

LEMMA B. *Let $g : X \to CP^m$ with $m > 0$ be as in Lemma* A. *Let CP^1 be a line in CP^m. Then $J_k(g^{-1}(CP^1), L)$ is trivial.*

The above also show if $m = 1$, then

$$T_X \approx g^*(H^2 \oplus \underbrace{H \oplus \cdots \oplus H}_{n-1 \text{ copies}})$$

where $\dim_C X = n$. This shows X is homogeneous if $m = 1$ and that one has a holomorphic one-dimensional foliation determined by $g^*(H^2)$ that is invariant under the biholomorphism group of X. Finally one uses the above foliation to show g is the product projection where $X \approx G/\Gamma \times CP^1$. A direct computation shows $J_k(G/\Gamma \times CP^1, g^*(H^k))$ is not trivial.

The rest of the cases mentioned in the introduction are straightforward consequences of the above two propositions.

II. Hartshorne's conjecture is true if $\dim_C X \leqq 2$. Thus the first nontrivial dimension for my conjecture is $\dim_C X = 3$, and in this case one can by Proposition II assume without loss of generality that X is projective. The first k for which the conjecture is unknown is $k = 2$. By a detailed analysis of various Hilbert polynomials one can show:

LEMMA. *If* $\dim_C X = 3$, *X is projective and $J_2(X, L)$ is trivial, then* $\dim_C H^0(X, L) = 10$ *and* $\dim_C H^0(X, T_X) = 15$. *Further X has ample tangent bundle.*

How much more like projective space can a manifold be? Nonetheless I can see no way of going much further and in fact I have doubts whether the conjecture can be true for $\dim_C X \geqq 4$.

BIBLIOGRAPHY

1. H. Grauert, *Über Modifikationen und exzeptionelle analytische Mengen*, Math. Ann. **146** (1962), 331–368. MR **25** #583.

2. P. Griffiths, *Hermitian differential geometry, Chern classes, and positive vector bundles*, Global Analysis (Papers in Honor of K. Kodaira), Univ. of Tokyo Press, Tokyo, 1969, pp. 185–251. MR **41** #2717.

3. R. Hartshorne, *Ample subvarieties of algebraic varieties*, Lecture Notes in Math., vol. 156, Springer-Verlag, Berlin and New York, 1970. MR **44** #211.

4. ———, *Ample vector bundles*, Inst. Hautes Études Sci. Publ. Math. No. 29 (1966), 63–94. MR **33** #1313.

5. S. Kobayashi and T. Ochiai, *Compact homogeneous manifolds complex with positive tangent bundle*, Differential Geometry (In Honor of Kentaro Yano), Kinokuniya, Tokyo, 1972, pp. 221–232. MR **48** #2432.

6. ———, *On complex manifolds with positive tangent bundles*, J. Math. Soc. Japan **22** (1970), 499–525. MR **43** #1231.

7. A. Kumpera and D. Spencer, *Lie equations*. Vol. I: *General theory*, Princeton Univ. Press, Princeton, N. J., 1972.

8. B. Malgrange, *Théorie analytique des équations différentielles*, Séminaire Bourbaki 329.

9. R. Remmert and K. Stein, *Eigentliche holomorphe Abbildungen*, Math. Z. **73** (1960), 159–189. MR **23** #A1840.

10. A. J. Sommese, *Compact complex manifolds possessing a line bundle with a trivial jet bundle*, Abh. Math. Sem. Univ. Hamburg (to appear).

11. K. Stein, *On factorization of holomorphic mappings*, Proc. Conf. Complex Analysis (Minneapolis, 1964), Springer, Berlin, 1965, pp. 1–7. MR **31** #2419.

THE INSTITUTE FOR ADVANCED STDUY

Proceedings of Symposia in Pure Mathematics
Volume 30, 1977

ON THE CLASSIFICATION OF HOLOMORPHICALLY CONVEX SPACES

VO VAN TAN

To my father, in honor of his 77th birthday

1. General status of the problem. All C-analytic spaces considered here are assumed to be countable at infinity and noncompact, unless the contrary is explicitly stated. Coh(X) will denote the category of analytic coherent sheaves on the space X.

One of our main goals in this paper is to study the distribution of compact analytic subvarieties in a given C-analytic space satisfying certain nice cohomological properties.

In that direction we get [10]

THEOREM 1. *If X is cohomologically q-complete, then X has no compact analytic subvariety of C-dim $\geqq q$.*

THEOREM 1′. *If X is cohomologically q-convex, then X admits only finitely many compact irreducible components of C-dim $\geqq q$.*

DEFINITION 1. For an integer $q \geqq 1$, X is said to be cohomologically q-complete (resp. cohomologically q-convex) if $H^\nu(X, \mathscr{F}) = 0$ (resp. $\dim_C H^\nu(X, \mathscr{F}) < \infty$) for all $\mathscr{F} \in \mathrm{Coh}(X)$ and for all $\nu \geqq q$.

The first result is quite satisfactory but the second one is not. What we really expect is,

Problem I. If X is cohomologically q-convex, does X admit a q-maximal compact analytic subvariety $S \subset X$?

DEFINITION 2. A compact analytic subvariety S in X is said to be q-maximal, if

(1) for any compact irreducible analytic subvariety $T \subset X$, with $\dim_C T \geqq q$, then necessarily $T \subset S$;

AMS (MOS) subject classifications (1970). Primary 32E05, 32F10, Secondary 14F25.

(2) $\dim_C S_x > 0$, for all $x \in S$.

The last condition, among other things, is to make our definition, for $q = 1$, coincide with the definition of maximal compact subvariety in the sense of Grauert **[3]**.

Problem I has a positive answer for $q = 1$ **[6]** (even a more precise one). For $q = n = \dim_C X$, Theorem 1′ provides an affirmative answer. Therefore, Problem I remains open for the case where $q > 1$ and $\dim_C X > 2$. In fact, a careful analysis of the proof for the case $q = 1$ shows that the holomorph-convexity played a vital role there. Moreover, if the cohomological 1-convexity gives the holomorph-convexity for free **[6]**, it is no longer the case for the cohomological q-convexity with $q > 1$. For example, $X := C^2\backslash\{0\}$ is cohomologically 2-convex, but certainly not holomorph-convex.

2. The holomorph-convexity. In view of these facts, we would like to propose the following

Problem II. Let X be a given holomorphically convex space. Suppose X is cohomologically q-convex; does X admit a q-maximal compact analytic subvariety S (with $1 < q < n = \dim_C X$)?

Our purpose here is twofold. First of all, we would like to solve Problem II and secondly try to classify the holomorph-convex spaces.

Before attacking Problem II, we would like to state some intrinsic properties of the holomorph-convexity which will be useful for us later **[4]**. A C-analytic space X is holomorph-convex if, for any infinite discrete set $E \subset X$, there exists a holomorph function f on X for which $f(E)$ is unbounded. Let X be a holomorph-convex space; then there exists a Remmert-Stein reduction Y, namely

(1) there exists a Stein space Y;

(2) a proper, surjective and holomorphic mapping $\pi : X \to Y$ with the properties that all its fibres are connected;

(3) $\Gamma(Y, O_Y) \simeq \Gamma(X, O_X)$.

Moreover, let $S := \{x \in X \mid x \text{ is not isolated in its fibre } \pi^{-1}\pi(x)\}$ be the degenerate set. Remmert proved that S is actually an analytic subvariety in X.

Besides the two well-known extreme cases of holomorph-convex spaces, namely Stein spaces and compact spaces, we would like to exhibit the intermediate one which will provide a clear picture for our next results. In the sequel, X will be a holomorph-convex space, Y its Remmert-Stein reduction and S its degenerate subvariety.

(a) Let X be the blowing up of C^n at the origin; then $Y = C^n$ and $S = P_m$ with $m = n - 1$.

(b) Let E be an infinite discrete set in C^n and let X be the monoidal transform of C^n with center E; then $Y = C^n$ and $S =$ disjoint union of infinitely many copies of P_m with $m = n - 1$.

(c) Let C^r be a submanifold of C^n with $r < n - 1$, and let X be the monoidal transform of C^n with center C^r; then $Y = C^n$ and $S = C^r \times P_m$ with $m = n - r - 1$.

(d) Let $X = C^n \times P_m$; then $Y = C^n$ and $S = X = C^n \times P_m$.

We are now in position to state our next result; see **[11]**.

THEOREM 2. *Problem* II *has an affirmative answer.*

Actually, in the special case where the holomorph-convex space X is nondegenerate, Theorem 2 is just a straightforward consequence of Theorem 1′.

DEFINITION 3. A holomorph-convex space X is said to be nondegenerate if the image $T := \pi(S)$ of its degenerate set S is a discrete set in Y, the Remmert-Stein reduction of X.

Our next step is toward the classification of holomorph-convex spaces, namely we would like to show that in this framework, the algebraic (cohomological) properties are completely characterized by the geometric properties (existence of compact analytic subvarieties).

First of all the converse of Theorem 1 is obviously false, even for $q = 1$, as is easily seen by the following counterexample:

$X := \mathbf{C}^2\backslash\{0\}$ does not have any compact analytic subvariety of $\mathbf{C}$-dim $\geqq 1$ but certainly X is not cohomologically 1-complete since $\dim_{\mathbf{C}} H^1(X, O_X) = \infty$. But we have the following version of Cartan's Theorem B [4].

(*) If X is holomorph-convex and has no compact analytic subvariety of $\mathbf{C}$-dim $\geqq 1$, then X is cohomologically 1-complete.

With this in mind, we obtain the following result [11].

THEOREM 3. *Let X be a given holomorph-convex space. If X does not have any compact analytic subvariety of* $\mathbf{C}$-dim $\geqq q$, *then X is cohomologically q-complete (with $1 < q < n = \dim_{\mathbf{C}} X$).*

Also, it is known [6] that,

(**) If X is holomorph-convex and admits a 1-maximal compact analytic subvariety then X is cohomologically 1-convex.

Similarly, we obtain [11]

THEOREM 4. *The converse of Theorem 2 also holds, with $1 < q < n$.*

Again here it is useful to take a look at the special case, namely the one of nondegenerate holomorph-convex spaces. Theorems 3 and 4 become transparent in view of the following [11]:

PROPOSITION. *Let X be a holomorph-convex space. X is nondegenerate iff $X = \bigcup_i X_i$ with $X_i \subset\subset X_{i+1}$, X_i are strongly pseudoconvex domains and (X_i, X) are Runge pairs.*

For the case where $q = n$, better results are obtained; see [7] or [8].

THEOREM 5. *Let X be a given* $\mathbf{C}$*-analytic space, with* $\dim_{\mathbf{C}} X = n$.

(a) *If X admits an n-maximal compact analytic subvariety S, then X is cohomologically n-convex.*

(b) *If $S = \emptyset$ then X is cohomologically n-complete.*

In view of Theorems 1 through 5, the classification problem for holomorph-convex spaces is achieved. Before mentioning the next two useful corollaries we would like to introduce the following

DEFINITION 4. $\mathrm{cd}(X) :=$ the smallest integer $\nu \geqq -1$ such that $H^i(X, \mathcal{F}) = 0$ for all $i > \nu$ and all $\mathcal{F} \in \mathrm{Coh}(X)$.

$\mathrm{fd}(X) :=$ the smallest integer $\nu \geqq 0$ such that $\dim_{\mathbf{C}} H^i(X, \mathcal{F}) < \infty$ for all $i \geqq \nu$ and all $\mathcal{F} \in \mathrm{Coh}(X)$.

COROLLARY 1. *Let X be an irreducible C-analytic space. X is compact iff* $\mathrm{cd}(X) = n = \dim_C X$.

COROLLARY 2. *Let X be a C-analytic space with* $\dim_C X = n$. *X is compact iff* $\mathrm{fd}(X) = 0$.

REMARK. The classification problem for holomorph-convex spaces has an analogue in algebraic geometry if one replaces C-analytic spaces by C-algebraic varieties, analytic coherent sheaves by algebraic coherent sheaves and holomorph-convex spaces by varieties proper over affine varieties. For more details see **[11]**.

3. The obstruction problem. Our previous investigation was the starting preparation for attacking

Problem III. Let X be a given C-analytic space without any compact analytic subvariety of C-dim $\geqq q$.

(a) If X is q-convex, is X then q-complete?

(b) If X is cohomologically q-convex, is X then cohomologically q-complete?

For the definition of q-convex spaces, we refer to **[1]**.

As can be seen from Theorem 1, the existence of compact analytic subvarieties of C-dim $\geqq q$ is a necessary obstruction for going from q-convexity (resp. cohomological q-convexity) to q-completeness (resp. cohomological q-completeness). Moreover, for $q = 1$, Problem III has positive answer **[6]**. Unfortunately, the answer is negative for $q > 1$, even in the nonsingular case. In fact, let us consider the following counterexample.

Let $X := P_4 \backslash Y$ where Y is a nonsingular compact and connected 2-dimensional submanifold in P_4. It is known **[2]** that X is 2-convex. Moreover, from the dimension argument, X does not have any compact analytic subvariety of C-dim $\geqq 2$. If X were cohomologically 2-complete,

Claim. The first betti number of Y, $b_1(Y) = 0$.

In fact if X is cohomologically 2-complete, the result of **[9]** tells us that $H_6(X, C) = 0$. Poincaré duality implies that $H_c^2(X, C) \simeq H_6(X, C) = 0$ since $\dim_R X = 8$. Then the following exact cohomology sequence with compact support

$$\cdots \to \underset{\substack{\wr\wr \\ 0}}{H^1(P_4, C)} \to H^1(Y, C) \to \underset{\substack{\wr\wr \\ 0}}{H_c^2(X, C)} \to \cdots$$

tells us that

$$H^1(Y, C) = 0.$$

But Hartshorne exhibited a nonsingular surface Y in P_4 with $b_1(Y) \neq 0$. More recently, Horrocks and Mumford **[5]** constructed C-2-dimensional tori T embedded in P_4; certainly $b_1(T) \neq 0$. For more details concerning these counterexamples, see **[10]**.

Of course, all these counterexamples are obviously not holomorph-convex. With a little more work, the results of §2 provide us an affirmative answer for Problem III(b). Within the framework of holomorph-convex spaces, namely **[11]**,

THEOREM 6. *Let X be a given holomorph-convex space with* $\dim_C X = n$, *and let* *us assume that X is cohomologically q-convex with its q-maximal compact analytic subvariety S $(1 < q < n)$.*

If $\dim_C S < p$ *then X is cohomologically p-complete* $(p \geqq q)$.

REMARK. For $q = n$, Theorem 5 provides a positive answer for our Problem III(b). As far as Problem III(a) is concerned, even in the framework of holomorph-convex spaces, the main difficulty centers at the singular points of the given $\boldsymbol{C}$-analytic space. See **[11]** for more discussion.

4. Prospects and comments. We would like to take this opportunity to mention some interesting open problems which have been at an impasse for more than a decade. In 1962, Andreotti and Grauert **[1]** obtained the following result:

THEOREM A-G. *Let X be a given $\boldsymbol{C}$-analytic space.*

(a) *If X is q-complete then X is cohomologically q-complete.*

(b) *If X is q-convex then X is cohomologically q-convex.*

Earlier [Bull. Soc. Math. France **97** (1959), pp. 341–350] Grauert stated the following:

CONJECTURE. *Let X be a given noncompact complex space with* $\dim_C X = n$; *then X is q-convex for some $q \leqq n$.*

For $q = 1$, the converse of Theorem A-G holds (Serre, Narasimhan **[4]**, **[6]**). Naturally, it raises the following

Problem IV. Does the converse of Theorem A-G hold for $q > 1$?

Also it would be interesting to consider the following weaker version of Problem IV, namely

Problem V. Let X be a given holomorph-convex space. Does the converse of Theorem A-G hold for $q > 1$?

Actually, Problem V is not as trivial as it looks, since in the following particular case a satisfactory answer for it is still unknown:

Let X be a 1-convex space. Certainly X is holomorph-convex. Let S be its 1-maximal compact analytic subvariety; then it is known that X is cohomologically q-complete with $q = \dim_C S + 1$.

Question. Is X then q-complete?

For more details and discussion see **[11]**.

REMARK. If Problem V has a positive answer, one will have a nice and complete characterization of holomorph-convex spaces (i) by their function-theoretic properties (q-convex functions), (ii) by their algebraic properties (sheaf-cohomology), and (iii) by geometric properties (existence of high dimensional compact analytic subvarieties).

Meanwhile a careful look at Theorem 1′ shows that the Grauert conjecture is not too accurate. Instead the following problem seems more plausible; see also **[10]**.

Problem VI. Let X be a given $\boldsymbol{C}$-analytic space with $\dim_C X = n$.

(a) If X has only finitely many compact n-irreducible components then X is n-convex.

(b) If X has no compact n-dimensional irreducible component then X is n-complete.

When $n = 1$, it is known that the answer for Problem VI is positive; see [4]. Furthermore, if X is nonsingular, an affirmative answer was given recently by Greene and Wu (*Whitney's imbedding theorem by solutions of elliptic equations and geometric consequences*, Proc. Sympos. Pure Math., vol. 27, Part II, Amer. Math. Soc., Providence, R. I., 1973, pp. 287–296). In view of Theorems 1 and I′, Problem VI is a special case of Problem IV, so it would be interesting to investigate first Problem VI before getting into Problem IV.

Finally, in some context, the holomorph-convex hypothesis is too restrictive, for example in Theorem 6. It is our hope to return to this question in the near future.

BIBLIOGRAPHY

1. A. Andreotti and H. Grauert, *Théorèmes de finitude pour la cohomologie des espaces complexes*, Bull. Soc. Math. France **90** (1962), 193–259. MR **27** #343.

2. W. Barth, *Der Abstand von einer algebraischen Mannigfaltigkeit im komplex-projektiven Räum*, Math. Ann. **187** (1970), 150–162. MR **42** #3080.

3. H. Grauert, *Über Modifikationen und exzeptionelle analytische Mengen*, Math. Ann. **146** (1962), 331–368. MR **25** #583.

4. R. Gunning and H. Rossi, *Analytic functions of several complex variables*, Prentice-Hall, Englewood Cliffs, N. J., 1965. MR **31** #4927.

5. G. Horrocks and D. Mumford, *A rank* 2 *vector bundle on* $\mathbf{P}^4$ *with* 15,000 *symmetries*, Topology **12** (1973).

6. R. Narasimhan, *The Levi problem for complex spaces*, 11, Math. Ann. **146** (1962), 195–216. MR **32** #229.

7. P. Ramis, G. Ruget and J. Verdier, *Dualité relative en géométrie analytique complexe*, Invent. Math. **13** (1971), 261–283. MR **46** #7553.

8. Y.-T. Siu, *Analytic sheaf cohomology groups of dimension n of n-dimensional complex spaces*, Trans. Amer. Math. Soc. **143** (1969), 77–94. MR **40** #5902.

9. G. Sorani and V. Villani, *q-complete spaces and cohomology*, Trans. Amer. Math. Soc. **125** (1966), 432–448. MR **34** #6154.

10. Vo Van Tan, *On the classification problem of q-convex spaces* (to appear).

11. ———, *On the geometry of holomorph-convex spaces* (to appear).

INSTITUTO MATEMATICO G. VITALE (MODENA)

Proceedings of Symposia in Pure Mathematics
Volume 30, 1977

POINCARÉ THETA SERIES AND L_1 COHOMOLOGY

R. O. WELLS, JR.* AND JOSEPH A. WOLF**

1. Introduction. Ninety-five years ago, Poincaré revolutionized the theory of automorphic forms by introducing the method of summing over a discontinuous group. In modern language and somewhat greater generality, one has

D: a bounded symmetric domain in $\boldsymbol{C}^n$;

$\boldsymbol{K}$: the canonical line bundle (of $(n, 0)$-forms) over D; and

Γ: a discontinuous group of analytic automorphisms of D.

One considers holomorphic sections φ of powers $\boldsymbol{K}^m \to D$, for example $(dz^1 \wedge \cdots \wedge dz^n)^m$, and forms the *Poincaré theta series*

$$\theta(\varphi) = \sum_{\gamma \in \Gamma} \gamma^*(\varphi) \equiv \sum_{\gamma \in \Gamma} \varphi \circ \gamma^{-1}.$$

$\boldsymbol{K}^m$ carries a natural Γ-invariant hermitian metric, and if m is sufficiently large ($m \geqq 2$ for the unit disc in $\boldsymbol{C}$), then $\boldsymbol{K}^m \to D$ has absolutely integrable holomorphic sections; in fact $(dz^1 \wedge \cdots \wedge dz^n)^m$ is L_1. When φ is L_1, the series $\theta(\varphi)$ is absolutely convergent, uniformly on compact subsets of D, and represents a Γ-invariant holomorphic section of $\boldsymbol{K}^m \to D$. The Γ-invariant holomorphic sections of $\boldsymbol{K}^m \to D$ are the *Γ-automorphic forms of weight m on D*. See Borel [4] for a systematic discussion.

Poincaré's construction is the primary source of automorphic forms on D. The automorphic forms of a given weight m form a finite-dimensional space $H^0_\Gamma(D; \mathcal{O}(\boldsymbol{K}^m))$. For m sufficiently large, the corresponding map of $\Gamma \backslash D$ is a quasi-projective embedding, i.e., the quotients of elements of $H^0_\Gamma(D; \mathcal{O}(\boldsymbol{K}^m))$ generate the function field of $\Gamma \backslash D$.

AMS (MOS) subject classifications (1970). Primary 32M10, 32N05, 32F10.

*Research partially supported by NSF grant MPS75–05270; Alexander von Humboldt Foundation awardee.

**Research partially supported by NSF grant MPS74–01477.

An important aspect of automorphic function theory in several variables is the special case

$$D = \{p \times p \text{ complex matrices } Z : Z = {}^tZ \text{ and } I - ZZ^* \gg 0\},$$

which is analytically equivalent to the "Siegel upper half-space"

$$H_p = \{p \times p \text{ complex matrices } Z : Z = {}^tZ \text{ and } \operatorname{Im} Z \gg 0\}$$

of degree p. It has complex dimension $p(p + 1)/2$, and is the space of normalized Riemann period matrices of degree p. For appropriate choice of Γ, the equivalence classes of period matrices of Riemann surfaces of genus p sit in $\Gamma\backslash D$.

When Griffiths studied periods of integrals on algebraic manifolds **[8]**, **[9]**, he saw that generally the corresponding period matrix domains D are not bounded symmetric domains. In fact **[20]**, they carry no nonconstant holomorphic functions. These period matrix domains belong to a well-understood **[12]**, **[20]** class of open homogeneous complex manifolds that we call *flag domains*. Here the first difficulty (see Schmid **[12]**, **[13]**) is that one cannot expect to find sections of line bundles, or even vector bundles, but must look to cohomology of degree $s = \dim_C Y$ where Y is a maximal compact subvariety of D. In particular there are no automorphic forms in the classical sense on D, and one is led to the *automorphic cohomology space*

$$H^s_\Gamma(D; \mathcal{O}(E)) = \{\Gamma\text{-invariant classes in } H^s(D; \mathcal{O}(E))\}$$

where $E \to D$ is a "nondegenerate" homogeneous holomorphic vector bundle.

At present, very little is known about automorphic cohomology, especially when $\Gamma\backslash D$ is noncompact. For example, even in the Griffiths period domain case one does not know whether $H^s_\Gamma(D; \mathcal{O}(E))$ is finite dimensional, nor does one know how to relate it to function theory on $\Gamma\backslash D$. Recently, however, we constructed absolutely integrable cohomology classes $\varphi \in H^s(D; \mathcal{O}(E))$ for a certain specific class of bundles $E \to D$, and we showed that the Poincaré series $\theta(\varphi) = \sum_{\gamma\in\Gamma} \gamma^*(\varphi)$ always converges to an automorphic cohomology class. That is what we describe below.

The detailed proof of the theorems discussed in this paper appear in **[18]**. Some of these results had been announced previously by one of us in a preliminary fashion in **[17]**.

2. Flag domains. A *complex flag manifold* is a compact complex homogeneous space $X = G_C/P$ where G_C is a complex semisimple Lie group and P is a parabolic subgroup. Fix a noncompact real form G of G_C. Then G has only finitely many orbits on X, so in particular there are open orbits. A *flag domain* is a (necessarily open) orbit $G(x) \subset X$ on which the isotropy subgroups of G are compact. Replacing P by a conjugate, the flag domains have the form $D = G(x_0) \cong G/V$ where $x_0 = 1 \cdot P$ and $V = G \cap P$ is compact. Then V contains a compact Cartan subgroup H of G, so it sits in a unique maximal compact subgroup K of G, and we have $Y = [K(x_0) \cong K/V$: maximal compact subvariety of $D]$. All this is classical **[20]**.

We now consider the "linear deformation space" $\pi : \mathcal{Y} \to M$ of Y, given as follows. M is the set of all gY, $g \in G_C$, such that $gY \subset D$, and $\mathcal{Y}$ is the disjoint union of these gY with $\pi(gY) = \{gY\}$. More precisely, let $L = \{g \in G_C : gY = Y\}$. Then L is a complex subgroup, $K_C \subsetneq L$, and we have

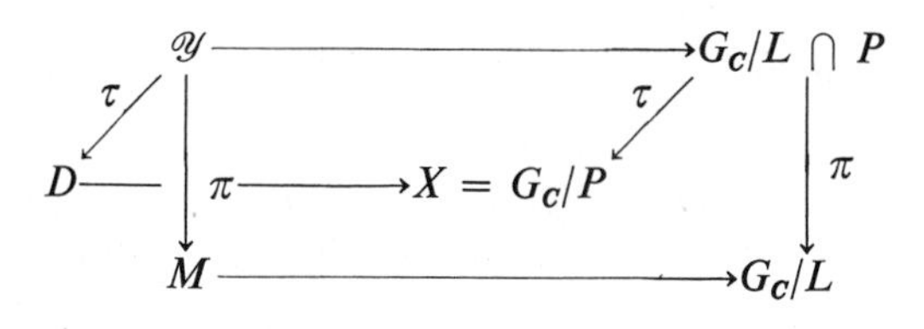

where the horizontal arrows are inclusions of open subsets. In particular $\pi: \mathcal{Y} \to M$ is a holomorphic mapping of maximal rank. We prove

THEOREM 1. *M is a Stein manifold.*

This had earlier been conjectured by Griffiths [8], and one of us [16] had checked the case $D = SO(2h, 1)/U(h)$. The principal tools in the proof are a clear understanding of the group L, Schmid's exhaustion function for D [12], the Andreotti-Norguet solution to the generalized Levi problem for analytic cycles on q-convex manifolds [1], [2], and the Docquier-Grauert exhaustion principle for Stein manifolds [7].

3. Homogeneous vector bundles. As above, $D = G(x_0) \cong G/V$ is a flag domain, $Y = K(x_0) \cong K/V$ is a maximal compact subvariety, and their dimensions are $n = \dim_C D$ and $s = \dim_C Y$.

If μ is a unitary representation of V then E_μ will denote the representation space, and $\boldsymbol{E}_\mu = G \times_\mu E_\mu \to G/V = D$ will denote the associated homogeneous hermitian C^∞ vector bundle. Any extension $\tilde{\mu}$ of μ to a holomorphic representation of P on E_μ defines a holomorphic vector bundle $\tilde{\boldsymbol{E}}_\mu \to G_C/P = X$ such that $\boldsymbol{E}_\mu = \tilde{\boldsymbol{E}}_\mu|_D$, and thus imposes a holomorphic vector bundle structure on $\boldsymbol{E}_\mu \to D$. If μ is irreducible, there is exactly one extension $\tilde{\mu}$, and so we may view $\boldsymbol{E}_\mu \to D$ as a G-homogeneous holomorphic vector bundle in a unique way.

Recall the compact Cartan subgroup H of G with $H \subset V \subset K$ and consider a positive $\mathfrak{h}_C$-root system Δ^+ on $\mathfrak{g}_C$ such that $\mathfrak{p} = \mathfrak{p}^r + \mathfrak{p}^n$ is the sum of reductive part and nilradical where, for some subset Φ of the simple roots,

$$\mathfrak{p}^r = \mathfrak{b}_C = \mathfrak{h}_C + \sum_{\langle\Phi\rangle} \mathfrak{g}_C^\beta + \mathfrak{g}_C^{-\beta} \quad \text{and} \quad \mathfrak{p}^n = \sum_{\Delta^+\setminus\langle\Phi\rangle} \mathfrak{g}_C^{-\alpha}.$$

Here $\langle\Phi\rangle = \{\alpha \in \Delta^+ : \alpha \text{ is a linear combination from } \Phi\}$ is the positive $\mathfrak{h}_C$-root system on $\mathfrak{b}_C$. In these orderings, we denote

μ_λ : irreducible representation of V with highest weight λ,
E_λ : representation space of μ_λ,
$\boldsymbol{E}_\lambda$: associated hermitian holomorphic vector bundle on D,
$\mathcal{E}_\lambda$: sheaf $\mathcal{O}(\boldsymbol{E}_\lambda)$ of germs of holomorphic sections.

If G_C is simply connected, which we may assume without loss of generality, and if $\Phi = \{\varphi_1, \cdots, \varphi_r\} \subset \{\varphi_1, \cdots, \varphi_l\} = \Psi$ is a simple system for $(\mathfrak{g}_C, \Delta^+)$, then the possibilities for λ are given by

$$\frac{2\langle\lambda, \varphi_i\rangle}{\langle\varphi_i, \varphi_i\rangle} \quad \begin{array}{l}\text{is an integer for } 1 \leqq i \leqq l \\ \text{and is } \geqq 0 \text{ for } 1 \leqq i \leqq r.\end{array}$$

Further, let Δ_K^+ denote the set of compact positive roots and Δ_S^+ the noncompact positive roots, so $\mathfrak{g} = \mathfrak{k} + \mathfrak{s}$ with

$$\mathfrak{k}_C = \mathfrak{h}_C + \sum_{\Delta_K^+} (\mathfrak{g}_C^\beta + \mathfrak{g}_C^{-\beta}) \quad \text{and} \quad \mathfrak{s}_C = \sum_{\Delta_S^+} (\mathfrak{g}_C^\gamma + \mathfrak{g}_C^{-\gamma}).$$

Finally define $2\rho = 2\rho_G = \sum_{\Delta^+} \gamma$, $2\rho_K = \sum_{\Delta_K^+} \beta$ and $2\rho_V = \sum_{\langle\Phi\rangle} \alpha$.

A homogeneous holomorphic vector bundle $E_\lambda \to D$ is *nondegenerate* if

$$\langle \lambda + \rho_K + \beta_1 + \cdots + \beta_l, \alpha \rangle > 0 \quad \text{for all } \alpha \in \langle \Phi \rangle \text{ and}$$
$$\langle \lambda + \rho_K + \beta_1 + \cdots + \beta_l, \gamma \rangle < 0 \quad \text{for all } \gamma \in \Delta_K^+ \backslash \langle \Phi \rangle$$

whenever $\{\beta_1, \cdots, \beta_l\} \subset \Delta_S^+$ are distinct. This is just what one needs to apply the Borel-Weil-Bott theorem **[6]** to conclude: The sheaf cohomology $H^q(Y, \mathcal{O}(E_\lambda \otimes \bigwedge^l N)) = 0$ for $0 \leqq q < s$ and all l, where $N \to Y$ is the holomorphic normal bundle of Y in D. Then a variation on Schmid's identity theorem **[12**, Corollary 6.5] says

Proposition 1. *If $E_\lambda \to D$ is nondegenerate then $H^q(D; \mathscr{E}_\lambda) = 0$ for $q \neq s$, and if $c \in H^s(D; \mathscr{E}_\lambda)$ with $c|_{gY} = 0$ for all $g \in G$ then $c = 0$. Further $H^s(D; \mathscr{E}_\lambda)$ is an infinite-dimensional Fréchet space on which G acts by a continuous representation.*

Recall the linear deformation space of §2. The maps $D \xleftarrow{\tau} \mathscr{Y} \xrightarrow{\pi} M$ are holomorphic, maximal rank and G-equivariant. First, that gives $F_\lambda = \tau^* E_\lambda \to \mathscr{Y}$ is a pull-back bundle. Second, it gives us $\pi_*^s \mathscr{F}_\lambda \to M$, sth direct image sheaf, where $\mathscr{F}_\lambda = \mathcal{O}(F_\lambda)$. Using the identity theorem one sees that $\tau^*: H^s(D; \mathscr{E}_\lambda) \to H^s(\mathscr{Y}; \mathscr{F}_\lambda)$ is a G-equivariant topological injection of Fréchet spaces. Since M is Stein, Cartan's Theorem B and the Grauert direct image theorem show that the edge homomorphism

$$e: H^s(\mathscr{Y}; \mathscr{F}_\lambda) \to H^0(M; \pi_*^s \mathscr{F}_\lambda)$$

of the Leray spectral sequence is a topological isomorphism. This establishes our principal representation theorem.

Theorem 2. *If $E_\lambda \to D$ is nondegenerate then $e \circ \tau^*$ is a G-equivariant topological injection*

$$\sigma: H^s(D; \mathscr{E}_\lambda) \to H^0(M; \pi_*^s \mathscr{F}_\lambda)$$

of Fréchet spaces.

We note that $\pi_*^s \mathscr{F}_\lambda \to M$ is locally free. In fact it is $\mathcal{O}(\tilde{E}_\lambda)$ where $\tilde{E}_\lambda \to M$ is the holomorphic vector bundle obtained by restriction from the G_C-homogeneous bundle $\tilde{E}'_\lambda \to G_C/L$ associated to the L-module $H^s(Y; \mathscr{E}_\lambda)$. Thus the theorem represents s-cohomology on the flag domain D by sections of a holomorphic vector bundle over the Stein manifold M.

The principal representation theorem is the exact statement of a theorem conjectured by Griffiths **[8]**, **[9]** and announced by one of us **[18]**.

4. Poincaré series. Since V is compact, the flag domain $D \cong G/V$ has a G-invariant hermitian metric, and so we can speak of the pointwise norm of differential forms with values in a hermitian vector bundle $E \to D$. That gives us the Lebesgue classes

$$\mathscr{E}_r^{p,q}(D; E) = \left\{ E\text{-valued } (p, q)\text{-forms } \varphi: \int_D \|\varphi(x)\|^r < \infty \right\}.$$

We say that a sheaf cohomology class $c \in H^q(D; \mathcal{O}(\boldsymbol{E}))$ is of Lebesgue class L_r if it has a Dolbeault representative in $\mathcal{E}_r^{0,q}(D; \boldsymbol{E})$, and $H_r^q(D; \mathcal{O}(\boldsymbol{E}))$ denotes the set of all such classes.

THEOREM 3. *Let $\boldsymbol{E}_\lambda \to D$ be nondegenerate, let $c \in H_1^s(D; \mathcal{E}_\lambda)$, and let Γ be a discrete subgroup of G. Then the Poincaré series $\theta(c) = \sum_{\gamma \in \Gamma} \gamma^* c$ converges, in the Fréchet space topology of $H^s(D; \mathcal{E}_\lambda)$, to a Γ-invariant class.*

A weaker version of this theorem was given by Griffiths in [8].

The idea is to use the principal representation theorem and reduce to

THEOREM 4. *Let $\boldsymbol{E}_\lambda \to D$ be nondegenerate, let $c \in H_1^s(D; \mathcal{E}_\lambda)$, let Γ be a discrete subgroup of G, and recall the Fréchet injection $\sigma : H^s(D; \mathcal{E}_\lambda) \to H^0(M; \pi_*^s \mathcal{F}_\lambda)$. The Poincaré series $\theta(\sigma(c)) = \sum_{\gamma \in \Gamma} \gamma^*(\sigma(c))$ converges in the Fréchet topology to a Γ-invariant section of $\pi_*^s \mathcal{F}_\lambda$.*

Theorem 4 is a variation on a result of Griffiths [8], and our proof follows the classical pattern, as amplified by Griffiths, but modified to take into account the nondegeneracy of $\boldsymbol{E}_\lambda$. The result is related to some theorems of Godement, Harish-Chandra and Borel (see [5, §9]) which are proved by methods of harmonic analysis on G. Those theorems apply to the case where

(1) c is K-finite, i.e., $\{k^*c : k \in K\}$ has finite-dimensional span, and

(2) c is $\mathfrak{Z}$-finite where $\mathfrak{Z}$ is the center of the enveloping algebra of $\mathfrak{g}_C$.

We will see below that $\mathfrak{Z}$-finiteness is not a serious restriction, but K-finiteness essentially says that c has finite Fourier series. At any rate, this gives convergence of $\theta(c)$, and also gives the result that $\theta(c)$ has a bounded Γ-invariant Dolbeault representative.

5. Square-integrable cohomology. In order to produce L_1 cohomology classes for the Poincaré series of Theorems 3 and 4, we must first digress and discuss L_2 cohomology and unitary representations.

If $\boldsymbol{E} \to D$ is a G-homogeneous hermitian holomorphic vector bundle, then one has the Kodaira-Hodge-Laplace operator $\bar{\square} = \bar\partial\bar\partial^* + \bar\partial^*\bar\partial$ on the spaces $\mathcal{E}^{p,q}(D; \boldsymbol{E})$ of smooth $\boldsymbol{E}$-valued (p, q)-forms. $\bar{\square}$ defines a selfadjoint operator $\bar{\square}$ on the Hilbert space completion of $\mathcal{E}_2^{p,q}(D; \boldsymbol{E})$, whose kernel

$$\mathcal{H}^{p,q}(D; \boldsymbol{E}) : L_2 \text{ harmonic } \boldsymbol{E}\text{-valued } (p, q)\text{-forms}$$

is a closed subspace consisting of C^∞ forms. That gives

$$\pi_\mu^q : \text{unitary representation of } G \text{ on } \mathcal{H}^{0,q}(D; \boldsymbol{E}_\mu).$$

Recently Schmid [14] settled the "Langlands conjecture," completing the identification of the π_μ^q as follows. Let $\Lambda' = \{\nu \in i\mathfrak{h}^* : e^\nu \text{ defined on } H \text{ and } \langle \nu, \alpha \rangle \neq 0 \text{ for all } \alpha \in \Delta^+\}$. Given $\nu \in \Lambda'$, Let

$$q(\nu) = |\{\alpha \in \Delta_K^+ : \langle \nu, \alpha \rangle < 0\}| + |\{\gamma \in \Delta_S^+ : \langle \nu, \gamma \rangle > 0\}|$$

and

$[\pi_\nu] = \omega(\nu)$: Harish-Chandra's discrete series representation class for G parametrized by ν (see [10]).

Then, if λ is the highest weight of μ,

(1) *if* $\lambda + \rho \notin \Lambda'$ *then every* $\mathcal{H}^{0,q}(D; \mathbf{E}_\mu) = 0$;
(2) *if* $\lambda + \rho \in \Lambda'$ *and* $q \neq q(\lambda + \rho)$ *then* $\mathcal{H}^{0,q}(D; \mathbf{E}_\mu) = 0$;
(3) *if* $\lambda + \rho \in \Lambda'$ *and* $q = q(\lambda + \rho)$ *then* $\pi^q_\lambda \in [\pi_{\lambda+\rho}]$.
One of the first consequences of this is

THEOREM 5. *If* $\mathbf{E}_\lambda \to D$ *is nondegenerate, then the natural surjective map*

$$\mathcal{H}^{0,s}(D; \mathbf{E}_\lambda) \ni \omega \mapsto (\textit{Dolbeault class}) \in H^s_2(D; \mathscr{E}_\lambda)$$

is injective. If $\lambda + \rho \in \Lambda'$ *with* $q(\lambda + \rho) = s$, *then* G *acts on the image by the discrete series representation* $[\pi_{\lambda+\rho}]$.

6. Absolutely integrable cohomology. An irreducible unitary representation π of G is *integrable* if the coefficient $f_{u,v}(g) = \langle u, \pi(g)v\rangle \in L_1(G)$ whenever u and v are K-finite. As $|f_{u,v}(g)| \leqq \|u\| \cdot \|v\|$, then $f_{u,v} \in L_2(G)$, so $[\pi]$ is in the discrete series.

We will say that the homogeneous holomorphic vector bundle $\mathbf{E}_\lambda \to D$ is L_1-*nonsingular* if $\lambda + \rho \in \Lambda'$, and $|\langle \lambda + \rho, \beta\rangle| > \frac{1}{2}\sum_{\alpha\in\Delta^+} |\langle \alpha, \beta\rangle|$ for all $\beta \in \Delta^+_S$. That is a necessary (Trombi and Varadarajan **[15]**) and sufficient (Hecht and Schmid **[11]**) condition for the discrete series class $[\pi_{\lambda+\rho}]$ to be integrable.

THEOREM 6. *Let* $\mathbf{E}_\lambda \to D$ *be nondegenerate and* L_1-*nonsingular with* $q(\lambda + \rho) = s$. *Then* G *acts on* $H^s_2(D; \mathscr{E}_\lambda)$ *by the integrable discrete series representation* $[\pi_{\lambda+\rho}]$, *and every* K-*finite class* $c \in H^s_2(D; \mathscr{E}_\lambda)$ *is absolutely integrable, i.e., is in* $H^s_1(D; \mathscr{E}_\lambda)$.

Since the K-finite elements are dense in the infinite-dimensional Hilbert space $H^s_2(D; \mathscr{E}_\lambda)$, this provides an abundance of L_1 cohomology classes that we can sum in Poincaré series to obtain automorphic cohomology.

The proof of Theorem 6 uses a direct image construction of Schmid **[12]** and follows a route suggested by him to one of us.

Fix $\mathbf{E}_\lambda \to D$ nondegenerate and denote $U_\lambda = H^s(Y; \mathscr{E}_\lambda)$ and $W_\lambda = U_\lambda \otimes \mathfrak{s}_C$. Then K acts irreducibly on U_λ with *lowest* weight $\nu = w(\lambda + \rho_K) - \rho_K$ for a certain element w of the Weyl group, and we have a K-invariant $W_\lambda = W^+_\lambda \oplus W^-_\lambda$ where $W^\pm_\lambda$ is a sum of K-modules of lowest weight $\nu \pm \beta$, $\beta \in \Delta^+_s$. Writing

$$\mathbf{U}_\lambda \to G/K, \quad \mathbf{W}^\pm_\lambda \to G/K \quad \text{and} \quad \mathbf{W} = \mathbf{W}^+ \oplus \mathbf{W}^- \to G/K$$

for the associated G-homogeneous vector bundles, we have an exact sequence of Fréchet space maps

$$0 \to H^s(D; \mathscr{E}_\lambda) \xrightarrow{\zeta} C^\infty(\mathbf{U}_\lambda) \xrightarrow{\mathscr{D}} C^\infty(\mathbf{W}^+_\lambda)$$

where $C^\infty(\,\cdot\,)$ denotes the Fréchet space of C^∞ sections viewed as a subspace of $C^\infty(G) \otimes U_\lambda$ or $C^\infty(G) \otimes W^+_\lambda$. It is given by

$$\zeta(c)\,(g) = (g^*c)|_Y \in U_\lambda$$

and

$$\mathscr{D}(F) = \text{projection}_{(W_\lambda \to W^+_\lambda)} \left\{ \sum_{\beta\in\Delta^+_s} e_\beta(F) \otimes e_{-\beta} \right\}$$

for a certain normalization of root vectors $e_\gamma \in \mathfrak{g}^\gamma_C$. In other words, the direct image map ζ is a G-equivariant Fréchet isomorphism of $H^s(D; \mathscr{E}_\lambda)$ onto the kernel $C^\infty(\mathbf{U}_\lambda)_\mathscr{D}$ of $\mathscr{D}$.

Using our knowledge of the representation of G on $H_2^s(D; \mathscr{E}_\lambda)$, one can follow square-integrability through the direct image map ζ and see that it maps $H_2^s(D; \mathscr{E}_\lambda)$ onto

$$^0L_2(U_\lambda)_{\mathscr{D}} : L_2 \text{ closure of } \{F \in C^\infty(U_\lambda)_{\mathscr{D}} : F \text{ is } L_2 \text{ and } \mathfrak{Z}\text{-finite}\}.$$

If $E_\lambda \to D$ is L_1-nonsingular with $q(\lambda + \rho) = s$, one can further see that *if* $F \in {}^0L_2(U_\lambda)_{\mathscr{D}}$ *is K-finite then* $F: G \to U_\lambda$ *is* L_1.

The identity theorem (Proposition 1) is proved by a careful examination of the order of vanishing of differential forms along the fibres of $D \to G/K$. Standard methods of harmonic analysis on semisimple Lie groups allow one to carry square-integrability through those considerations and obtain both Theorem 5 and the above characterization of $\zeta \cdot H_2^s(D; \mathscr{E}_\lambda)$. To carry absolute integrability we make use of an estimate as follows.

LEMMA. *Let* $f \in L_p(G)$ *where* $1 \leqq p \leqq 2$. *Let* ξ *belong to the universal enveloping algebra* $\mathscr{G}$ *so that both* f *and* $\xi(f)$ *are* $\mathfrak{Z}$*-finite, left K-finite and* L_2. *Then* $\xi(f) \in L_p(G)$.

Using the lemma, we obtain the L_1 version of the technique used to prove the identity theorem, and that tells us that *if* $F \in {}^0L_2(U_\lambda)_{\mathscr{D}}$ *is K-finite and* L_1 *then* $\zeta^{-1}(F) \in H_1^s(D; \mathscr{E}_\lambda)$. Theorem 6 follows.

7. Some questions. Some obvious questions come to mind at this point.

(1) Which Poincaré series $\theta(c)$ are nonzero?

(2) Is $H_p^s(D; \mathscr{E}_\lambda)$ finite dimensional, as in the classical cases?

(3) What is the dimension of the space of Poincaré series arising from a given $E_\lambda \to D$? How does that space compare with the full automorphic cohomology space $H_p^s(D; \mathscr{E}_\lambda)$?

(4) How does one obtain quasi-projective embeddings from automorphic cohomology?

(5) Can one construct meromorphic functions on $\Gamma\backslash D$ using holomorphic arc components of boundary orbits **[20]** in the way that Bailey and Borel **[3]** use boundary components **[19]**? How would such Eisenstein series be related to our Poincaré series?

REFERENCES

1. A. Andreotti and F. Norguet, *Problème de Lévi et convexité holomorphe pour les classes de cohomologie*, Ann. Scuola Norm. Sup. Pisa (3) **20** (1966), 197–241. MR **33** #7583.

2. ———, *La convexité holomorphe dans l'espace analytique des cycles d'une variété algébrique*, Ann. Scuola Norm. Sup. Pisa (3) **21** (1967), 31–82. MR **39** #477.

3. W. L. Baily, Jr. and A. Borel, *Compactification of arithmetic quotients of bounded symmetric domains*, Ann. of Math. (2) **84** (1966), 442–528. MR **35** #6870.

4. A. Borel, *Les fonctions automorphes de plusieurs variables complexes*, Bull. Soc. Math. France **80** (1952), 167–182. MR **14**, 1077.

5. ———, *Introduction to automorphic forms*, Algebraic Groups and Discontinuous Subgroups, Proc. Sympos. Pure Math., vol. 9, Amer. Math. Soc., Providence, R. I., 1966, pp. 199–210. MR **34** #7465.

6. R. Bott, *Homogeneous vector bundles*, Ann. of Math. (2) **66** (1957), 203–248. MR **19**, 681.

7. F. Docquier and H. Grauert, *Levisches Problem und Rungescher Satz für Teilgebeite Steinscher Mannigfaltigkeiten*, Math. Ann. **140** (1960), 94–123. MR **26** #6435.

8. P. A. Griffiths, *Periods of integrals on algebraic manifolds*. I. *Construction and properties of the modular varieties*, Amer. J. Math. **90** (1968), 568–626. MR **37** #5215.

9. ———, *Periods of integrals on algebraic manifolds: Summary of main results and discussion of open problems*, Bull. Amer. Math. Soc. **76** (1970), 228–296. MR **41** #3470.

10. Harish-Chandra, *Discrete series for semisimple Lie groups*. II. *Explicit determination of the characters*, Acta Math. **116** (1966), 1–111. MR **36** #2745.

11. H. Hecht and W. Schmid, *On integrable representations of a semisimple Lie group*, Math. Ann. (to appear).

12. W. Schmid, *Homogeneous complex manifolds and representations of semisimple Lie groups*, Thesis, Univ. of California at Berkeley, 1967.

13. ———, *On a conjecture of Langlands*, Ann. of Math. (2) **93** (1971), 1–42. MR **44** #4149.

14. ———, *L^2-cohomology and discrete series*, Ann. of Math. (2) **103** (1976), 373–394.

15. P. C. Trombi and V. S. Varadarajan, *Asymptotic behavior of eigenfunctions on a semisimple Lie group: the discrete spectrum*, Acta Math. **129** (1972), 237–280.

16. R. O. Wells, Jr., *Parametrizing the compact submanifolds of a period matrix domain by a Stein manifold*, Sympos. on Several Complex Variables (Park City, Utah, 1970), Lecture Notes in Math., vol. 184, Springer-Verlag, Berlin and New York, 1971, pp. 121–150. MR **46** #7555.

17. ———, *Automorphic cohomology on homogeneous complex manifolds*, Rice Univ. Studies **59** (1973), 147–155.

18. R. O. Wells, Jr. and J. A. Wolf, *Poincaré series and automorphic cohomology on flag domains*, Ann. of Math. (to appear).

19. J. A. Wolf and A. Korányi, *Generalized Cayley transformations of bounded symmetric domains*, Amer. J. Math. **87** (1965), 899–939. MR **33** #229.

20. J. A. Wolf, *The action of a real semisimple group on a complex flag manifold*. I. *Orbit structure and holomorphic arc components*, Bull. Amer. Math. Soc. **75** (1969), 1121–1237. MR **40** #4477.

RICE UNIVERSITY
UNIVERSITÄT GOTTINGEN

UNIVERSITY OF CALIFORNIA, BERKELEY
HEBREW UNIVERSITY

PRINCIPAL LECTURE

R. E. GREENE AND H. WU

ANALYSIS ON NONCOMPACT KÄHLER MANIFOLDS

Proceedings of Symposia in Pure Mathematics
Volume 30, 1977

ANALYSIS ON NONCOMPACT KÄHLER MANIFOLDS

R. E. GREENE AND H. WU*

The main concern of this article is the relationship between the Riemannian structure and the complex structure on noncompact Kähler manifolds. Restricting attention to Kähler manifolds has the effect of insuring a closer relationship between Riemannian structure and function theory than that which exists on an arbitrary Hermitian manifold. Although the present knowledge of the subject is somewhat fragmentary, its scope is quite vast. In a sense, one is here dealing with "applied several complex variables". On the one hand, one tries to apply the existing theory of complex manifolds to answer concrete geometric questions. On the other, these geometric questions in turn raise open problems in complex analysis itself. In order to give some idea of these applications and their associated problems in complex analysis, we shall address ourselves here to two specific questions:

(I) When do nonconstant holomorphic functions exist on noncompact complex manifolds?

(II) What is the function theory of Stein manifolds diffeomorphic to C^n?

These two questions are seemingly unrelated but, as we shall see, one naturally leads to the other. First, some general comments are in order: Calabi and Eckmann have shown in [2] that a complex manifold can be diffeomorphic to C^n yet not possess any nonconstant holomorphic functions. This shows that, to answer (I), one needs to go beyond the topological level. The introduction of a Kähler metric is geometrically natural. The general utility of the Kähler assumption will become apparent later; an immediate consequence of this assumption is the elimination of the Calabi-Eckmann examples from consideration, as those manifolds admit no Kähler metric. Regarding (II), one can interpret the uniformization theorem for

---

AMS (MOS) subject classifications (1970). Primary 32C10, 53C55; Secondary 32F05, 32A99.
*Work of both authors partially supported by the National Science Foundation.

Riemann surfaces as the complete answer to this question in the case of complex dimension one. In the general case, there is a long way to go before this question can be answered in any reasonable way.

Before turning to the results and problems centering around (I) and (II), we give in §1 an intuitive discussion of the relevant geometric concepts, with the intention of making this paper nearly self-contained as far as geometry is concerned. Then in §2 we give known theorems and in §3 conjectures and open questions. In §4 we outline the proofs of Theorems 1A and 1B (of §2). Finally, in §5, we discuss the proof of Theorem 1D. We hope that these proof-outlines will give an adequate idea of some of the technical aspects of the subject (cf. **[11]** in this respect). Some of the theorems and most of the geometric concepts discussed in this article are actually valid for Riemannian manifolds and harmonic functions rather than just for Kähler manifolds and holomorphic functions. We have purposely stated almost everything in terms of the latter purely for the sake of clarity.

1. Geometric preliminaries. (Cf. §1 of **[11]**). Let M be a complex manifold of (complex) dimension n. A *Hermitian metric* G on M is a C^∞ assignment to each $p \in M$ a Hermitian inner product $G(p)$ on the complex tangent space M_p^C. If local coordinates are chosen around p, G can be expressed as

$$G = \sum_{i,j} G_{i\bar{j}} dz^i \otimes d\bar{z}^j,$$

where $\{G_{i\bar{j}}(p)\}$ is a positive definite Hermitian matrix for each p. G is called a *Kähler metric* (and M is then called a *Kähler manifold*) iff the associated 2-form

$$\Omega \equiv \sum_{i,j} G_{i\bar{j}} dz^i \wedge d\bar{z}^j$$

is closed, i.e., $d\Omega = 0$. The form Ω and hence this condition can be checked to be independent of the choice of local coordinates. The most important examples of Kähler metrics are the *flat metric* $\sum_{i=1}^n dz^i \otimes d\bar{z}^i$ on C^n, and the *Fubini-Study metric* on complex projective space P_nC. The latter can be described in one of two ways:

(1) $P_nC = U(n+1)/U(n) \times U(1)$ where $U(l)$ denotes the unitary group in l-variables. To define a Hermitian metric on P_nC, it suffices to pick a $p \in P_nC$ and to pick a Hermitian inner product $G(p)$ on the complex tangent space at p which is invariant under the isotropy action of $U(n) \times U(1)$; translation of $G(p)$ to other points of P_nC by $U(n+1)$ then defines the Hermitian metric G completely. Since C^n sits in P_nC as an open set, we simply let p be the origin of C^n and let $G(p)$ be the usual inner product of C^n.

(2) If homogeneous coordinates $\{z^0, \cdots, z^n\}$ are used for P_nC, then we define a Hermitian metric G by:

$$G = \left(\sum_i z^i\bar{z}^i\right)^{-2}\left\{\left(\sum_i z^i\bar{z}^i\right)\left(\sum_i dz^i \otimes d\bar{z}^i\right) - \left(\sum_i \bar{z}^i dz^i\right) \otimes \left(\sum_i z^i d\bar{z}^i\right)\right\}.$$

One must work a little to verify that the two definitions coincide and that G is in fact a Kähler metric. This is the Fubini-Study metric.

Kähler manifolds are *hereditary* in the sense that a submanifold N of a Kähler manifold M is itself a Kähler manifold. Indeed, if G is the Kähler metric on M, then the restriction of G to N is a Kähler metric on N. Since Stein manifolds may be regarded as submanifolds on $\boldsymbol{C}^n$, all Stein manifolds are Kähler manifolds. One may anticipate what is to follow by immediately asking which Kähler manifolds are also Stein manifolds. Partial answers are given in Theorem 1 of §2, but a full characterization is unknown.

Algebraic varieties are compact complex submanifolds of $P_n C$, so all smooth algebraic varieties are Kähler manifolds. This fact is of course of decisive importance in the transcendental theory of algebraic geometry. By and large, we shall avoid compact Kähler manifolds in this article because they do not carry any nonconstant holomorphic functions (recall question (I)).

We proceed to discuss some Riemannian geometry. A *Riemannian metric* on a real manifold N is just a C^∞ assignment to each point of N a real inner product. Now the real part of a Hermitian inner product on a complex vector space is a real inner product on the underlying real vector space. Thus, if G is a Kähler metric on M, Re G (= the real part of G) defines a Riemannian metric on the underlying real manifold M. One can therefore start doing Riemannian geometry on Kähler manifolds. For the understanding of the theorems and conjectures in this article, there is no need to worry about the technical fine points of Riemannian geometry. However, an informal discussion of the concept of sectional curvature would be helpful. This we now provide.

First we must define geodesics. On the Riemannian manifold M, let us agree once and for all to look only at curves with tangent vectors of constant length. Let I be an interval in R and let $\gamma: I \to M$ be a curve. Denoting the length of γ by $L(\gamma)$, we define for two points $p, q \in M$ the distance $d(p, q) \equiv \inf L(\gamma)$, where the infimum is taken over all γ joining p to q. It is a nontrivial fact that (M, d) then becomes a metric space. M is called a *complete Riemannian manifold* (or the Riemannian metric is said to be *complete*) iff (M, d) is a complete metric space. A curve $\gamma: I \to M$ is called a *geodesic* iff it has the property that for any two points p, q on γ sufficiently close together, $d(p, q)$ equals the length of that portion of γ joining p to q. In other words, geodesics are curves which locally minimize length. Geodesics are always C^∞ curves. On $\boldsymbol{R}^2$ with the usual flat metric, the geodesics are straight line segments. On a sphere of radius a around the origin in $\boldsymbol{R}^3$, again with the usual metric, the geodesics are the great circles, i.e., the intersections of the sphere with planes through the origin, or equivalently, the circles of radius a. Geodesics enjoy two properties: (a) Given $p \in M$ and $v \in M_p$ (= (real) tangent space to M at p), there is a unique geodesic passing through p and tangent to v, i.e., having tangent vector at p equal to v. (b) If V is a k-dimensional linear subspace of M_p, then the set of all geodesics through p tangent to (some vector of) V forms a k-dimensional submanifold of M in a neighborhood of p.

Now, to define sectional curvature: It is a function K which assigns to each 2-dimensional subspace of each tangent space of M a real number. Thus $K: G_2(M) \to \boldsymbol{R}$, where $G_2(M)$ is the Grassmann bundle of 2-planes in M. If $\dim_{\boldsymbol{R}} M = 2$, then K is just a real-valued function on M; in this case, K is called either the *Gaussian curvature* or more simply, the *curvature*. Gaussian curvature is easier to describe, so we shall define it first.

Let N be a 2-dimensional Riemannian manifold and let $q \in N$. By the *r-disc around* q, we mean the set of points in N of distance at most r from q. Fixing $p \in N$, we shall define $K(p)$ (the curvature of N at p) precisely. Let r be a small positive number and let $A(r)$ be the area of the r-disc around p. In $\boldsymbol{R}^2$, let $A_0(r)$ be the area of the r-disc around 0; thus $A_0(r) = \pi r^2$. By definition,

$$K(p) \equiv \lim_{r \to 0} 12 \frac{A_0(r) - A(r)}{r^2 A_0(r)}.$$

For our purpose, the precise formula is hardly as important as the simple observation that

$$K(p) \geqq 0, \quad \text{if } A_0(r) \geqq A(r),$$
$$K(p) \leqq 0, \quad \text{if } A_0(r) \leqq A(r),$$

for all small $r > 0$. Thus, taking $\boldsymbol{R}^2$ as our model space, the curvature of N is nothing but a comparison between the area of the small r-discs of N and that of the model. For instance, $\boldsymbol{R}^2$ has zero curvature. To elucidate this further, note that $A_0(r) \geqq A(r)$ for all small $r > 0$ iff the geodesics issuing from p are spreading apart from each other at a rate not greater than that of the straight lines through the origin in $\boldsymbol{R}^2$. There is a corresponding interpretation for $A_0(r) \leqq A(r)$. Using the observation above, we can say that curvature $\geqq 0$ (resp. $\leqq 0$) roughly means that geodesics through a point are crowding together (resp. diverging from each other) compared to geodesics, i.e., straight lines, in $\boldsymbol{R}^2$.

Among surfaces in $\boldsymbol{R}^3$, the sphere of radius a has constant positive curvature $1/a^2$; this can be computed using the above formula for K and elementary calculus. The surfaces of constant negative curvature are not so easily described, but the hyperboloid and the catenoid (the surface obtained by revolving the graph of $y = \cosh x$ around the x-axis) may be offered as generic examples of surfaces of negative curvature. Note that the latter surfaces look like a saddle at each of its points, and they are easily seen to possess the property that $A(r) > A_0(r)$.

We can now define the sectional curvature K of a general Riemannian manifold: Fix a 2-dimensional linear subspace $P \subset M_p$ for some $p \in M$. Let N be the 2-dimensional Riemannian manifold in a neighborhood of p formed by the geodesics through p tangent to P. By definition, the *sectional curvature of* $P = K(P) =$ the Gaussian curvature of N at p. (For the equivalence of this definition of sectional curvature and the more usual—but less intuitively meaningful—definition in terms of the Riemannian curvature tensor, see [**18**, p. 64 ff.] for instance.)

It follows from the definition of sectional curvature that the interpretation of Gaussian curvature in terms of the rate of divergence of geodesics compared to the rate of divergence of straight lines in $\boldsymbol{R}^2$ holds as an interpretation of sectional curvature in general also. A specific example of this interpretation is the following situation: Let M be a complete simply connected Riemannian manifold with nonpositive sectional curvature, i.e., $K(P) \leqq 0$ for all P. Let C_1 and C_2 be geodesics through a point $p \in M$. Let p_1 be the point obtained by going in the positive direction along C_1 arc-length r_1 from p, and p_2 be the point obtained by going in the positive direction along C_2 arc-length r_2 from p. (The completeness of M implies that p_1 and p_2 are defined for all nonnegative r_1 and r_2.) Then

$$d(p_1, p_2) \geqq [r_1^2 + r_2^2 - 2r_1 r_2 \cos \alpha]^{1/2}$$

where $\alpha =$ the angle between the positive directions of C_1 and C_2 at p. In other words, $d(p_1, p_2)$ is at least as great as it would be if M were $\boldsymbol{R}^n$. This result, which is a classical theorem of Riemannian geometry (see, e.g., [**18**, p. 73] or [**24**, p. 103]) will be used in §4 in the proof of Theorem 1A of §2.

To round off the picture, we return to Kähler manifolds and adopt the convention that a Kähler manifold (M, G) is complete, has positive sectional curvature, etc., iff the associated Riemannian manifold $(M, \operatorname{Re} G)$ has the corresponding property. In this language, the *Poincaré metric*

$$(4/(1 - |z|^2)^2)\, dz \otimes d\bar{z}$$

on the unit disc $\Delta \equiv \{|z| < 1\} \subset \boldsymbol{C}$ converts Δ into a complete Kähler manifold of constant negative curvature -1. This metric plays an important role in geometric function theory.

Having gone through the definition of sectional curvature, we must reluctantly point out that professional geometers employ four other related curvature functions on a Kähler manifold: (a) holomorphic bisectional curvature, (b) holomorphic sectional curvature, (c) Ricci curvature, (d) scalar curvature. We shall use primarily the first and the third in this article, and in addition a fleeting acquaintance with them would suffice. The most important fact to keep in mind is that both the holomorphic bisectional curvature and the Ricci curvature are merely functions obtained from the sectional curvature by some combinations. Using the notation $>$ to denote "contains strictly more information than," we can state: sectional curvature $>$ holomorphic bisectional curvature $>$ Ricci curvature. For example, on a Kähler manifold M:

M has positive (nonnegative, etc.) sectional curvature $\Rightarrow$

M has positive (nonnegative, etc.) holomorphic bisectional curvature $\Rightarrow$

M has positive (nonnegative, etc.) Ricci curvature.

The reverse implications are false in general. Incidentally, in the language of Griffiths [**17**], the Kähler manifold having positive holomorphic bisectional curvature implies that the holomorphic tangent bundle of M is positive.

Without attempting to maintain a self-contained exposition at this point, we give for completeness the technical definitions of the curvature functions (a), (b), and (c) (the scalar curvature (d) will not be used in this paper). Fix a point $p \in M$ and let $\mathscr{C}$ denote the set of 1-dimensional complex subspaces of the tangent space M_p. Note that if J denotes the structure tensor of M at p, $\mathscr{C}$ is just the set of all 2-dimensional real subspaces of M_p which are invariant under J. Let K be the curvature function as before. The *holomorphic sectional curvature* function is just the restriction of K to $\mathscr{C}$. The *holomorphic bisectional curvature* is the function $H : \mathscr{C} \times \mathscr{C} \to \boldsymbol{R}$ such that if $h, h' \in \mathscr{C}$,

$$H(h, h') = K(\operatorname{span}\{a, a'\}) + K(\operatorname{span}\{a, Ja'\}),$$

where $0 \neq a \in h$, $0 \neq a' \in h'$, and the definition can be shown to be independent of the choices of a and a'. Next, let S be the unit sphere in M_p. Then the *Ricci curvature* is the function $\operatorname{Ric} : S \to \boldsymbol{R}$ such that: if $e_1 \in S$, and $\{e_1, Je_1, \cdots, e_n, Je_n\}$ is an orthonormal basis of M_p, then

$$\operatorname{Ric}(e_1) = \sum_{i=1}^{n} H(h_1, h_i),$$

where $h_i = \operatorname{span}\{e_i, Je_i\}$, $\forall\, i$.

Complex differential geometry is presently suffering from the lack of a unified and systematic exposition. This is all the more unfortunate because not only is the formalism formidable, but there are actually two distinct approaches to the subject which differ in numerical factors and conceptual outlooks; the end results are always qualitatively the same of course. At the moment, we can only offer as references: Kobayashi and Nomizu [24] and Kobayashi [22] for the "real approach," and Griffiths [17], Wells [31] and Wu [32] for the "complex approach."

2. Known results. Our first concern is to produce Kähler manifolds that will carry nonconstant holomorphic functions; see question (I) of the opening paragraph. The assumptions to be made are strong, but they are justified by the strong conclusions.

THEOREM 1 ([9], [14], [15]). *Let M be a complete Kähler manifold. Then M is a Stein manifold if any one of the following holds:*

(A) *M is simply connected and the sectional curvature $\leqq 0$.*

(B) *M is noncompact, the sectional curvature $\geqq 0$ and moreover > 0 outside a compact set.*

(C) *M is noncompact, the sectional curvature $\geqq 0$, and the holomorphic bisectional curvature > 0.*

(D) *M is noncompact, the Ricci curvature > 0, the sectional curvature $\geqq 0$, and the canonical bundle is trivial.*

It may illuminate the various assumptions of this theorem to consider some counterexamples to the statements obtained by weakening the assumptions in various ways. Here are some complete Kähler manifolds which are not Stein manifolds. The numbering corresponds to that of the preceding theorem.

(A) $P_nC \times C^k$, $n > 0$, $k \geqq 0$ with the product metric of the Fubini-Study metric on P_nC and the standard metric on C^k. $\pi_1(P_nC \times C^k) = 0$ and the metric is complete but the sectional curvature is not $\leqq 0$ everywhere. $S \times C^k$, $k \geqq 0$, S compact Riemann surface of genus $\geqq 1$ with a Hermitian (hence Kähler) metric[1] of Gaussian curvature $\leqq 0$, with the product metric of the preceding metric on S and the standard metric on C^k. Then sectional curvature $\leqq 0$ on $S \times C^k$ and metric is complete, but $\pi_1 \neq 0$.

(B) $P_nC \times C^k$, $n > 0$, $k > 0$, with the product of the Fubini-Study metric on P_nC and the standard metric on C^k. The sectional curvature $\geqq 0$ on $P_nC \times C^k$. But for any point $p \in P_nC \times C^k$ there is a 2-plane $P \in (P_nC \times C^k)_p$ with sectional curvature of $P = 0$; namely, take $P =$ any 2-plane spanned by a vector tangent to P_nC and a vector tangent to C^k.

(C) The example listed for (B). Sectional curvature $\geqq 0$. But for any point p there exists a pair of (J-invariant) 2-planes $h, h' \in \mathscr{C}$ such that the holomorphic bisectional curvature $H(h, h') = 0$; namely, take any $h \in \mathscr{C}$ tangent to P_nC and any $h' \in \mathscr{C}$ tangent to C^k.

[1] Such metrics always exist. If genus $S = 1$, then $S = C/$(a group of translations) so S inherits from the metric of C a metric of 0 curvature. If genus $S \geqq 2$, then $S =$ the unit disc $D/$(a group of linear fractional transformations). Linear fractional transformations which preserve the unit disc act as isometries relative to the Poincaré metric. Thus S inherits a metric of constant negative curvature from the Poincaré metric on D.

(D) $S \times \boldsymbol{C}^k$, $k > 0$, S a Riemann surface of genus 1 with zero curvature metric, $S \times \boldsymbol{C}^k$ with the product of the metric on S and the standard metric on $\boldsymbol{C}^k$. Then $S \times \boldsymbol{C}^k$ has sectional curvature $\geqq 0$ (actually $\equiv 0$) and has a topologically (even an analytically) trivial canonical bundle, but the Ricci curvature is $\equiv 0$, not > 0.

The construction of an appropriate counterexample for (D) with the canonical bundle restriction removed is somewhat more subtle than the examples in the previous list. For this construction, we first show how to put a complete Hermitian (and hence Kähler) metric of positive curvature on $\boldsymbol{C}$: There are certainly complete Riemannian metrics on $\boldsymbol{R}^2$ with positive curvature; for instance, the metric induced by embedding $\boldsymbol{R}^2$ as a paraboloid of revolution $(x, y) \to (x, y, x^2 + y^2)$ is one such metric. By a standard result of Riemann surface theory, there is a complex structure on $\boldsymbol{R}^2$ relative to which this metric is (the real part of) a Hermitian metric. A theorem of Blanc and Fialla (generalized by Huber **[20]**) shows that the resulting Riemann surface is (biholomorphic to) $\boldsymbol{C}$; Huber's general theorem is that an open Riemann surface is parabolic if it admits a complete Hermitian metric with the property that there is a (negative) constant A such that $\int_\Omega$ Gaussian curvature $\geqq A$ for any region Ω with compact closure. If $P_n\boldsymbol{C} \times \boldsymbol{C}$ is given the product of the Fubini-Study metric on $P_n\boldsymbol{C}$ and the complete positive curvature Hermitian metric on $\boldsymbol{C}$ just indicated, then $P_n\boldsymbol{C} \times \boldsymbol{C}$ has sectional curvature $\geqq 0$ and Ricci curvature > 0. But the canonical bundle K of $P_n\boldsymbol{C} \times \boldsymbol{C}$ is of course not topologically trivial. (To check this nontriviality directly note that, if $i : P_n\boldsymbol{C} \to P_n\boldsymbol{C} \times \boldsymbol{C}$ is the map $i(p) = p \times 0$ then $i^*K_{P_n\boldsymbol{C}\times\boldsymbol{C}} \cong K_{P_n\boldsymbol{C}}$ so that topological triviality of $K_{P_n\boldsymbol{C}\times\boldsymbol{C}}$ would imply topological triviality of $K_{P_n\boldsymbol{C}}$, contradicting $c_1(K_{P_n\boldsymbol{C}}) \neq 0$.)

The role of completeness in Theorem 1 is an essential one, for if M is a complete Kähler manifold with $\dim_{\boldsymbol{C}} M \geqq 2$ satisfying one of (A)—(D), then $M - \{p\}$ for any $p \in M$ also satisfies the hypotheses of Theorem 1 excluding completeness, but $M - \{p\}$ cannot be a Stein manifold. In fact, if M is any complex manifold with $\dim_{\boldsymbol{C}} M \geqq 2$ and $p \in M$, then $M - \{p\}$ is not holomorphically convex because of Hartogs' phenomenon. A particularly conspicuous example of this situation is $P_n\boldsymbol{C} - \{p\}$, $p \in P_n\boldsymbol{C}$, $n \geqq 2$; the extension principle of Hartogs implies that every holomorphic function on $P_n\boldsymbol{C} - \{p\}$ extends holomorphically to $P_n\boldsymbol{C}$ and hence is constant, even though $P_n\boldsymbol{C} - \{p\}$ has a well-behaved Kähler metric of positive sectional curvature, albeit not a complete one. Similar observations apply to the role of completeness in later theorems and there are usually obvious specific examples showing this necessity of assuming completeness. A second consideration, which is less specific but is still compelling in practice, is that almost all the results of global differential geometry hold only with completeness assumptions. Thus to search for theorems with geometric content but no completeness assumptions is likely to prove futile.

The assumption that the metric be a Kähler, not just a Hermitian, one was justified philosophically in the introduction. But its specific justification by counterexamples is not easy. Klembeck has recently shown by an example in **[21]** that a (non-Kähler) complete Hermitian manifold satisfying (A) of Theorem 1 need not be a Stein manifold. But similar examples showing the necessity of the Kähler condition in (B), (C), and (D) are not known.

Although (B)—(D) look superficially similar, their proofs call for widely divergent techniques. The proof of (B) (and of (A)) is outlined in §4; it uses primarily geometric techniques. The proof of (D) requires, in addition to the geometric results, the L^2 method of Kohn and Hörmander; the proof of (D) is discussed in §5 (cf. [**11**]). The proof of (C) is based on a recent result of G. Elencwajg [**6**]. Recall that a complex manifold X is called 0-*convex* iff there is a C^∞ exhaustion function ϕ on X (i.e., a proper function $\phi : X \to [0, +\infty)$) which is strictly plurisubharmonic outside a compact set. It is known that X is 0-convex iff X is obtained from a Stein space by blowing up a finite number of points (Andreotti and Grauert [**1**], Narasimhan [**27**]). A domain D in X is called *locally pseudoconvex* iff $\forall\, x \in \partial D$, there exists a neighborhood U_x of x in X such that $D \cap U_x$ is a Stein manifold. The theorem alluded to is the following.

THEOREM OF ELENCWAJG ([**6**]). *Let X be a Kähler manifold with positive holomorphic bisectional curvature and let $D \subset\subset X$ be locally pseudoconvex. Then D is 0-convex.*

To return to Theorem 1, we note that the manifolds satisfying (A) or (B) are diffeomorphic to C^n. For (A) this is the classical theorem of Cartan-Hadamard (see Kobayashi and Nomizu [**24**] for instance), and for (B) this is the recent result of Cheeger and Gromoll [**3**]. From a geometric standpoint then, one sees immediately how question (I) naturally leads to question (II) posed at the beginning of this article.

Once nonconstant holomorphic functions are known to exist, it would be desirable to ask for some with special analytic properties. The most obvious such property is membership in L^p $(1 \leqq p \leqq \infty)$. The following series of theorems give an indication of what is presently known.

THEOREM 2 ([**9**], [**12**]). (A) *Let M be an n-dimensional, complete, simply connected Kähler manifold with nonpositive sectional curvature. Let f be a holomorphic function on M which is not identically zero, and let B_r denote the points of distance $\leqq r$ from a fixed point $p_0 \in M$. Then there exists a constant $\alpha > 0$ which is independent of r, such that $\forall\, p \geqq 1$,*

$$\int_{B_r} |f|^p \geqq \alpha r^{2n}.$$

(B) *Let M be a noncompact complete Kähler manifold with nonnegative sectional curvature. Then there is no holomorphic q-form in L^p for $q \geqq 0$, $1 \leqq p < \infty$, except the zero form.*

There are more precise estimates of the integral in case (A) (cf. [**9**]), and there are stronger versions of (B) in case $q = 0$, i.e., for holomorphic functions (see [**12**]). The following related result of S.-T. Yau also excludes holomorphic functions from L^p $(p < \infty)$; it is more general than Theorem 2 because it makes no curvature assumptions and allows p to be less than 1, but it is also more restrictive because it does not yield information either about q-forms for $q > 0$, or about constants, or about the optimal rate of growth of the relevant integrals.

THEOREM 3 ([**34**]). *Let M be a noncompact complete Kähler manifold. Then there are no nonconstant holomorphic functions in L^p for $0 < p < \infty$.*

The case of L^∞ is much more delicate. At the moment, all theorems in this direction are nonexistence theorems. For example, an immediate consequence of

Theorem 3 is that on a complete Kähler manifold with finite volume, there are no bounded holomorphic functions except constants. These assumptions are satisfied by manifolds of the type D/Γ, where D is a symmetric bounded domain equipped with the usual metric and Γ is an arithmetic subgroup. Of course, in this case the absence of nonconstant bounded holomorphic functions (in fact, of any nonconstant holomorphic functions) also follows from compactification theory. But complete manifolds generally tend to have infinite volume, and in such cases a lot more work is needed to obtain information about L^∞. What is known is summarized in the following two theorems.

THEOREM 4 (YAU **[35]**, **[36]**). *Let M be a complete Kähler manifold with Ricci curvature $\geqq 0$. Then the only holomorphic functions in L^∞ are constants.*

THEOREM 5 (**[9]**, **[14]**). *Let M be a complete, simply connected Kähler manifold with nonpositive sectional curvature. Suppose for some $p_0 \in M$ and for all p outside some compact set, there exists a constant $(-A) < 0$ such that sectional curvature$(p) \geqq -A/\rho^{2+\varepsilon}$, where ε is some positive number and ρ is the distance of p to p_0. Then there are no bounded holomorphic functions on M except constants.*

The preceding inequality is to be interpreted as being valid for the sectional curvature of any 2-plane at p; this convention will be in force in all subsequent inequalities of a similar nature. Both theorems are basically global studies of solutions of elliptic operators, but their proofs are entirely different in character. The assumptions in Theorem 4 are self-explanatory, but those of Theorem 5 deserve some comment. Recall that for $\mathbf{C}^n$ equipped with the standard (flat) metric, the sectional curvature is zero. Theorem 5 is therefore a generalization of Liouville's theorem that a bounded holomorphic function on $\mathbf{C}^n$ must be a constant. The inequality in Theorem 5 is necessary for the validity of the conclusion: we have seen that the Poincaré metric on the unit disc Δ is a complete Kähler metric of nonpositive curvature, in fact of curvature -1. Δ equipped with this metric satisfies every assumption of Theorem 5 except the inequality, which is fortunately not satisfied, because Δ carries many bounded nonconstant holomorphic functions. We can go further. For n an integer $\geqq 3$, the following Kähler metrics on Δ are related to the Poincaré metric:

$$G_n = (4/(1 - |z|^2)^n)\, dz \otimes d\bar{z}.$$

These metrics are all complete and they all have the property that outside some compact set containing 0,

$$\text{curvature}(p) \leqq -A_n/\rho^2$$

where $(-A_n)$ is some negative constant depending on n, and ρ is the distance of p from 0 with respect to G_n. Moreover, G_n has negative curvature everywhere. This shows two things: (a) The exponent $2 + \varepsilon$ in Theorem 5 is best possible. (b) One should study Kähler manifolds satisfying the condition:

$$\text{sectional curvature}(p) \leqq -A/\rho^2$$

outside a compact set. Before we can state the known result in this case, we must briefly define the Kobayashi metric on a complex manifold (see Kobayashi **[22]**).

Let us denote the distance function associated with the Poincaré metric on the unit disc Δ by σ. An elementary integration shows that $\forall\, z \in \Delta$,

$$\sigma(0,z) = \frac{1}{2} \log \frac{1 + |z|}{1 - |z|}.$$

Now, given a complex manifold M and two points $p, q \in M$, a nonnegative number $\tilde{k}(p, q)$ can be defined by $\tilde{k}(p, q) = \inf \sigma(0, z)$ where the infimum is taken over all $z \in \Delta$ with the property: There is a holomorphic mapping $f : \Delta \to M$ such that $f(0) = p$, $f(q) = z$. For arbitrary $p, q \in M$, consider sequences of points $p_0 = p$, $p_1, \cdots, p_{l+1} = q$. The *Kobayashi metric* k is then by definition the function $k : M \times M \to [0, \infty)$ such that

$$k(p, q) \equiv \inf \sum_{i=0}^{l} \tilde{k}(p_i, p_{i+1}),$$

where the infimum is taken over all such finite sequences $\{p_0, \cdots, p_{l+1}\}$, l arbitrary. In general k is a metric in the sense of general topology except $k(p, q)$ may be zero even when $p \neq q$. For instance, the Kobayashi metric of $\boldsymbol{C}^n$ identically vanishes, as it is easy to see. If k is an honest metric (i.e., if $k(p, q) = 0$ implies that $p = q$), then M is called a *hyperbolic manifold*, and if k is a complete metric M is called a *complete hyperbolic manifold.* Examples of (complete) hyperbolic manifolds are (symmetric) bounded domains in $\boldsymbol{C}^n$. With these definitions, we can now give the complement to Theorem 5.

THEOREM 6. *Let M be a complete simply connected Kähler manifold with nonpositive sectional curvature. Suppose that with respect to some point $p_0 \in M$, there exists a negative constant $(-A)$ such that*

$$\textit{sectional curvature}(p) \leqq -A/\rho^2$$

for all p outside some compact set, where ρ denotes the distance of p to p_0. Then M is a complete hyperbolic manifold.

(We note in passing a related result: the following two inequalities on the holomorphic sectional curvature of a Hermitian manifold are sufficient to guarantee hyperbolicity:

(i) holomorphic sectional curvature < 0,

(ii) holomorphic sectional curvature$(p) \leqq -A/\rho^2$ outside a compact set.)

In the case $\dim M = 1$, Theorem 6 was first proved by Paul Yang in his thesis **[33]** using the concept of extremal length; of course the theorem then reads that M is the disc (uniformization theorem). The proof of the general case of Theorem 6 makes crucial use of this result of Yang.

There is one more result related to Theorem 6. We have already implicitly pointed out that, in terms of the Kobayashi metric, $\boldsymbol{C}^n$ is the antithesis of a hyperbolic manifold. With this in mind, we have:

THEOREM 7 (**[9]**). *A Kähler metric with nonpositive Ricci curvature on $\boldsymbol{C}^n$ must satisfy*

$$\limsup_{|z|\to\infty} |z|^2 \cdot (\textit{Ricci curvature}(z)) > -\infty,$$

where $|z|^2 \equiv \sum_{i=1}^n z^i \bar{z}^i$.

The exponent 2 in this theorem is the best possible. Moreover, the theorem is even valid for pseudo-Hermitian metrics; see [**9**, II] for a definition of the latter.

We emphasize that in Theorems 5, 6, and 7 the main weight of the hypotheses other than nonpositivity of the curvature lies in its rate of decay "near infinity." This suggests the philosophical outlook that, for simply connected complete Kähler manifolds of nonpositive sectional curvature, the behavior of the manifold near infinity alone should suffice to determine most of the function theory. This is the guiding spirit in the formulation of most of the conjectures in the next section. On the other hand, this philosophy is certainly not valid for general Stein manifolds. Indeed, the famous Myrberg example in Riemann surface theory (cf. Nakai and Sario [**26**, p. 93]) shows that two open Riemann surfaces (thus 1-dimensional Stein manifolds) can be diffeomorphic and biholomorphic outside respective compact sets, yet nonconstant bounded holomorphic functions can exist on one but not on the other.

3. Open questions. With the theorems of the preceding section as background, we are now in a position to pose some unsolved problems. This particular choice of problems has been dictated by the criteria of simplicity of statement and relative accessibility to analysts. Technical geometric problems have been entirely left out.

Let us therefore consider two different sets of assumptions on a complete Kähler manifold M:

(a) M is noncompact, has nonnegative sectional curvature, and has positive sectional curvature outside a compact set.

(b) M is simply connected and has nonpositive sectional curvature.

It has already been remarked that, in both cases, M is topologically diffeomorphic to C^n and complex analytically a Stein manifold (Theorem 1). We are therefore considering question (II) posed at the beginning of this article.

CONJECTURE 1. *In case* (a), *M is biholomorphic to C^n.*

It may be more enlightening to break this into two subconjectures:

CONJECTURE 1a. *In case* (a), *the Kobayashi metric identically vanishes.*

CONJECTURE 1b. *If a Stein manifold is diffeomorphic to C^n and has identically vanishing Kobayashi metric, then it is biholomorphic to C^n.*

A question related to Conjecture 1a was also raised by Kobayashi in his excellent recent survey article [**23**, Problem C1]. Conjecture 1b is purely a problem in complex analysis; implicit in this conjecture is the belief that the complex structure of C^n admits no nontrivial deformation. It would be interesting to verify it just for domains of holomorphy in C^n. Although neither conjecture seems accessible at present, 1b may well turn out to be the easier of the two. Now we turn to case (b). Fix a $p_0 \in M$ and as usual let ρ denote the distance from a point p to p_0.

CONJECTURE 2. *In case* (b), *assume further that outside a compact set, there is an* $\varepsilon < 0$ *such that*

$$\textit{sectional curvature}(p) \geqq -A/\rho^{2+\varepsilon}$$

for some constant $(-A) < 0$. *Then* M *is biholomorphic to* $\mathbf{C}^n$.

This conjecture is inspired by Theorem 5. Again, it seems to us advantageous to try first to prove that the Kobayashi metric vanishes identically in this case, and then try to resolve Conjecture 1b. To the extent that Theorem 5 complements Theorem 6, this conjecture complements the following.

CONJECTURE 3. *In case* (b), *assume further that outside a compact set*

$$\textit{sectional curvature}(p) \leqq -A/\rho^2$$

for some negative constant $-A$. *Then* M *is a Stein manifold relative to bounded holomorphic functions, i.e., bounded holomorphic functions separate points, give local coordinates, and make* M *convex.* (*Here by convexity of* M *relative to bounded holomorphic functions is meant that for every compact set* K *in* M *the set* $\{p \in M \mid |f(p)| \leqq \sup f$ *for every bounded holomorphic function* $f: M \to \mathbf{C}\}$ *should be compact.*)

In view of Theorem 6, it suffices to resolve:

CONJECTURE 3′. *If a Stein manifold is diffeomorphic to* $\mathbf{C}^n$ *and is complete hyperbolic, then* M *is a Stein manifold relative to bounded holomorphic functions.*

If this is true, then in particular bounded holomorphic functions separate points. If the bounded holomorphic functions on a complex manifold N separate points, N is called *Caratheodory hyperbolic* (Kobayashi **[22]**, **[23]**). It is easy to see that Caratheodory hyperbolicity implies hyperbolicity (cf. Kobayashi **[22]**). Conjecture 3′ then naturally leads to a related conjecture:

CONJECTURE 3″. *For Stein manifolds diffeomorphic to* $\mathbf{C}^n$, *hyperbolicity and Caratheodory hyperbolicity are equivalent.*

The preceding three conjectures demand many bounded holomorphic functions by imposing rather stringent assumptions on the curvature. Were we content to have just one nonconstant bounded holomorphic function, much weaker assumptions should be sufficient:

CONJECTURE 4. *In case* (b), *assume that*

$$\textit{Ricci curvature}(p) \leqq -A/(1 + \rho^2)$$

holds everywhere on M, *where* $(-A)$ *is some negative constant. Then* M *carries nonconstant bounded holomorphic functions.*

We have shown that Hermitian manifolds satisfying just the Ricci curvature hypothesis of Conjecture 4 are measure hyperbolic (see Kobayashi **[22]** for the definition of measure hyperbolicity); it is not necessary for the result to assume sectional curvature $\leqq 0$ or that the metric be Kähler.

Instead of proliferating conjectures in the same vein, we shall content ourselves with pointing out that as soon as the existence of nonconstant bounded holomorphic functions has been established, one can immediately ask questions about the compactifications of such manifolds, the boundary behavior of bounded holomorphic functions (Fatou's theorem), etc. For manifolds satisfying the assumptions of Conjecture 3, the Eberlein-O'Neill compactification [5] seems to be a reasonable candidate, but nothing is known at the moment. Needless to say, the ultimate object of studying bounded holomorphic functions on abstract manifolds is to answer

Question 5. Are the manifolds of Conjecture 3 biholomorphic to a bounded domain in C^n?

There are other equally basic problems of a slightly different nature, so we shall round off this section by mentioning three explicitly. We began this article by asking when nonconstant holomorphic functions can exist on Kähler manifolds, and we provided some answers by making fairly drastic assumptions (Theorem 1). To a large extent, this failure to ease up on the curvature assumptions is a reflection on the lack of good theorems in complex analysis which can guarantee the existence of one nonconstant holomorphic function on a manifold without going to the extreme of making it a Stein manifold. An interesting example in this connection has been given by Grauert [8]: there exists a (noncompact) complex manifold which is not holomorphically convex but which has an exhaustion function which is strictly plurisubharmonic except on a subvariety of codimension 2. This example suggests that general characterization of manifolds which are holomorphically convex without being Stein manifolds is probably a difficult problem. But in a suitable geometric setting the problem may be more easily approachable. In the context of the present article, we specifically pose the following problem.

Question 6. If M is a complete noncompact Kähler manifold with nonnegative sectional curvature and positive Ricci curvature, is M holomorphically convex?

For integers $k \geqq 1$, $l \geqq 0$, the manifolds $C^k \times P_l C$ can be given Kähler metrics satisfying all the hypotheses; these manifolds are of course holomorphically convex. From Theorem 1D, we also know that a manifold satisfying the hypotheses of Question 6 would be a Stein manifold if its canonical bundle is topologically trivial. Without this assumption on the canonical bundle, what is known is that the manifold M carries nonconstant meromorphic functions [14, Theorem 9], $H^i(M, \mathcal{O}) = 0 \; \forall \, i \geqq 1$ where $\mathcal{O}$ is the structure sheaf [14, Theorem 8], M admits a continuous plurisubharmonic exhaustion function to be called ϕ (see §4 following), and the sublevel sets of ϕ, $M_a \equiv \{p \in M: \phi(p) \leqq a\}$, have the property that each (M, M_a) is a Runge pair $\forall \, a \in R$ ([9, III, Theorem 7]; cf. also [11], Theorem 3]). Are these enough to answer Question 6?

Going back to a more primitive level, the Calabi-Eckmann example [2] mentioned earlier is a complex manifold which admits no Kähler metric. We now ask:

Question 7. If a complete Kähler manifold is diffeomorphic to C^n, does it carry nonconstant holomorphic functions?

Finally, the question was raised in §1 of a geometric characterization of Stein manifolds. There is an obvious candidate and it is best to motivate it by two simple remarks: (i) a closed submanifold of a complete Riemannian manifold is complete (elementary general topology), and (ii) the holomorphic bisectional curvature of a Kähler manifold decreases on complex submanifolds (Griffiths [17], Kobayashi [22]). Since a Stein manifold may be regarded as a closed complex submanifold of some C^n (which possesses a complete Kähler metric of zero sectional curvature), every Stein manifold is a complete Kähler manifold with nonpositive holomorphic bisectional curvature. We now conjecture the converse.

CONJECTURE 8. *A noncompact complex manifold is a Stein manifold iff it has no compact subvarieties of positive dimension and admits a complete Kähler metric of nonpositive holomorphic bisectional curvature.*

4. Convex plurisubharmonic functions on Kähler manifolds. The purpose of this section and the next one is to outline the proofs of parts (A), (B), and (D) of Theorem 1. The considerations involved in these proofs fall into two parts: The first of these parts is the relationship between convexity and plurisubharmonicity on Kähler manifolds and the differential geometric conditions under which such functions exist; results of this type are discussed in this section and it is shown here how these results imply Theorems 1A and 1B. The second part, needed in order to establish Theorem 1D, is a differential geometric version of the L^2 theory for the $\bar\partial$ operator. In §5, this recasting of the $\bar\partial$ operator analysis is discussed, and the outline of the proof of Theorem 1D is completed.

The central notion in this section is that of a convex function, and the main problem to be overcome is that of approximating arbitrary convex functions by C^∞ ones. Convex functions are defined in the following way: Let M be a Riemannian manifold and $f: M \to \boldsymbol{R}$ a function. Then f is *convex* iff, for any geodesic $\gamma: I \to M$, $f \circ \gamma$ is a convex function of one variable, that is, for all $t_1, t_2 \in I$,

$$f(\gamma(\lambda t_1 + (1 - \lambda)t_2)) \leqq \lambda f(\gamma(t_1)) + (1 - \lambda) f(\gamma(t_2))$$

for all $\lambda \in [0, 1]$. A pleasant feature of convex functions is that they are automatically Lipschitz continuous on compact sets.

If $M = C^n$ with the standard flat metric, then convexity of a function f simply means that restricted to each straight line the graph of f is convex upward, like that of e^x, for instance. An analytically expressed example of a convex function on C^n is $f = \sum_i z^i \bar z^i$ (summation over any nonempty subset of $\{1, \cdots, n\}$).

On any Riemannian manifold M, a C^2 function $f: M \to \boldsymbol{R}$ is convex if and only if $(f \circ \gamma)'' \geqq 0$ for any geodesic γ. From this observation, a natural notion of strict convexity follows: f is *strictly convex* iff $(f \circ \gamma)'' > 0$ for every geodesic γ which is not a constant curve. For functions which are not necessarily C^2, this definition is generalized as follows: $f: M \to \boldsymbol{R}$ is *strictly convex* iff it is locally the sum of a convex function and a C^2 strictly convex function, that is, iff for every point $p \in M$ there are functions ϕ and ψ defined in a neighborhood U of p such that ϕ is convex on U, ψ is C^2 and strictly convex on U, and $\phi + \psi = f|U$. A strictly convex function is convex, since convexity is a local property and strict convexity immediately implies convexity locally.

The definition of strict convexity of functions which are not necessarily C^2 is analogous to the definition of strict plurisubharmonicity of functions which are not necessarily C^2. The relationship between convexity and plurisubharmonicity is more than just an analogy. We shall show later in this section that on Kähler manifolds convexity implies plurisubharmonicity. This fact is well known and easily established for C^n, but it is a more subtle result for general Kähler manifolds. In the special case of functions which are C^2, a quantitative version of the implication is given by the following lemma:

PLURISUBHARMONICITY LEMMA. *If M is a Kähler manifold and $f: M \to \boldsymbol{R}$ is a C^2 function then for any point $p \in M$ and any vector $V \in M_p$ (= the real tangent space of M at p),*

$$L_f(V, V) \geqq 2\|V\|^2 \inf_{X \in M_p; \|X\|=1} D_f^2(X, X),$$

where L_f = the Levi form of

$$f = 4 \sum_{i,j} \frac{\partial^2 f}{\partial z_i \partial \bar{z}_j} dz_i \otimes d\bar{z}_j$$

and $D_f^2(X, X)$ = the second derivative of f along the geodesic through p whose tangent vector at p is X. In particular, if f is convex, then $L_f \geqq 0$ and so f is plurisubharmonic; and if f is strictly convex, then $L_f > 0$ and so f is strictly plurisubharmonic.

PROOF. We may assume, since the case $V = 0$ is obvious, that $V = \partial/\partial x_1|_p$ where $z_1 = x_1 + iy_1, \cdots, z_n = x_n + iy_n$ is a holomorphic coordinate system in a neighborhood of p. Then

$$L_f(V, V) = 4 \left.\frac{\partial^2 f}{\partial z_1 \partial \bar{z}_1}\right|_p = \left.\frac{\partial^2 f}{\partial x_1^2}\right|_p + \left.\frac{\partial^2 f}{\partial y_1^2}\right|_p.$$

Now

$$\left.\frac{\partial^2 f}{\partial x_1^2}\right|_p = D_f^2\left(\left.\frac{\partial}{\partial x_1}\right|_p, \left.\frac{\partial}{\partial x_1}\right|_p\right) + \left(D_{\partial/\partial x_1} \frac{\partial}{\partial x_1}\right)_p f$$

and

$$\left.\frac{\partial^2 f}{\partial y_1^2}\right|_p = D_f^2\left(\left.\frac{\partial}{\partial y_1}\right|_p, \left.\frac{\partial}{\partial y_1}\right|_p\right) + \left(D_{\partial/\partial y_1} \frac{\partial}{\partial y_1}\right)_p f$$

where D is the Riemannian covariant derivative on M. Since $\partial/\partial y_1 = J(\partial/\partial x_1)$ and $DJ = 0$,

$$\begin{aligned} D_{\partial/\partial y_1} \frac{\partial}{\partial y_1} &= D_{\partial/\partial y_1} J\left(\frac{\partial}{\partial x_1}\right) = JD_{\partial/\partial y_1} \frac{\partial}{\partial x_1} = JD_{\partial/\partial x_1} \frac{\partial}{\partial y_1} \\ &= JD_{\partial/\partial x_1} J\left(\frac{\partial}{\partial x_1}\right) = J^2 D_{\partial/\partial x_1} \frac{\partial}{\partial x_1} = -D_{\partial/\partial x_1} \frac{\partial}{\partial x_1}. \end{aligned}$$

Thus

$$\begin{aligned} L_f(V, V) &= \left.\frac{\partial^2 f}{\partial x_1^2}\right|_p + \left.\frac{\partial^2 f}{\partial y_1^2}\right|_p \\ &= D_f^2\left(\left.\frac{\partial}{\partial x_1}\right|_p, \left.\frac{\partial}{\partial x_1}\right|_p\right) + D_f^2\left(\left.\frac{\partial}{\partial y_1}\right|_p, \left.\frac{\partial}{\partial y_1}\right|_p\right). \end{aligned}$$

Since $D_f^2(\lambda V, \lambda V) = \lambda^2 D_f^2(V, V)$ for any $\lambda \in \boldsymbol{R}$,

$$D_f^2\left(\left.\frac{\partial}{\partial x_1}\right|_p, \left.\frac{\partial}{\partial x_1}\right|_p\right) \geqq \left\|\left(\left.\frac{\partial}{\partial x_1}\right|_p\right)\right\|^2 \inf = \|V\|^2 \inf$$

where inf stands for the infimum inf $D_f^2(X, X)$ over all $X \in M_p$ with $\|X\| = 1$. Similarly, since J is an isometry so that $\|\partial/\partial y_1\| = \|\partial/\partial x_1\|$,

$$D_f^2\left(\frac{\partial}{\partial y_1}\bigg|_p, \frac{\partial}{\partial y_1}\bigg|_p\right) \geqq \|V\|^2 \inf.$$

The inequality of the lemma thus follows. □

The plurisubharmonicity lemma implies that a Kähler manifold on which there exists a C^∞ strictly convex exhaustion function is a Stein manifold, because then that C^∞ exhaustion function is strictly plurisubharmonic. The implication establishes Theorem 1A as a consequence of the following purely differential geometric result:

> If M is a complete Riemannian manifold of nonnegative sectional curvature then there is a C^∞ strictly convex exhaustion function on M; in fact for any $p \in M$ the function d_p^2: $M \to \boldsymbol{R}$ defined by $d_p^2(q)$ = the square of the Riemannian distance from p to q is such an exhaustion function.

Since this result involves differential geometric considerations only, its proof has been relegated to an appendix. Incidentally, it is precisely because the plurisubharmonicity lemma may fail if the metric of M is a Hermitian, but not Kähler, metric that the Kähler hypothesis is needed in Theorem 1A, since as indicated the existence of a strictly convex exhaustion in no way depends on the Kähler assumption (or even the assumption that the metric is Hermitian). Precise conditions under which a Hermitian metric satisfies the plurisubharmonicity lemma are given in **[21]**.

Convex and strictly convex functions arise frequently in differential geometry, but the fortunate situations such as the one just encountered in which C^∞ strictly convex functions arise in a natural way occur much less frequently. In particular, the geometrically natural convex functions which appear on manifolds of nonnegative curvature [3] are in general not C^∞ or even C^1 (this lack of smoothness comes from the fact that these functions are constructed from the Riemannian distance function, which is not in general C^1). However, in the case of positive curvature one does obtain strict convexity:

THEOREM **[12]**. *If M is a complete noncompact Riemannian with positive sectional curvature then there is a strictly convex exhaustion function on M. If M is a complete Riemannian manifold with sectional curvature positive outside some compact subset, then there is an exhaustion function on M which is strictly convex outside some compact subset of M.*

The proof of this result in **[12]** is based on the technique used in [3] to show that on a complete noncompact Riemannian manifold of nonnegative curvature there is a convex exhaustion function.

To approach Theorem 1B as well as for other purposes, it would clearly be desirable to know how to approximate convex functions by C^∞ convex functions. On $\boldsymbol{C}^n$, it is well known and easy to see that the usual convolution smoothing technique supplies such approximations. For general Riemannian manifolds on Kähler manifolds, it is not known whether arbitrary convex functions can be approximated in neighborhoods of a compact set by C^∞ functions which are convex in a neighborhood of the compact set. However, the following somewhat weaker result in this direction is sufficient for most purposes:

APPROXIMATION LEMMA [10]. *Let K be a compact subset of a Kähler manifold and let f be a convex function defined in a neighborhood of K. Given $-\delta < 0$ there exists a family of C^∞ functions f_ε, $0 < \varepsilon \leqq \varepsilon_0$ for some positive ε_0 defined on a neighborhood of K (which may be chosen independently of δ), such that*

(i) *$f_\varepsilon \downarrow f$ uniformly on K as $\varepsilon \downarrow 0$.*

(ii) *Each f_ε is $(-\frac{1}{2}\delta)$-convex in K in the sense that at any point in K of any geodesic $\gamma: I \to M$ with unit tangent vectors $(f \circ \gamma)'' \geqq -\delta/2$.*

(iii) *Each f_ε is $(-\delta)$-plurisubharmonic on K in the sense that all the eigenvalues of the Levi form of each f_ε are $\geqq -\delta$ in K.*

Note that this lemma implies the result for Kähler manifolds (well known already for C^n) that a convex function is plurisubharmonic. Indeed, given a convex function f on a Kähler manifold M, it suffices to show that f is plurisubharmonic in a neighborhood of each point $p \in M$. For that purpose let ϕ be a C^∞ strictly plurisubharmonic function defined in a neighborhood of p. If for each $k = 1, 2, 3, \cdots$, $f + \phi/k$ is plurisubharmonic on a neighborhood of p (the neighborhood being independent of k) then f is plurisubharmonic on a neighborhood of p since the uniform-on-compact-subsets limit of plurisubharmonic functions is plurisubharmonic. But for any fixed k if one takes $\delta > 0$ sufficiently small then the eigenvalues of the Levi form of $f_\varepsilon + \phi/k$ (δ and f_ε as in the approximation lemma) are positive. Also $f_\varepsilon + \phi/k \downarrow f + \phi/k$ as $\varepsilon \downarrow 0$. Thus $f + \phi/k$ is plurisubharmonic for each k and the plurisubharmonicity of f near p follows.

From the fact that convexity implies plurisubharmonicity on Kähler manifolds, it follows easily that strict convexity implies strict plurisubharmonicity there. For if f is strictly convex then locally $f = \phi + \psi$, ϕ convex and ψ C^2 strictly convex. But then ϕ is plurisubharmonic by the previous paragraph and ψ is strictly plurisubharmonic by the plurisubharmonicity lemma so that f is strictly plurisubharmonic.

Here are a few comments on the proof of the approximation lemma: Part (iii) follows from part (ii) by the plurisubharmonicity lemma. For the definition of the functions f_ε, a convolution smoothing process is employed as in the case $M = C^n$, but it is necessarily more sophisticated than the one encountered in C^n because there is no natural way to translate functions around on a Kähler manifold. This convolution involves the Riemannian structure of the manifold closely. Once the convolution process is set up, the proof of (i) is easy, but the proof of (ii) is by no means trivial. (See [10] for details.)

As noted earlier, every complete noncompact Riemannian manifold of nonnegative sectional curvature has a (continuous) convex exhaustion function [3] and thus according to the previous paragraph a continuous plurisubharmonic exhaustion function. In general, this plurisubharmonic function need not be strictly plurisubharmonic anywhere, so that no useful information can yet be deduced. (Recall the celebrated example of Grauert showing that a pseudoconvex manifold need not have any nonconstant holomorphic functions; see Narasimhan [28].) This observation explains why in Theorem 1, parts (B) — (D), additional assumptions have to be imposed.

If the manifold satisfies the hypotheses of Theorem 1B, then as noted earlier there is an exhaustion function on it which is strictly convex outside some compact set. One of the virtues of strict convexity is that the approximation result is entirely satisfactory for strictly convex functions:

APPROXIMATION LEMMA II [**15**]. *Let f be a strictly convex function defined in a neighborhood of a compact set K in a Kähler manifold. Then there exists a family of C^∞ functions f_ε defined in a fixed neighborhood of K, $0 < \varepsilon \leqq \varepsilon_0$ for some fixed ε_0, such that*

(i) *$f_\varepsilon \downarrow f$ uniformly on K as $\varepsilon \downarrow 0$.*
(ii) *Each f_ε is strictly convex.*
(iii) *Each f_ε is strictly plurisubharmonic.*

The proof of this is merely a technical modification of that of the approximation lemma, and (iii) is again a consequence of (ii). Armed with this improvement, we would like to claim immediately that:

(*) On a manifold M satisfying the hypotheses of Theorem 1B, there is a C^∞ exhaustion function which is strictly plurisubharmonic outside a compact set K_0.

But we are ahead of ourselves, for the following reason: If $M\backslash K_0$ were compact, then intuitively we would replace $f|M\backslash K_0$ by one such C^∞ function f_ε and extend f_ε in a C^∞ manner to be an exhaustion function on all of M; this extension is then C^∞ and strictly plurisubharmonic on $M\backslash K_0$. Unfortunately, $M\backslash K_0$ is anything but compact. This means that, to arrive at the desired goal, an additional ingredient to the second approximation lemma is needed; namely, how to pass from approximation on compact sets to global approximations.

There are two ways to accomplish this. One is the method of Richberg, who proved directly in [**29**] that a continuous strictly plurisubharmonic function on an arbitrary complex manifold can be approximated in the *C^0-fine topology* by a C^∞ function of the same type. Recall that density in the C^0-fine topology means: Given a continuous strictly plurisubharmonic function f and an arbitrary positive continuous function δ, there exists a C^∞ strictly plurisubharmonic function g such that $|f-g| < \delta$ on the whole manifold. In particular, we can approximate continuous strictly plurisubharmonic functions by C^∞ ones uniformly on the whole manifold. With the help of this result, the assertion (*) follows because we have already seen that strictly convex functions are strictly plurisubharmonic. We may thus replace the original exhaustion function f on M by a C^∞ function f such that $f_\varepsilon|K_0$ is arbitrary, and that $f_\varepsilon|M\backslash K_0$ is a strictly plurisubharmonic function that uniformly approximates $f|M\backslash K_0$. The latter property guarantees that f_ε is in fact an exhaustion function.

This theorem of Richberg was proved in [**29**] by a step-by-step alteration of the original function, one coordinate patch at a time. Another method which we now outline will lead to the same conclusion (*) by directly exploiting strict convexity. This method is particularly suitable for dealing with functions of interest in geometry, e.g., strictly subharmonic functions on a Riemannian manifold. So let $\mathscr{S}$ be a subsheaf of the sheaf of germs of continuous functions on a manifold N. Note that $\mathscr{S}$ is assumed to be a subsheaf in the topological sense only: It is not assumed to be closed under the algebraic operations on the sheaf of germs of continuous functions. Consider the following possible properties of $\mathcal{S}$.

(1) C^∞ *stability*. On any compact subset K of N a sufficiently small C^∞ perturbation of a section of $\mathcal{S}|K$ is a section of $\mathcal{S}|K$.

(2) *Maximum closure*. The maximum of two germs in $\mathcal{S}_p$, $p \in M$, is in $\mathcal{S}_p$.

(3) *Semilocal approximation*. On compact sets, C^∞ sections C^0-approximate a given section, and uniformly approximate it up to k derivatives on every closed subset where it is C^k, $1 \leqq k \leqq \infty$.

GLOBAL APPROXIMATION THEOREM (**[13]**, **[15]**, **[16]**). *If N is a paracompact manifold and if $\mathcal{S}$ (a subsheaf of the sheaf of germs of continuous functions on N) satisfies* (1), (2), *and* (3), *then the sections of $\mathcal{S}$ which are C^∞ functions on N are dense in the space of all sections of $\mathcal{S}$ in the C^0-fine topology.*

By virtue of the second approximation lemma and standard elementary arguments, strictly convex functions on a Kähler manifold enjoy properties (1)—(3). The preceding theorem then implies:

(**) On the Kähler manifold M under discussion, there is a C^∞ exhaustion function ϕ which is strictly convex outside K_0 (and hence strictly plurisubharmonic outside K_0).

The proof of Theorem 1B can now be completed quite simply. We first observe that if M satisfies the hypotheses of Theorem 1B, then M has no compact positive-dimensional subvarieties. For suppose it does; then they give rise to nonzero positive-dimensional homology classes because M is a Kähler manifold. But we have already observed (in §2) that M is diffeomorphic to C^n, a contradiction. Next, let M_a denote the sublevel set $\{p \in M: \phi(p) \leqq a\}$, where ϕ is as in (**). Without loss of generality, we may assume $K_0 = M_0$. Now $\forall\, a > 0$, M_a is a strongly pseudoconvex domain and hence holomorphically convex, by Grauert's solution of the Levi problem. Since M_a contains no compact positive-dimensional subvarieties, M_a is actually a Stein manifold. But $M = \bigcup_{a>0} M_a$, so M is the increasing union of a one-parameter family of Stein manifolds. By a theorem of Docquier-Grauert **[4]**, M is itself a Stein manifold. Q.E.D.

It remains to make a few concluding remarks about the preceding proof. First of all, strictly plurisubharmonic functions on a complex manifold can be shown to satisfy the hypotheses of the global approximation theorem and Richberg's theorem is thereby recovered. But this presents no major simplification over Richberg's original proof. Second, it is not apparent here that there is any advantage in having the strict convexity of ϕ in (**) rather than just strict plurisubharmonicity as in (*). Actually, for the geometric analysis of M, this point is decisive (see **[15]**). Third, one could have by-passed either Richberg's theorem or the global approximation theorem in the latter part of the proof if one is willing to invoke Narasimhan's solution of Levi's problem **[27]**. There, he uses only continuous functions to define strongly pseudoconvex domains, and it is relatively easy to prove that a (continuous) strictly convex function on a Kähler manifold is strictly plurisubharmonic in his sense. However, our presentation has the virtue of avoiding a difficult proof (**[27]** is concerned with spaces) while doing something that is illustrative of the basic techniques in the subject. Last, it is to be observed that the overall nonnegativity of the sectional curvature was only needed to conclude that the topology of the Kähler manifold M is trivial. The proof as it stands essentially suffices for the following theorem (see [15] for a detailed justification).

Let M be a noncompact, complete Hermitian manifold. Assume that, outside a compact set, the metric is Kählerian and has positive sectional curvature. Then M is 0-convex and hence is obtained from a Stein space by blowing up a finite number of points.

5. The $\bar{\partial}$ operator on Kähler manifolds. The first goal of this section is to demonstrate certain properties of the $\bar{\partial}$ operator on noncompact Kähler manifolds, specifically on those satisfying the hypotheses of Theorem 1D. The analysis here of the $\bar{\partial}$ operator is based on the L^2 method developed by Kohn and Hörmander (see [19]) and also by Andreotti and Vesentini (cf. [30]), which we shall reformulate in a different geometric version. Using this and some of the results in §4, we shall show that on a manifold satisfying the hypotheses of Theorem 1D the equation $\bar{\partial}f = \phi$ has a solution f for any $C^\infty(0, 1)$ form ϕ with $\bar{\partial}\phi = 0$ (see [11] for a closely related and in some aspects more detailed discussion of similar results). As an aside, we shall discuss the fact that the estimates used to prove this result can also be used to establish the Kodaira vanishing theorem (for compact Kähler manifolds), being in fact essentially the same estimates used by Kodaira in his original proof (see [25], for instance). This fact shows the underlying unity of the compact and noncompact cases. The second goal of the section is to show how the solvability of $\bar{\partial}f = \phi$ on manifolds satisfying the hypotheses of Theorem 1D implies Theorem 1D itself, i.e., that the manifold is a Stein manifold.

Expressed in very general terms, the L^2 method consists of considering the operator $\bar{\partial}$ as a densely defined operator on certain Hilbert spaces and studying its behavior by functional analysis. In this setting, the solution of $\bar{\partial}f = \phi$ when $\bar{\partial}\phi = 0$ is obtained if an estimate of a certain type can be obtained. We shall describe the L^2 method more specifically in several parts: (a) the Hilbert spaces; (b) the $\bar{\partial}$ operator and its adjoint and the density of C^∞ forms of compact support (notation: C_0^∞ forms) in the graph norm; (c) estimates for C_0^∞ forms, which will by virtue of (b) imply Hilbert space estimates; (d) the final form of the estimates and the solution of $\bar{\partial}$. Since the functional analysis involved in (d) is virtually the same in the present setting as in the case of domains in euclidean space, it will not be discussed in detail here: for a thorough treatment in the absence of differential geometric considerations see [19], and for a discussion of the modifications appropriate to a differential geometric setting see [11].

(a) *The Hilbert space.* The Riemannian metric of a Riemannian manifold M induces for each $p \in M$ an inner product on the dual space M_p^* of the real tangent space M_p, namely the dual of the inner product on M_p. Specifically if $X_1, \cdots, X_n$ is an orthonormal basis of M_p and $\omega_1, \cdots, \omega_n$ the dual basis for M_p^* then $\omega_1, \cdots, \omega_n$ is defined to be an orthonormal basis for M_p^*. The inner product thus determined on M_p^* is independent of the choice of $X_1, \cdots, X_n$ among orthonormal bases of M_p. This inner product on M_p^* in turn induces inner products on the space $\Lambda^r M_p^*$, $r = 1, 2, \cdots$, of r-forms at p: If $\omega_1, \cdots, \omega_n$ is an orthonormal basis of M_p^* then the inner product on $\Lambda^r M_p^*$ is determined by making $\omega_{i_1} \wedge \cdots \wedge \omega_{i_r}$, $1 \leqq i_1 < \cdots < i_r \leqq n$, an orthonormal basis for $\Lambda^r M_p^*$. This inner product will be denoted by g, the point p and the value of r being made clear from the context. The complex linear extension of g to $\Lambda_C^r M_p^*$ = the complex-valued r-forms at p will also be denoted by g. On $\Lambda_C^r M_p^*$, there is then a natural Hermitian (positive definite) inner product: $\alpha, \beta \to g(\alpha, \bar{\beta})$, where $^-$ denotes conjugation. We now define the inner product of C_0^∞ complex-valued r-forms ω, ψ on M to be $\langle \omega, \psi \rangle = \int_M g(\omega, \bar{\psi})$, where integration is relative to the measure induced from the Riemannian metric on M. This is a positive definite Hermitian inner product on the C_0^∞ r-forms on M.

Now suppose that $\tau : M \to \boldsymbol{R}$ is a continuous function.
Define

$$\langle \omega, \psi \rangle_\tau = \int_M g(\omega, \bar{\psi}) e^{-\tau}.$$

This is again a positive definite Hermitian inner product on the C_0^∞ r-forms on M. We shall be interested here in the restrictions of these inner products to complex-valued r-forms of type $(0, r)$. In particular, the functional analysis of $\bar{\partial}$ will be considered in Hilbert spaces $L_{0,r}^2(\tau) =^{\text{def}}$ the Hilbert space completion of the C_0^∞ $(0, r)$-forms relative to the Hermitian inner product $\langle\ ,\ \rangle_\tau$.

(b) *The $\bar{\partial}$ operator and graph norm density of C_0^∞.* The operator $\bar{\partial}$ maps the C_0^∞ forms of type $(0, r)$ to the C_0^∞ forms of type $(0, r+1)$. It is thus a densely defined operator on $L_{0,r}^2(\tau)$. It is closed and its closure has a densely defined adjoint δ_τ'': $L_{0,r+1}^2(\tau) \to L_{0,r}^2(\tau)$. The $''$ occurs in this notation because of the conventions: δ = adjoint of d, δ' = adjoint of ∂, and δ'' = adjoint of $\bar{\partial}$. If τ is a C^1 function (in our applications, τ will in fact be C^∞) then δ_τ'' is actually the closure of a differential operator $\delta_\tau'' = e^\tau(\delta'' e^{-\tau})$ where $\delta'' = \delta_0'' =$ the formal adjoint of $\bar{\partial} = - * \partial *$.

For elements $\omega \in L_{0,r}^2(\tau) \cap$ domain $\bar{\partial} \cap$ domain δ_τ'', there is a graph norm

$$(\|\omega\|^2 + \|\bar{\partial}\omega\|^2 + \|\delta_\tau''\omega\|^2)^{1/2}$$

where $\|\ \|$ is the L^2-norm on the Hilbert spaces. The density assertion which will be used to pass from estimates on C_0^∞ forms to Hilbert space estimates is:

DENSITY LEMMA. *If M is a complete Kähler manifold, then the C_0^∞ forms of type $(0, r)$ are dense in $L_{0,r}^2(\tau) \cap$ domain $\bar{\partial} \cap$ domain δ_τ'' in the graph norm.*

This fact was first noted explicitly by Andreotti and Vesentini **[30]**: in the approach of **[19]** its role was taken by a density statement involving not a complete Kähler metric but rather the use of three different weight factors of the form $e^{-\tau}$ on the spaces of $r-1$, r, and $r+1$ forms.

An informal description of the role of completeness in the density statement may be illuminating to the outline of a formal proof to be given shortly: Given a form $\omega \in L_{0,r}^2(\tau) \cap$ domain $\bar{\partial} \cap$ domain δ_τ'', one wishes to approximate ω in the graph norm by a form of compact support. Once this approximation is accomplished then the approximation of the compact-support form by a C_0^∞ form follows by using the standard partition-of-unity and convolution smoothing process (Friedrichs' mollifier technique). So one would like in particular to show that ω can be approximated in the graph norm by a form $\mathscr{E}_K\omega$ where $\mathscr{E}_K$ is a function which is $\equiv 1$ on a (large) compact set K and $\equiv 0$ outside a somewhat larger compact set. However, first derivatives of $\mathscr{E}_K$ occur in the expression for the graph norm of $\omega - \mathscr{E}_K\omega$, so one needs control of the first derivatives of $\mathscr{E}_K$, i.e., the gradient of $\mathscr{E}_K$, as the set K is made larger. A difficulty arises if M is not complete. If M is for instance a bounded domain in C^n with the metric of C^n restricted, M being thus not complete, when K is made large the first derivatives of $\mathscr{E}_K$ will also have to be large to enable $\mathscr{E}_K$ to reach 0 by the boundary of M, the boundary being very near K when K is large in the finite-diameter domain M. Of course, not every noncomplete manifold has a boundary in this literal sense (of being a proper open subset of a larger manifold) but the general principle still applies. However, if M is complete and noncompact, then the distance from any compact set to "infinity" (the analogue of the boundary) is infinite in the following precise sense: If M is complete, then for every compact set K in M and every real number r the set $\{p \in M \mid \text{distance from } p \text{ to some point of } K \leqq r\}$ is compact and in particular if M is noncompact, this set is a proper subset on M. The "infinite distance to infinity" in this sense makes it possible to construct function $\mathscr{E}_K$ with uniformly (independently of K) bounded gradient vectors.

To make the previous general observations precise it is convenient to use the following result of Gaffney [7]: If a Riemannian manifold M is complete then there exists a C^∞ exhaustion function $\eta: M \to \boldsymbol{R}$ with the length of the gradient of η bounded on M. (The converse also holds, as is easily established from standard Riemannian-geometric characterizations of completeness. But we shall not need the converse here.) Let η be such an exhaustion function and let $b: \boldsymbol{R} \to \boldsymbol{R}$ be a C^∞ function with $b(x) = 1$ for $x \leqq 0$, $0 \leqq b(x) \leqq 1$ for $0 \leqq x \leqq 1$ and $b(x) = 0$ for $x \geqq 1$. Define $\mathscr{E}_r: M \to \boldsymbol{R}$ by $\mathscr{E}_r(p) = b(\eta(p) - r)$ so that $\mathscr{E}_r = 1$ on $\eta^{-1}((-\infty, r])$ and $= 0$ outside of $\eta^{-1}((-\infty, r+1])$. Then $\mathscr{E}_r\omega \to \omega$ in the graph norm as $r \to +\infty$ for all $\omega \in L^2_{0,r}(\tau) \cap \text{domain } \bar\partial \cap \text{domain } \delta''_r$. To check this latter assertion, note first that $\mathscr{E}_r\omega \to^{L^2(\tau)} \omega$ as $r \to +\infty$ since η is an exhaustion function. Also $\bar\partial(\mathscr{E}_r\omega - \omega) = (\bar\partial\mathscr{E}_r)\,\omega + \mathscr{E}_r(\bar\partial\omega) - \bar\partial\omega$ and $\mathscr{E}_r(\bar\partial\omega) - \bar\partial\omega \to^{L^2(\tau)} 0$ as $r \to +\infty$ since $\bar\partial\omega \in L^2_{0,r+1}(\tau)$. Moreover, the $L^2(\tau)$ norm $\|(\bar\partial\mathscr{E}_r)\omega\|$ satisfies, since the support of $\bar\partial\mathscr{E}_r \subset \eta^{-1}([r, r+1])$,

$$\|(\bar\partial\mathscr{E}_r)\omega\| \leqq C\|\omega\chi_r\|$$

where $\chi_r =$ the characteristic function of $\eta^{-1}([r, r+1])$ and C is a constant independent of r: this estimate follows from the fact that the gradient of $\mathscr{E}_r$, and hence the size of $\bar\partial\mathscr{E}_r$, is bounded independently of r. (Here one makes a local coordinate calculation to check that a bound on gradient $\mathscr{E}_r$ implies one on $\bar\partial\mathscr{E}_r$.) But $\|\omega\chi_r\| \to 0$ as $r \to +\infty$ so $\bar\partial(\mathscr{E}_r\omega) - \omega \to^{L^2(\tau)} 0$ as $r \to +\infty$. A similar analysis applies to δ''_τ. Thus the approximation by compact-support forms is established, and the further approximation by C_0^∞ follows by a standard process, as already noted.

(c) *The estimates for C_0^∞ forms.* Since for the purpose of proving Theorem 1D we need to solve $\bar\partial f = \phi$ only when ϕ is a (0, 1) form, we shall discuss in detail only the estimate needed for that case. But a similar method applies to the establishment of the appropriate estimates for the (0, r) case for all $r > 0$.

Now let τ be a C^∞ function and ω a C_0^∞ (0, 1) form. Write $\square_\tau$ for the complex Laplacian relative to the weight factor τ, i.e., $\square_\tau = \bar\partial\delta''_\tau + \delta''_\tau\bar\partial$. Then for any C_0^∞ (0, 1) form ω

$$\langle\square_\tau\omega, \omega\rangle_\tau = \langle\delta''_\tau\omega, \delta''_\tau\omega\rangle_\tau + \langle\bar\partial\omega, \bar\partial\omega\rangle_\tau,$$

since δ''_τ is the adjoint of $\bar\partial$ relative to the inner product $\langle\;,\;\rangle_\tau$ on C_0^∞ forms. The computation of $\square_\tau\omega$ and thence $\langle\square_\tau\omega, \omega\rangle_\tau$ is a standard result in complex differential geometry. In fact it is a special case of the computation of the complex Laplacian for forms with values in a holomorphic line bundle with a Hermitian metric. We shall describe the result of this computation in the special case of $\Box_\tau$ first, and we shall then indicate briefly how it is derived from the general case.

The information we need about the complex Laplacian $\Box_\tau$ can be represented schematically as follows: If ϕ is a C^∞ $(0, r)$ form then

$$g(\Box_\tau\phi, \bar{\phi}) = \text{Ricci curvature term} + \text{Levi form term} + \text{covariant derivative term.}$$

Moreover, if ϕ has compact support,

$$\int_M (\text{covariant derivative term})\, e^{-\tau} \geqq 0.$$

Thus for a C_0^∞ $(0, r)$ form ϕ

$$\langle \Box_\tau\phi, \phi\rangle_\tau \geqq \int_M (\text{Ricci curvature term})e^{-\tau} + \int_M (\text{Levi form term})e^{-\tau}.$$

To give this schematic representation specific content, we need to set up some notation. We shall give the specific content only in the case of (0, 1) forms; the $(0, r)$ case is quite similar but is notationally more complicated, and it will not be needed here.

Let p be any point of M. We shall give the value of the Ricci curvature and bundle curvature terms at p: Let $X_1,\cdots, X_n$, $n = \dim_C M$, be an orthonormal set of vectors in the real tangent space M_p with the property that $X_1, JX_1, \cdots, X_n, JX_n$ is an orthonormal basis for M_p. Then let $\theta_1,\cdots,\theta_{2n}$ be the (real) basis of M_p^* dual to $X_1, JX_1, \cdots, X_n, JX_n$. The forms $\omega_j = (\theta_{2j-1} + \sqrt{-1}\,\theta_{2j})/\sqrt{2}$, $j = 1,\cdots, n$, are of type (1, 0); their conjugates $\bar{\omega}_j = (\theta_{2j-1} - \sqrt{-1}\,\theta_{2j})/\sqrt{2}$ are of type (0, 1), and the forms $\omega_1, \cdots, \omega_n, \bar{\omega}_1, \cdots, \bar{\omega}_n$ are a complex basis for $M_p^* \otimes \boldsymbol{C} = \Lambda_C^1 M_p^*$. Then for any (0, 1) form ϕ, define the components ϕ_j of ϕ at p by requiring that $\phi = \sum_{j=1}^n \phi_j\bar{\omega}_j(p)$. In this notation, the Ricci curvature term in the $g(\Box_\tau\phi, \bar{\phi})$ formula is (at p) $\sum_{j=1}^n |\phi_j|^2\, R(X_j, X_j)$ where $R(X_j, X_j)$ = the Ricci curvature in the direction X_j. The notation $R(X_j, X_j)$ is used because the Ricci curvature in fact arises from a quadratic form on M_p. The Levi form term is (at p)

$$\frac{1}{2}\left\{\sum_{j,k=1}^n \phi_j\bar{\phi}_k L_\tau\left(\frac{1}{2}(X_j - \sqrt{-1}\,JX_j), \frac{1'}{2}(X_k + \sqrt{-1}\,JX_k)\right)\right\},$$

where L_τ is as usual the Hermitian form

$$4\sum_{k,l=1}^n \frac{\partial^2\tau}{\partial z_k \partial\bar{z}_l}\, dz_k \otimes d\bar{z}_l$$

for any holomorphic coordinate system $(z_1,\cdots, z_n)$. (Incidentally, this expression of the Levi form term corrects an error on pp. 154–155 of **[11]**.)

Define functions $c_R : M \to \boldsymbol{R}$ and $c_\tau : M \to \boldsymbol{R}$ by

$$c_R(p) = \min_{X\in M_p; \|X\|=1} R(X, X), \qquad c_\tau(p) = \tfrac{1}{2} \min_{X\in M_p; \|X\|=1} L_\tau(X, X).$$

The functions c_R and c_τ are continuous. Moreover, one obtains from the previous inequality on $\langle \square_\tau \phi, \bar{\phi} \rangle_\tau$ and the explicit forms of the Ricci curvature and Levi form terms that

$$\langle \square_\tau \phi, \phi \rangle_\tau \geqq \int_M (c_R + c_\tau) g(\phi, \bar{\phi}) e^{-\tau}.$$

This is the estimate which will be used to complete the solution of $\bar{\partial} f = \phi$ when $\bar{\partial}\phi = 0$ and M satisfies the hypotheses of Theorem 1D.

We shall now turn aside from the discussion directed toward the proof of Theorem 1D in order to discuss the result of the computation of the complex Laplacian for line bundle valued forms, the resulting estimate, and its applications to compact Kähler manifolds. This discussion is not needed in order to prove Theorem 1D; the considerations involved in that proof recommence with part (d) of this section.

Let $E \to M$ be a holomorphic line bundle with a C^∞ Hermitian metric h. There is then a metric induced on r forms with values in E: At a point p, one sets $g_h(\phi e_1, \psi e_2) = g(\phi, \bar{\psi}) h(e_1, e_2)$ where $\phi, \psi \in \Lambda_C^r M_p^*$ and $e_1, e_2 \in E_p$, the fibre at p. Then if s_1 and s_2 are sections of E and ϕ and ψ are C_0^∞ C-valued forms, one defines

$$\langle \phi s_1, \psi s_2 \rangle_h = \int_M g_h(\phi s_1, \psi s_2).$$

Since E has holomorphic transition functions, $\bar{\partial}$ operates on E-valued forms in an invariantly defined way: If s is a local holomorphic section of E and ϕ is a C^∞ r-form, then $\bar{\partial}(\phi s) = (\bar{\partial}\phi)s$. Associated to $\bar{\partial}$ acting on E-valued forms, there is a formal adjoint denoted by δ_h'' which satisfies as usual

$$\langle \bar{\partial}\phi, \psi \rangle_h = \langle \phi, \delta_h'' \psi \rangle_h$$

for all C_0^∞ E-valued forms ϕ and ψ. The Laplacian $\square_h$ for E-valued forms is of course defined to be $\bar{\partial}\delta_h'' + \delta_h''\bar{\partial}$.

Note that if $E = M \times C$, the trivial line bundle, then the ideas just given reduce to the previous ones associated to $\langle\ ,\ \rangle$ if one takes the Hermitian metric h on E to be that determined by $h(1, 1) = e^{-\tau}$ where 1 denotes the constant section of value 1. Namely, then $\langle\ ,\ \rangle_h = \langle\ ,\ \rangle_\tau$, $\delta_h'' = \delta_\tau''$, and $\square_h = \square_\tau$.

The schematic formula for $g_h(\square_h \phi, \phi)$, ϕ an E-valued $(0, r)$ form, $r > 0$, is

$$g_h(\square_h \phi, \phi) = \text{Ricci curvature term} + \text{bundle curvature term} + \text{covariant derivative term}.$$

As before, one always has $\int_M$ covariant derivative term $\geqq 0$ if ϕ is of compact support. (Note: The role of the weight factor $e^{-\tau}$ has here been taken over by the metric h; thus our schematic formula for $g_h(\square_h \phi, \phi)$ reduces directly in the previous special case to the schematic formula for $g(\square_\tau \phi, \bar{\phi})$ *with both sides multiplied by* $e^{-\tau}$.) So one obtains the inequality

$$\langle \square_h \phi, \phi \rangle_h \geqq \int_M \text{Ricci curvature term} + \int_M \text{bundle curvature term}.$$

We now give the specific meaning of the terms of this formula when ϕ is an E-

valued (0, 1) form. The Ricci curvature term here has virtually the same meaning as before: If $\phi = \phi_0 e$, $e \in E_p$, at p, then the present Ricci curvature term = the former Ricci curvature term for ϕ_0 multiplied by $h(e, e)$. The term now called the bundle curvature term is closely related to the Levi form term in the $\square_\tau$ formula. Specifically, let s be a holomorphic nonvanishing section of E defined in a neighborhood of p and let $\phi = \phi_0 s$ in that neighborhood of p. Then the bundle curvature term at $p = \{$the Levi form term for ϕ_0 with $\tau = -\log h(s, s)\} \times h(s, s)$. Thus locally the bundle curvature term is computable from the Levi form term. In view of these specifications, one sees immediately that the $g_h(\square_h\phi, \phi)$ reduces to the $g(\square_\tau\phi, \bar{\phi})$ formula multiplied by $e^{-\tau}$ when E is trivial with $e^{-\tau}$ as its metric.

The $\square_h$ inequality, the special case of which will yield a cohomology vanishing theorem in the case of noncompact manifolds (see (d) following) also yields such a theorem in the case of compact manifolds. In the latter case, the result is the Kodaira vanishing theorem. The calculation of $\square_h$ was in fact the method used by Kodaira to prove his vanishing theorem. This proof was in turn a generalization to the complex case of Bochner's proof that a compact Riemannian manifold of positive Ricci curvature has first Betti number = 0: This proof is obtained by using the corresponding calculation of the real Laplacian on 1-forms to show that a harmonic 1-form must vanish identically.

Assume for the remainder of this paragraph that M is a compact Kähler manifold. Suppose that E is a Hermitian holomorphic line bundle with the property that for a particular $r > 0$ the sum of the Ricci curvature and the bundle curvature terms in the $g_h(\square_h\phi, \phi)$ formula is positive definite in the $(0, r)$ form ϕ at each point of M (these terms are quadratic forms in ϕ for any $r > 0$). Then since M is compact, there is an $\varepsilon > 0$ such that

$$\langle \square_h\phi, \phi\rangle_h \geqq \varepsilon\langle\phi, \phi\rangle_h.$$

Thus if $\square_h\phi \equiv 0$ then $\langle\phi, \phi\rangle_h = 0$ so $\phi \equiv 0$. In other words, every $\square_h$-harmonic $(0, r)$ form is identically 0. Hodge theory then implies that $H^r(M, \mathcal{O}(E)) = 0$ for that value of r ($\mathcal{O}(E)$ = sheaf of germs of holomorphic sections of E). To formulate a familiar condition under which the sum (Ricci curvature term) + (bundle curvature term) is positive definite, one computes that this sum is positive definite for all $r > 0$ provided that the bundle curvature of the tensor product $K^* \otimes E$ is positive definite, where K^* = the dual of K, K being the bundle of $(n, 0)$ forms with the metric obtained from the metric of M, and the bundle $K^* \otimes E$ is given the metric which is the tensor product of the metrics of K^* and E. Any metric on $K^* \otimes E$ has the form: (given metric of K^*) $\otimes$ (some metric on E). So there is a metric on E such that $K^* \otimes E$ has positive definite bundle curvature if and only if there exists a metric on $K^* \otimes E$ such that with this metric $K^* \otimes E$ has positive definite bundle curvature. A holomorphic line bundle which admits a Hermitian metric with an associated bundle curvature which is positive definite is called a positive line bundle. In this terminology, the previous remarks lead to the Kodaira vanishing theorem: If M is a compact Kähler manifold and if $E \to M$ is a holomorphic line bundle with the property that $K^* \otimes E$ is a positive line bundle, then, for all $r > 0$, $H^r(M, \mathcal{O}(E)) = 0$.

For a precise formula for $\square_h$, one can take a look at **[25]**.

(d) *The final form of the estimates and the functional analysis.* Under the hypotheses of Theorem 1D, $R(X, X) > 0$ for all unit vectors X. Hence the continuous function c_R (defined in part (c)) is everywhere positive. Now one would like to choose for τ a C^∞ exhaustion function which would make $c_\tau \geqq 0$ or at least make $c_R + c_\tau > 0$. It has been noted already that there is on M a convex (and hence plurisubharmonic) exhaustion function. But this function is not in general C^∞. However, "nearly convex" C^∞ exhaustion functions, which will serve the purpose indicated, can be obtained as a corollary of the global approximation theorem of §4:

COROLLARY OF THE GLOBAL APPROXIMATION THEOREM. *Suppose M is a noncompact Riemannian manifold on which there is a convex exhaustion function σ. Then for any positive continuous function $\varepsilon : M \to \boldsymbol{R}$ there is a C^∞ function $\tau : M \to \boldsymbol{R}$ such that*

(a) $|\tau - \sigma| < \varepsilon$ *everywhere on M;*

(b) *for any point $p \in M$ and any unit vector $X \in M_p$, $D_\tau^2(X, X) > -\varepsilon(p)$.*

Here $D_\tau^2(X, X)$ is as before (§4) the second derivative at p of τ along the geodesic through p with tangent X. We shall postpone the deduction of this corollary from the global approximation theorem itself until we have shown how to use the corollary to obtain the solution of $\bar\partial f = \phi$ on manifolds satisfying the hypotheses of Theorem 1D.

So suppose now that M satisfies the hypotheses of Theorem 1D and that ϕ is a $C^\infty(0,1)$ form on M such that $\bar\partial\phi = 0$. To solve $\bar\partial f = \phi$, we first need to pick a weight factor $e^{-\tau}$ with $\tau \in C^\infty$ such that $\phi \in L_{0,1}^2(\tau)$. For this purpose, let σ be a convex exhaustion function on M, the existence of which was noted earlier. And let $\chi : \boldsymbol{R} \to \boldsymbol{R}$ be a C^∞ nonnegative strictly increasing convex function. Then $\chi \circ \sigma : M \to \boldsymbol{R}$ is an exhaustion function which is again convex, as is easily seen. Furthermore, if χ is chosen to increase sufficiently rapidly as its argument $\to +\infty$, then ϕ will be in $L_{0,1}^2(\chi \circ \sigma)$. $\chi \circ \sigma$ is still not in general C^∞. Now choose, in the corollary of the global approximation theorem, the function ε by $\varepsilon(p) = \min(1, c_R(p)/4)$ and the function σ there to be the present $\chi \circ \sigma$. (The function c_R was defined in part (c) of this section.) Then the resulting C^∞ function τ has the property that $\phi \in L_{0,1}^2(\tau)$ since $|\tau - \chi \circ \sigma| < 1$ everywhere. Furthermore, by the plurisubharmonicity lemma of §4, the function $c_\tau : M \to \boldsymbol{R}$ corresponding to this choice of τ satisfies $c_R(p) + c_\tau(p) > \frac{1}{2} c_R(p)$ for all $p \in M$ (in more detail: $c_\tau(p) = \frac{1}{2}$ minimum eigenvalue of $L_\tau|_p \geqq$ minimum of $D_\tau^2(X, X)$ for $\|X\| = 1$, $X \in M_p$, by plurisubharmonicity lemma, $> -c_R(p)/4$). By having chosen the function χ to increase rapidly enough, one may also arrange that the function τ satisfies the condition, which is needed for technical reasons (see [**11**, p. 159]):

$$\int_M g(\phi, \bar\phi) \frac{e^{-\tau}}{c_R} < \infty.$$

Since $c_R + c_\tau > \frac{1}{2} c_R$ the estimate on $\langle \Box_\tau \psi, \psi \rangle_\tau$ yields when ψ is a $C_0^\infty(0, 1)$ form:

$$\langle \Box_\tau \psi, \psi \rangle_\tau \geqq \int_M \tfrac{1}{2} c_R g(\psi, \bar\psi) e^{-\tau}$$

or equivalently,

$$\langle \bar\partial\phi, \bar\partial\phi\rangle_\tau + \langle \delta''_\tau\phi, \delta''_\tau\phi\rangle_\tau \geqq \int_M \tfrac{1}{2} c_R g(\phi, \bar\phi) e^{-\tau}.$$

The graph norm density lemma of part (b) implies that this second form of the estimate holds for all $\phi \in L^2_{0,1}(\tau) \cap$ domain $\bar\partial \cap$ domain δ''_τ.

Now it cannot happen that c_R is bounded away from 0 on M, for it is a theorem of Riemannian geometry (Myers' theorem) that a complete Riemannian manifold whose Ricci curvature is positive and bounded away from 0 is compact. Thus one cannot deduce directly from the estimate just given the usual estimate for the $\bar\partial$ problem:

$$\langle \bar\partial\phi, \bar\partial\phi\rangle_\tau + \langle \delta''_\tau\phi, \delta''_\tau\phi\rangle_\tau \geqq C \langle \phi, \phi\rangle_\tau$$

for some positive constant C and all $\phi \in L^2(\tau) \cap$ domain $\bar\partial \cap$ domain δ''_τ. But a minor refinement of the functional analysis involved (**[11]** and **[19]**) shows that the estimate that we have established for $\phi \in L^2_{0,1}(\tau) \cap$ domain $\bar\partial \cap$ domain δ''_τ is sufficient, in the presence of the condition $\int g(\phi, \bar\phi)\, e^{-\tau}/c_R < +\infty$, to imply that there is a (locally L^2) solution of $\bar\partial f = \phi$. Standard elliptic regularity theory then implies that in fact f is a C^∞ function with $\bar\partial f = \phi$.

The previous paragraph is the end of our discussion of the L^2 method, and its concluding statements are what we shall use to finish the proof of Theorem 1D. We first prove a lemma:

Strictly Plurisubharmonic Function Lemma. *If M is a complete Kähler manifold with topologically trivial canonical bundle K (= the bundle of $(n, 0)$ forms, $n = \dim_C M$) and with the equation $\bar\partial f = \phi$ having a C^∞ solution f for every $C^\infty(0, 1)$ form ϕ such that $\bar\partial\phi = 0$, then there exists a C^∞ strictly plurisubharmonic function on M.*

The proof of the lemma is a global version of the classical construction of a local potential for a Kähler metric: As indicated earlier, there is a quadratic form on M_p, the Ricci form R, determined by $R(X, X) =$ the Ricci curvature of the direction X for $X \in M_p$, $\|X\| = 1$. The existence of this form is a standard result of Riemannian geometry. Let $\tilde R$ be the 2-form $X, Y \to^{\tilde R} R(JX, Y)$. Then $\tilde R$ is a real form of type (1, 1). It is closed and, up to a constant factor, represents in $H^2(M; \boldsymbol{R})$ the (first) Chern class of K^* (=dual of the canonical bundle). Since K and hence K^* are topologically trivial, $\tilde R$ represents 0 in $H^2(M; \boldsymbol{R})$, i.e., R is d-exact (by de Rham's theorem). Let $\tilde R = d\theta$, θ is a real 1-form. The form θ has a unique C^∞ decomposition $\theta = \theta_1 + \theta_2$ where θ_1 is type (1, 0), θ_2 is type (0, 1) and $\theta_2 = \bar\theta_1$: In holomorphic coordinates $z_1, \cdots, z_n$ the decomposition is determined by

$$\begin{aligned} dx_j &= \tfrac{1}{2}(dx_j + \sqrt{-1}\, dy_j) + \tfrac{1}{2}(dx_j - \sqrt{-1}\, dy_j), \\ dy_j &= -\tfrac{1}{2}\sqrt{-1}(dx_j + \sqrt{-1}\, dy_j) + \tfrac{1}{2}\sqrt{-1}(dx_j - \sqrt{-1}\, dy_j), \end{aligned}$$

$z_j = x_j + \sqrt{-1}\; y_j$, and local existence plus uniqueness implies global existence. Since $d\theta = \tilde R$ is of type (1, 1), $\partial\theta_1 = 0 = \bar\partial\theta_2$. Thus by the solvability of the $\bar\partial$ equation, $\theta_2 = \bar\partial f$ for some C^∞ function f. Then $\theta_1 = \partial\bar f$ and

$$\tilde R = (\partial + \bar\partial)(\partial\bar f + \bar\partial f) = \partial\bar\partial f + \bar\partial\partial\bar f = \bar\partial(f - \bar f).$$

The positive definiteness of the form R is equivalent to the strict plurisubharmonicity of $-\sqrt{-1}(f - \bar f)$ where $\tilde R = \partial\bar\partial(f - \bar f)$. □

We can now complete the proof of Theorem 1D: On a manifold M satisfying the hypotheses of that theorem, there is by the lemma just proved a C^∞ strictly plurisubharmonic function, say $g: M \to \boldsymbol{R}$. The function e^g is positive and again strictly plurisubharmonic. Let σ be a convex exhaustion function on M (existence of such having been previously noted), and let $\varepsilon: M \to \boldsymbol{R}$ be the positive continuous function determined by

$$\varepsilon(p) = \min\left(1, \frac{1}{4} \min_{X \in M_p; \|X\|=1} L_{e^g}(X, X)\right).$$

With this σ and ε, apply the corollary of the global approximation theorem. The sum $\tau + e^g$ of the resulting C^∞ function τ and e^g is an exhaustion function since $|\sigma - \tau| < 1$ and $e^g \geqq 0$. Moreover it is C^∞ and its Levi form $L_{\tau+e^g}$ satisfies for $X \in M_p$, $\|X\| = 1$:

$$\begin{aligned} L_{\tau+e^g}(X, X) &= L_\tau(X, X) + L_{e^g}(X, X) \\ &\geqq -2\varepsilon(p) + 4\varepsilon(p) = 2\varepsilon(p) > 0 \end{aligned}$$

since $\frac{1}{4} \min L_{e^g}(X, X) \geqq \varepsilon(p)$ and $L_\tau(X, X) \geqq -2\varepsilon(p)$ by the plurisubharmonicity lemma and (b) of the conclusion of the corollary. Thus $\tau + e^g$ is a C^∞ strictly plurisubharmonic exhaustion function on M. M is therefore a Stein manifold.

There is one last point left to nail down: the deduction of the corollary from the global approximation theorem. For this purpose let $\eta: M \to \boldsymbol{R}$ be a continuous function and define the subsheaf $\mathscr{S}$ consisting of germs of η-convex functions as follows: A germ f of a continuous function is in $\mathscr{S}$, i.e., is η-convex, iff f in a neighborhood of p can be expressed as the sum $f_1 + f_2$ where f_1 is convex and f_2 is C^∞ and satisfies

$$\min_{X \in M_p, \|X\|=1} D^2_{f_2}(X, X) > \eta(p).$$

Note that in this sense the germs of 0-convex functions are exactly the germs of strictly convex functions in the previously defined sense. For a fixed function $\eta: M \to \boldsymbol{R}$, the subsheaf $\mathscr{S}$ clearly satisfies the C^∞ stability property and is easily checked to satisfy the maximum closure property. That it satisfies the semilocal approximation property is obtained by a minor technical modification of the approximation lemma of §4 (and **[10]**; see also **[15]**). Thus $\mathscr{S}$ satisfies the hypotheses of the global approximation theorem for any continuous $\eta: M \to \boldsymbol{R}$. To obtain the corollary, take $\eta = -\varepsilon$: The corollary follows from the final observation that a convex function is $(-\varepsilon)$-convex for any positive continuous function $\varepsilon: M \to \boldsymbol{R}$.

Appendix. The strict convexity of the square of the distance on a complete simply connected manifold of nonpositive curvature. In this appendix, we shall prove the indicated strict convexity result, which was used in §4 to prove Theorem 1A. We take as starting point in this proof the result in §1 comparing the Riemannian distance on complete simply connected manifolds of nonpositive curvature with the distance function on euclidean space. (For a treatment of the subject which includes the proof of this comparison result see **[24]**.)

We need the following standard theorem of Riemannian geometry: If p, q are points of a complete Riemannian manifold, then there is a (not necessarily unique) geodesic from p to q such that the distance from p to q along this geodesic = the Riemannian distance from p to q. Such a geodesic is called a *minimizing geodesic* from p to q.

Now suppose that M is a complete simply connected manifold of nonpositive curvature. Let p, q be points of M and let C be a geodesic through q with the parameter of C = (oriented) arc length along C so that in particular $C(0) = q$. Let C_1 be a minimizing geodesic from p to q. And let α = the angle between the positive direction of C at q and the q-to-p direction of the geodesic C_1. Then by the comparison result of §1:

$$\text{distance}^2(p, C(t)) \geqq t^2 + \text{distance}^2(p, q) - 2t(\cos\alpha)(\text{distance}(p, q)).$$

If one knew that the function d_p^2 ($x \xrightarrow{d_p^2} \text{distance}^2(p, x)$, $x \in M$) were C^∞ then elementary calculus considerations applied to the inequality just given would show that the second derivative of d_p^2 along C at $q \geqq 2$.

Thus once it is shown that d_p^2 is C^∞, the proof of the strict convexity of d_p^2 will be complete. (Note: The function d_p^2 is not C^∞ on all Riemannian manifolds; its C^∞ character in the present case is a rather special property.) It is convenient at this point to introduce another Riemannian geometry concept, the exponential map $\exp_p$: $M_p \to M$. This map is defined as follows: $\exp_p(0) = p$, if $X \in M_p$, $X \neq 0$, then $\exp_p(X)$ = the point obtained by going arc length $\|X\|$ from p along the positive direction of the geodesic whose tangent at p is X. On any complete Riemannian manifold, $\exp_p$ is defined on all of M_p. Furthermore the existence of minimizing geodesics between any two points of a complete Riemannian manifold implies that $\exp_p(M_p) = M$ when M is complete. Also (on any Riemannian manifold) the mapping $\exp_p$ is C^∞ because of the C^∞ dependence of geodesics on their initial point and tangent at that point. However, in general the mapping $\exp_p$ is not 1-1: A simple example in which $\exp_p$ is not 1-1 for any p is the 2-sphere (= $P_1\boldsymbol{C}$) with its usual metric of constant (positive) curvature, or, for a noncompact case, $P_1\boldsymbol{C} \times \boldsymbol{C}$ with the product metric. But on complete simply connected Riemannian manifolds of nonpositive curvature the exponential maps are not only 1-1 but are in fact diffeomorphisms, for the following reason:

As noted already, $\exp_p$ is onto. Also, in this case $\exp_p$ is 1-1 because the comparison result of §1 implies that

$$\text{distance}(\exp_p X_1, \exp_p X_2) \geqq \|X_1 - X_2\|$$

for all $X_1, X_2 \in M_p$. This inequality also shows that $\exp_p$ is nonsingular: For any distance nondecreasing map between Riemannian manifolds is easily checked to be necessarily nonsingular, and the inequality just given can be rephrased as saying that $\exp_p$ is distance nondecreasing from M_p to M. (Here M_p is also considered as a Riemannian manifold with metric obtained from its inner product and thus with the Riemannian distance from X_1 to $X_2 = \|X_1 - X_2\|$.) Hence $\exp_p$ is indeed a diffeomorphism.

Now $d_p^2: M \to \boldsymbol{R}$ is the same function as $q \to \|\exp_p^{-1}(q)\|^2$, because $\exp_p$ being a diffeomorphism implies that there is only one geodesic from p to q which must consequently be a minimizing geodesic. Since $\exp_p^{-1}: M \to M_p$ is C^∞ and $X \to \|X\|^2$ is a C^∞ function on M_p, d_p^2 is C^∞.

In summary: If M is a complete simply connected Riemannian manifold of nonpositive curvature, then for any $p \in M$

(a) $\exp_p : M_p \to M$ is a C^∞ diffeomorphism;

(b) d_p^2 is a C^∞ function and its second derivative along any arc-length parameter geodesic is $\geqq 2$ so that d_p^2 is a C^∞ strictly convex function.

It also follows from the previous discussion that d_p^2 is an exhaustion function since

$$\{q \in M \mid d_p^2(q) \leqq \alpha\} = \exp_p\{X \in M_p \mid \|X\|^2 \leqq \alpha\},$$

which is compact, being the image of a compact set under a continuous, in fact C^∞, mapping. Actually, on any complete Riemannian manifold d_p^2 is an exhaustion function, but this general fact is less obvious than its particular present instance.

It can be shown that any Riemannian manifold on which there is a strictly convex exhaustion function is diffeomorphic to euclidean space (the exhaustion need not be assumed to be C^∞ since the global approximation theorem applies: see [**15**]). But the diffeomorphism may not be obtainable from $\exp_p$, $p \in M$. Thus complete simply connected manifolds of nonpositive curvature have many properties not shared by Riemannian manifolds in general, nor even by other Riemannian manifolds diffeomorphic to euclidean space.

ADDED IN PROOF (July 2, 1976). It has come to our attention that a problem related to Conjecture 8 on p. 82 of this paper has also been raised by S. Kobayashi in [**23**, Problem C.5, p. 402].

The following developments related to this paper have taken place in the meantime:

(A) The role of bisectional curvature in complex analysis and the theorem of Elencwajg quoted on p. 76 have both been clarified by the recent works of the authors. In particular, Theorems 1B and 1C on p. 74 are now known to be special cases of a more general theorem. See R. E. Greene and H. Wu, *On Kähler manifolds of positive bisectional curvature and a theorem of Hartogs*, Abh. Math. Sem. Univ. Hamburg, Kähler Jubilee Volume (to appear), and H. Wu, *An elementary method in the study of nonnegative curvature* (to appear).

(B) Theorem 5 on p. 77 has been substantially strengthened by the authors. The details will appear in a forthcoming paper, *On the nonexistence of nonconstant bounded harmonic functions on certain Riemannian manifolds.*

(C) In connection with Conjecture 1 on p. 79, P. Klembeck has recently constructed a complete Kähler metric of positive sectional curvature on $\mathbf{C}^n$. See his paper, *A complete Kähler metric of positive curvature on* $\mathbf{C}^n$, preprint, Princeton University.

(D) Conjecture 2 on p. 79 has been affirmatively resolved by Y.-T. Siu and S.-T. Yau in their papers: *On the structure of complete simply connected Kähler manifolds with nonpositive curvature*, Proc. Nat. Acad. Sci. U.S.A. (to appear) and *Complete Kähler manifolds with nonpositive curvature of faster than quadratic decay*, Ann. of Math. (to appear).

(E) Concerning the remarks following Question 6 on p. 81, A. Sommese has made the observation that for the existence of nonconstant meromorphic functions on a Kähler manifold M, it suffices to assume M is complete and possesses a line bundle whose curvature is nonnegative and positive at one point. See his paper, *Addendum to "Criteria for quasi-projectivity"*, Math. Ann. (to appear).

Bibliography

1. A. Andreotti and H. Grauert, *Théorèmes de finitude pour la cohomologie des espaces complexes*, Bull. Soc. Math. France **90** (1962), 193–259. MR **27** #343.
2. E. Calabi and B. Eckmann, *A class of compact complex manifolds which are not algebraic*, Ann. of Math. (2) **58** (1953), 494–500. MR **15**, 244.
3. J. Cheeger and D. Gromoll, *On the structure of complete manifolds of nonnegative curvature*, Ann. of Math. (2) **96** (1972), 413–443. MR **46** #8121.
4. F. Docquier and H. Grauert, *Levisches Problem und Rungescher Satz für Teilgebiete Steinscher Mannifaltigkeiten*, Math. Ann. **140** (1960), 94–123. MR **26** #6435.
5. P. Eberlein and B. O'Neill, *Visibility manifolds*, Pacific J. Math. **46** (1973), 45–109. MR **49** #1421.
6. G. Elencwajg, *Pseudo-convexité locale dans les variétiés Kähleriennes*, Ann. Inst. Fourier **25** (1975), 295–314.
7. M. P. Gaffney, *The conservation property of the heat equation on Riemannian manifolds*, Comm. Pure Appl. Math. **12** (1959), 1–11. MR **21** #892.
8. H. Grauert, *Bermerkenswerte pseudoconvexe Mannifaltigkeiten*, Math. Z. **81** (1963), 377–391. MR **29** #6054.
9. R. E. Greene and H. Wu, *Curvature and complex analysis*. I, II, III, Bull. Amer. Math. Soc. **77** (1971), 1045–1049; ibid. **78** (1972), 866–870; **79** (1973), 606–608. MR **44** #473; **45** #7657; **47** #4188.
10. ———, *On the subharmonicity and plurisubharmonicity of geodesically convex functions*, Indiana Univ. Math. J. **22** (1973), 641–653.
11. ———, *A theorem in complex geometric function theory*, Value Distribution Theory, Part A (Proc. Tulane Univ. Program on Value-Distribution Theory in Complex Analysis and Related Topics in Differential Geometry, 1972/73), Dekker, New York, 1974, pp. 145–167. MR **50** #5021.
12. ———, *Integrals of subharmonic functions on manifolds on nonnegative curvature*, Invent. Math. **27** (1974), 265–298.
13. ———, *Approximation theorems, C^∞ convex exhaustions and manifolds of positive curvature*, Bull. Amer. Math. Soc. **81** (1975), 101–104.
14. ———, *Some function-theoretic properties of noncompact Kähler manifolds*, Proc. Sympos. Pure Math., vol. 27, part II, Amer. Math. Soc., Providence, R. I., 1975, pp. 33–41.
15. ———, *C^∞ convex functions and manifolds of positive curvature*, Acta Math. (to appear).
16. ———, *C^∞ approximations of convex, subharmonic and plurisubharmonic functions* (to appear).
17. P. A. Griffiths, *Hermitian differential geometry, Chern classes, and positive vector bundles*, Global Analysis (Papers in Honor of K. Kodaira), Univ. of Tokyo Press, Tokyo, 1969, pp. 185–251. MR **41** #2717.
18. S. Helgason, *Differential geometry and symmetric spaces*, Pure and Appl. Math., vol. 12, Academic Press, New York, 1962. MR **26** #2986.
19. L. Hörmander, *An introduction to complex analysis in several variables*, Van Nostrand, Princeton, N. J., 1966. MR **34** #2933.
20. A. Huber, *On subharmonic functions and differential geometry in the large*, Comment. Math. Helv. **32** (1957), 13–72.
21. P. F. Klembeck, *Function theory on complete open Hermitian manifolds*, Thesis, University of California, Los Angeles, 1975.
22. S. Kobayashi, *Hyperbolic manifolds and holomorphic mappings*, Dekker, New York, 1970. MR **43** #3503.
23. ———, *Intrinsic distances, measures and geometric function theory*, Bull. Amer. Math. Soc. **82** (1976), 357–416.
24. S. Kobayashi and K. Nomizu, *Foundations of differential geometry*. Vol. II, Interscience Tracts in Pure and Appl. Math., no. 15, part II, Interscience, New York, 1969. MR **38** #6501.

25. J. Morrow and K. Kodaira, *Complex manifolds*, Holt, Rinehart and Winston, New York, 1971. MR **46** #2080.

26. M. Nakai and L. Sario, *Classification theory of Riemann surfaces*, Die Grundlehren der math. Wissenschaften, Band 164, Springer-Verlag, Berlin and New York, 1970. MR **41** #8660.

27. R. Narasimhan, *The Levi problem for complex spaces*. II, Math. Ann. **146** (1962), 195–216. MR **32** #229.

28. ———, *The Levi problem in the theory of functions of several complex variables*, Proc. Internat. Congress Math. (Stockholm, 1962), Inst. Mittag-Leffler, Djursholm, 1963, pp. 385–388. MR **31** #371.

29. R. Rochberg, *Stetige streng pseudoconvexe Functionen*, Math. Ann. **175** (1968), 257–286. MR **36** #5386.

30. E. Vesentini, *Lectures on Levi convexity of complex manifolds and cohomology vanishing theorems*, Tata Inst. Fund. Res. Lectures on Math., no. 39, Tata Institute of Fundamental Research, Bombay, India, 1967. MR **38** #342.

31. R. O. Wells, Jr., *Differential analysis on complex manifolds*, Prentice-Hall, Englewood Cliffs, N. J., 1973.

32. H. Wu, *Normal families of holomorphic mappings*, Acta Math. **119** (1967), 193–233. MR **37** #468.

33. P. Yang, *Some theorems in complex differential geometry*, Thesis, University of California, Berkeley, 1974; See also: *Curvature of complex submanifolds of* C^n, J. Differential Geometry (to appear).

34. S.-T. Yau, *Some function-theoretic properties of complete Riemannian manifolds and their applications to geometry*, Indiana Univ. Math. J. (to appear).

35. ———, *Harmonic functions on complete Riemannian manifolds*, Comm. Pure Appl. Math. **28** (1975), 201–228.

36. ———, *A general Schwarz lemma for Kähler manifolds*, 1975 Summer Institute.

University of California, Los Angeles

University of California, Berkeley

SEMINAR SERIES

DIFFERENTIAL GEOMETRY AND COMPLEX ANALYSIS

Proceedings of Symposia in Pure Mathematics
Volume 30, 1977

MOVING FRAMES IN HERMITIAN GEOMETRY

SIMONE DOLBEAULT

Frenet frames are useful tools to study holomorphic curves in projective space [6]. Following a paper by S. S. Chern, M. J. Cowen and A. L. Vitter III [3], we shall try to outline E. Cartan's method of moving frames for Hermitian geometry and characterize those Hermitian manifolds possessing Frenet frames along every holomorphic curve by curvature conditions. Then, we shall consider a particular case of Kähler manifolds and examine Riemannian manifolds. As an example, we use Frenet frames to obtain easily classical results about holomorphic curves in complex projective space.

1. E. Cartan's moving frame in Hermitian geometry. Let M be a Hermitian manifold, i.e. a complex manifold of complex dimension n, with a metric

$$H = \langle \cdot \rangle = \sum_{1 \leq k \leq n} \omega_k \bar{\omega}_k,$$

where $\omega_1, \cdots, \omega_n$ are forms of type (1, 0) determined up to a unitary transformation, constituting a basis of the cotangent space T_m^*M at $m \in M$. Let TM be the tangent bundle and $(e_1, \cdots, e_n)$ the unitary basis of T_mM dual to $(\omega_1, \cdots, \omega_n)$; then $(m; e_1, \cdots, e_n)$ is a *unitary frame* at m. Given a connection $\omega = (\omega_{ik})$ on TM, then

$$De_i = \sum_{1 \leq k \leq n} \omega_{ik} e_k \qquad (i = 1, \cdots, n),$$

and the *torsion matrix* τ is determined by

$$\tau_i = d\omega_i - \sum_{1 \leq k \leq n} \omega_k \wedge \omega_{ki} \qquad (i = 1, \cdots, n).$$

It is well known [2], [5], that there is on TM a unique connection, called the *Hermitian connection*, such that the following two conditions are fulfilled:
$\omega_{ik} + \bar{\omega}_{ki} = 0$;
τ is of type (2, 0).

AMS (MOS) subject classifications (1970). Primary 53C25, 53C55.

REMARK. The first condition means that the connection preserves H, i.e., $D\langle e_i, e_k\rangle = 0$; the second is equivalent to: ω is of type (1, 0) [2].

Let ω be that Hermitian connection; then the curvature matrix Ω is of type (1, 1) and its components satisfy

$$\Omega_{ik} = d\omega_{ik} - \sum_{1\leqq j\leqq n} \omega_{ij} \wedge \omega_{jk}, \qquad \Omega_{ik} + \bar{\Omega}_{ik} = 0.$$

For each $m \in M$, X, $Y \in T_mM$, the curvature forms Ω_{ik} define a linear transformation $\Omega_m(X, \bar{Y}): T_mM \to T_mM$ determined by

$$\langle \Omega_m(X, \bar{Y})\, e_i, e_k \rangle = \Omega_{ik}(X, \bar{Y}).$$

Introducing the curvature tensor $\Omega_{ik} = \sum_{lj} R_{iklj}\, \omega_l \wedge \bar{\omega}_j$ and taking two orthonormal vectors X, $Y \in T_mM$, the *holomorphic sectional curvature* is $R(m; X, Y) = \sum_{ikjl} R_{ikjl}\, X_iY_kX_jY_l$; it does not depend on the choice of the basis $(e_1, \cdots, e_n)$ at m [5].

2. Frenet frames along holomorphic curves. A holomorphic curve is a holomorphic map $f: Z \to M$ from a Riemann surface into M. Let ζ and z_i $(i = 1, \cdots, n)$ be local coordinates of Z and M respectively. Then f is given by $z_i = z_i(\zeta)$ and its tangent vector $f'(0) \in T_{f(0)}M$ is given by $\sum_{l=i=n} (dz_j/d\zeta)\, \partial/\partial z_i = \zeta^\rho V$ where ρ is a nonnegative integer and V a nonzero vector.

DEFINITIONS. A *unitary frame for M along f* is a set e_i: $Z \to TM$ such that $(e_1(\zeta), \cdots, e_n(\zeta))$ is a unitary basis for $T_{f(\zeta)}M$ for each $\zeta \in Z$.

Given the Hermitian connection ω on TM, *a Frenet frame along f* is a unitary frame along f such that

$$e_1 = \langle V, V\rangle^{-1/2}\, V,$$
$$f^*(\omega_{ik}) = 0, \qquad \begin{cases} i = 1, & k > 2, \\ 1 < i < n, & |k - i| \neq 0, 1. \end{cases}$$

THEOREM. *Let M be a Hermitian manifold of dimension n. Consider the following conditions*:

(A) $\forall\, m \in M$, $\forall\, X \in T_mM$, $\Omega_m(X, \bar{X})\, X = a(X)\, X$, *where* $a(X) \in \boldsymbol{R}$.

(B) $\forall\, m \in M$, $\forall\, X, Y \in T_mM$, $X \perp Y$, $\Omega_m(X, \bar{X})Y = b(X)Y$, *where* $b(X) \in \boldsymbol{R}$.

Then M has a Frenet frame along every holomorphic curve iff one of the three following conditions is fulfilled:

(i) $n = 2$,

(ii) $n = 3$ *and* (A) *holds,*

(iii) $n = 4$ *and both* (A) *and* (B) *hold.*

IDEA OF THE PROOF [3]. Along f, we can take $f^*(\omega_1) = h(\zeta)d\zeta$, $f^*(\omega_k) = 0$, $2 \leqq k \leqq n$; the fact that τ is of type (2, 0) implies $f^*(\tau_i) = 0$ $(i = 1, \cdots, n)$; hence it follows after, eventually, a change of unitary frame that

$$f^*(\omega_{lk}) = \begin{cases} h_1(\zeta)d\zeta, & k = 2, \\ 0, & k > 2. \end{cases}$$

This proves (i).

For $n \geqq 3$, exterior differentiation shows that $h_1(\zeta)$ satisfies a differential equation whose integrability condition is (A); this proves (ii).

For $n \geqq 4$, an analogous procedure for all the $f^*(\omega_{kj})$ leads to the unique new condition (B), so that (iii) is proved.

Particular case. We consider Hermitian manifolds whose torsion is null, i.e., Kähler manifolds; then the result is simpler:

COROLLARY. *A Kähler manifold of dimension $n \geqq 3$ has a Frenet frame along every holomorphic curve iff it has constant holomorphic sectional curvature* [**3**].

REMARK. The case where M is a Riemannian manifold of dimension n and class C^r $(r \geqq 4)$ can be studied in a similar manner; then since $(e_1, \cdots, e_n)$ is an orthonormal basis of $T_m M$, the Riemannian connection implies the following conditions:

$$\omega_{ik} + \omega_{ki} = 0, \quad \tau = 0, \quad \Omega_{ik} + \Omega_{ki} = 0;$$

and it is easy to see that, if M admits a Frenet frame along every curve, then all the components of its curvature tensor are null, except, perhaps, R_{ikik} $(i, k = 1, \cdots, n)$.

3. Holomorphic curves in complex projective space [4]. For $M = \boldsymbol{P}^n$, we write $\theta = f^*(\omega_1) = h d\zeta$, $\theta_{ik} = f^*(\omega_{ik})$; Z has a conformal metric $h_0 = \theta\bar{\theta}$ and a Kähler form $\hat{H}_0 = \frac{1}{2}\sqrt{-1}\,\theta \wedge \bar{\theta}$; then there exists a unique connection φ such that $d\theta = \theta \wedge \varphi$, $\varphi + \bar{\varphi} = 0$ (explicitly one finds $\varphi = -\partial \log h + \bar{\partial} \log h$); then the Ricci form is defined by $\operatorname{Ric} H_0 = \frac{1}{2}\sqrt{-1}\, d\varphi = \sqrt{-1}\, \partial\bar{\partial} \log h$; this is equivalent to $\operatorname{Ric} H_0 = -2K\hat{H}_0$, where K is the gaussian curvature of the metric.

In a similar way, let us write, for $k = 1, \cdots, n-1$,

$$\hat{H}_k = \frac{\sqrt{-1}}{2}\,\theta_{k\,k+1} \wedge \bar{\theta}_{k\,k+1},$$

$\varphi_k = \theta_{kk} - \theta_{k+1\,k+1}$ and $\operatorname{Ric} H_k = \sqrt{-1}\, d\varphi_k/2$. Then one finds:

SECOND MAIN THEOREM.

$$\operatorname{Ric} H_0 = -2\hat{H}_0 + \hat{H}_1,$$
$$\operatorname{Ric} H_k = \hat{H}_{k-1} - 2\hat{H}_k + \hat{H}_{k+1} \qquad (k = 1, \cdots, n-1).$$

An immediate application is the theorem of Blaschke: The Poincaré metric $(1 - |\zeta|^2)^{-2} d\zeta\, d\bar{\zeta}$ on the unit disc cannot be obtained by an isometric imbedding into $\boldsymbol{P}^n$.

Another application is the following theorem of Calabi [**1**]: A nondegenerate holomorphic curve is uniquely determined up to rigid motion by its first fundamental form H_0.

REFERENCES

1. E. Calabi, *Isometric imbeddings of complex manifolds*, Ann. of Math. (2) **58** (1953), 1–23. MR **15**, 160.

2. S. S. Chern, *Complex manifolds without potential theory*, Van Nostrand Math. Studies, no. 15, Van Nostrand, Princeton, N. J., 1967. MR **37** #940.

3. S. S. Chern, M. J. Cowen and A. L. Vitter III, *Frenet frames along holomorphic curves*, Value Distribution Theory, part A, Dekker, New York, 1974, pp. 191–203. MR **50** #13616.

4. P. Griffiths, *On Cartan's method of Lie groups and moving frames as applied to uniqueness and existence questions in differential geometry*, Duke Math. J. (1974), 775–814.

5. S. Kobayashi and K. Nomizu, *Foundations of differential geometry*. Vols. I, II, Interscience Tracts in Pure and Appl. Math., no. 15, Interscience, New York, 1963, 1966. MR **27** #2945; **38** #6501.

6. H. Wu, *The equidistribution theory of holomorphic curves*, Ann. of Math. Studies, no. 64, Princeton Univ. Press, Princeton, N. J.; Univ. of Tokyo Press, Tokyo, 1970. MR **42** #7951.

UNIVERSITÉ DE POITIERS

Proceedings of Symposia in Pure Mathematics
Volume 30, 1977

LOCAL INVARIANTS OF REAL AND COMPLEX RIEMANNIAN MANIFOLDS

PETER B. GILKEY*

Introduction. In this note, we discuss some recent results concerning the characteristic classes of a Riemannian manifold. In the first section, we discuss the real case where our results are complete; in the second section we discuss the complex case. For complex manifolds we have some partial results, but there are still several open problems. Our work was motivated by the following question which was proposed by I.M. Singer: Let M^m denote a compact m-dimensional Riemannian manifold without boundary. Let |dvol| denote the Riemannian measure and let $\chi(M)$ denote the Euler characteristic of M. The Chern-Gauss-Bonnet-von Dyck theorem [1] states:

$$\chi(M) = \int_M E_m(G)\,|\,\mathrm{dvol}\,|$$

where $E_m(G)$ is an invariant which depends polynomially on the first and second derivatives of the metric tensor G. Singer's question is the following: Let $P(G)$ be any invariant which depends polynomially on the derivatives of the metric and suppose that

$$P(G)(M) = \int_M P(G)\,|\,\mathrm{dvol}\,|$$

is independent of the metric G for every M. Is there a constant c such that $P(G)(M) = c\chi(M)$? In other words, is the only diffeomorphism invariant in the category of unoriented Riemannian manifolds which is obtained by integrating a local formula in the derivatives of the metric the Euler characteristic?

In this form, the problem was first solved by E. Miller [6]. Miller used techniques from cobordism to establish this result for a wider class of invariants of the metric.

AMS (MOS) subject classifications (1970). Primary 32C10; Secondary 53A55, 53B20, 58G10.

*Research partially supported by NSF grant MPS72-04357.

We prove a local version of this result using the techniques of differential geometry. We show under the assumptions above that

$$P(G) = cE_m + \delta Q(G)$$

where $Q(G)$ is a 1-form valued invariant which depends polynomially on the derivatives of the metric tensor; δ is the formal adjoint of exterior differentiation d. This local version implies the global integrated version and was conjectured by Kulkarni [5]. The Euler class is just one example of a characteristic class; we will discuss similar theorems for the other characteristic classes in the remainder of this paper.

1. Let M be a real Riemannian manifold of dimension m which is isometrically embedded in a Riemannian manifold N of dimension $m + r$. M is compact, but N need not be. Let V be the normal bundle of the embedding; we consider invariance relative to the following four groups: let $\nu = \nu_1 \times \nu_2$. $\nu_1 = SO(m)$ if M is oriented and $O(m)$ otherwise; $\nu_2 = SO(r)$ if V is oriented and $O(r)$ otherwise. Let $X = (x_1, \cdots, x_{m+r})$ be a coordinate system defined in a neighborhood U of x_0 in N. X is a ν-coordinate system iff:

(a) $U \cap M = \{x\colon x_j(x) = 0 \text{ for } m + 1 \leqq j \leqq m + r\}$.

(b) If $\nu_1 = SO(m)$, $(\partial/\partial x_1, \cdots, \partial/\partial x_m)$ is a frame for TM which induces the orientation of M.

(c) If $\nu_2 = SO(r)$, the projection of $(\partial/\partial x_{m+1}, \cdots, \partial/\partial x_{m+r})$ to V defines a frame which induces the orientation of V.

Let $\alpha = (\alpha(1), \cdots, \alpha(m + r))$ be a multi-index. If X is a ν-coordinate system, let $d_{X,\alpha} = (\partial/\partial x_1)^{\alpha(1)} \cdots (\partial/\partial x_{m+r})^{\alpha(m+r)}$. We adopt the notational convention that indices $i, j, \cdots$ run from 1 through m; indices $u, v, \cdots$ run from 1 through $m + r$. We sum over repeated indices.

We introduce formal variables $g_{uv/\alpha} = g_{vu/\alpha}$, for the derivatives of the metric. If $\alpha = (0, \cdots, 0)$, $g_{uv/\alpha} = g_{uv}$, let

$$g = \det(g_{uv})^{1/2}, \qquad g_M = \det(g_{ij})^{1/2}.$$

Let $\mathscr{P}$ denote the polynomial algebra in the $\{g_{uv/\alpha}, g, g^{-1}, g_M, g_M^{-1}\}$ variables subject to the relations $g^2 = \det(g_{uv})$ and $g_M^2 = \det(g_{ij})$. Let $I = (i_1, \cdots, i_p)$ for $1 \leqq i_1 < \cdots < i_p \leqq m$. Let $|I| = p$ and $dx^I = dx_{i_1} \wedge \cdots \wedge dx_{i_p} \in \bigwedge^p(T^*M)$. P is a p-form valued polynomial in the derivatives of the metric if P has the form $P = \sum_{|I|=p} P_I\, dx^I$ for $P_I \in \mathscr{P}$.

Let X be a ν-coordinate system defined near $x_0 \in M$. Define

$$\begin{aligned} g(X,G)(x_0) &= \det(G(\partial/\partial x_u, \partial/\partial x_v)(x_0))^{1/2}, \\ g_M(X,G)(x_0) &= \det(G(\partial/\partial x_i, \partial/\partial x_j)(x_0))^{1/2}, \\ g_{uv/\alpha}(X,G)(x_0) &= d_{X,\alpha}(G(\partial/\partial x_u, \partial/\partial x_v))(x_0). \end{aligned}$$

If P is as above, define $P(X, G)(x_0) \in \bigwedge^p(T^*M)$ by evaluation. If $P(X, G)(x_0) = P(Y, G)(x_0)$ for any two ν-coordinate systems X and Y and for any G, P is ν-invariant. Let $P(G)(x_0)$ denote this common value and let $\mathscr{P}_{\nu,p}$ denote the vector space of all such ν-invariants.

The two natural differential operators d and δ induce maps:

$$d\colon \mathscr{P}_{\nu,p} \to \mathscr{P}_{\nu,p+1} \quad \text{and} \quad \delta\colon \mathscr{P}_{\nu,p} \to \mathscr{P}_{\nu,p-1}.$$

$d^2 = \delta^2 = 0$. If M is oriented and $P \in \mathscr{P}_{\nu,m}$, let $P(G)(M) = \int_M P(G)$. Dually, if $P \in \mathscr{P}_{\nu,0}$ let $P(G)(M) = \int_M P(G)\,|\text{dvol}|$.

The Pontrjagin forms of $T(M)$ are $O(m)$ invariants of the metric on M. If $r \geqq 2$, there are additional characteristic classes which arise from the normal bundle V. Let W be an s-dimensional real vector bundle over M with a Riemannian connection ∇. The algebra generated by the Pontrjagin forms are the $O(s)$ characteristic forms of the connection. If s is even and W is oriented, we can define an additional invariant, the Euler form $E(\nabla)$. If s is odd, let $E(\nabla) = 0$. This is an $SO(s)$-invariant s-form valued polynomial in the curvature tensor of the connection. $E(\nabla)^2$ is a Pontrjagin form. The algebra generated by the Pontrjagin forms and $E(\nabla)$ are the $SO(s)$-characteristic forms of the connection.

Let ∇_N denote the Levi-Civita connection on the tangent space $T(N)$ over M. This induces a connection ∇_V on V. Let ∇_M denote the Levi-Civita connection of the restriction of the metric to M. If P is a ν_1-characteristic p-form of the connection ∇_M and Q is a ν_2-characteristic q-form of the connection ∇_V, $PQ \in \mathscr{P}_{\nu,p+q}$. The vector space generated by all such products is the space of ν-characteristic forms.

Let $T^m = S^1 \times \cdots \times S^1$ denote the m-dimensional torus. Identify T^m with $T^m \times 0$ in $T^m \times R^r$ to define the standard embedding. We proved [2]:

THEOREM 1.1. *Let $P \in \mathscr{P}_{\nu,p}$ with $dP = 0$. If $p = m$, we assume $P(G)(T^m) = 0$ for every G on $T^m \times R^r$. Then there exists $Q \in \mathscr{P}_{\nu,p-1}$ and a ν-characteristic form $R \in \mathscr{P}_{\nu,p}$ such that $P = dQ + R$.*

COROLLARY 1.2. *Let $P \in \mathscr{P}_{\nu,p}$.*

(a) *If $dP(G) = 0$ for all G, $P(G)$ induces a map from metrics to the pth de Rham cohomology of M. If the cohomology class represented by $P(G)$ is independent of G for all G, this class is a ν-characteristic class.*

(b) *Let $p = 0$ and $\nu_1 = O(m)$. If $P(G)(M)$ is independent of G for all G, there is a constant c such that $P(G)(M) = c\chi(M)$.*

We generalized these results to the case that both M and N are pseudo-Riemannian manifolds [2], [3]. The bilinear form G on N and its restriction to M are assumed to be nonsingular. Let $O(p, q)$ be the group of linear maps of R^{p+q} which preserve a nondegenerate bilinear form of type (p, q); $SO(p, q)$ is the subgroup of orientation-preserving maps. Invariance is defined in relation to the groups $\nu = \nu_1 \times \nu_2$ where $\nu_1 = O(p_1, q_1)$ or $SO(p_1, q_1)$ for $p_1 + q_1 = m$ and $\nu_2 = O(p_2, q_2)$ or $SO(p_2, q_2)$ for $p_2 + q_2 = r$. We define the spaces $\mathscr{P}_{\nu,p}$ similarly and both Theorem 1.1 and Corollary 1.2 hold for these groups as well.

We can further generalize these results to the class of invariants which are C^∞ functions of the metric and its derivatives up to some finite order. Let $P(g_{uv/\alpha})$ be a C^∞ function of the variables $\{g_{uv/\alpha}\}$ for $|\alpha| \leqq k(P)$ and where g_{uv} is a nondegenerate bilinear form of type (p, q). We say that P is ν-invariant if $P(X, G) = P(Y, G)$ for any two ν-coordinate systems X and Y and for all G. Let $\mathscr{C}^\infty_{\nu,p}$ denote the vector space of all such ν-invariants. We show [4] that Theorem 1.1 generalizes to:

THEOREM 1.3. *Let $P \in \mathscr{C}^\infty_{\nu,p}$ with $dP = 0$. If $p = m$, we assume $P(G)(T^m) = 0$ for every G on $T^m \times R^r$. Then there exists $Q \in \mathscr{C}^\infty_{\nu,p-1}$ and a ν-characteristic form $R \in \mathscr{P}_{\nu,p}$ such that $P = dQ + R$. If P is real-analytic, Q can be chosen to be real-analytic.*

The map $P \to Q$ can be chosen to be linear and continuous in the C^∞ topology. The analogue of Corollary 1.2 *is still true for the spaces* $\mathscr{C}^\infty_{\nu,p}$.

2. The results for complex manifolds are not complete. Let M be a compact holomorphic manifold of real dimension $2m$, let G be a Hermitian metric on M, and let ∇ be the holomorphic connection on the complex tangent space $T_c(M)$. The algebra generated by the Chern forms of the connection ∇ are the $U(m)$ characteristic forms of the metric. X is a $U(m)$ coordinate system iff X is a system of holomorphic coordinates for M. We define the spaces $\mathscr{P}_{U(m),p}$ and $\mathscr{C}^\infty_{U(m),p}$ as in the real case. If P is a $U(m)$ characteristic p-form, $P \in \mathscr{P}_{U(m),p}$, let T^{2m} denote the complex torus which is defined by the lattice of Gaussian integers. The techniques used in the proof of Theorem 1.2 can be used to show

THEOREM 2.1. *Let* $P \in \mathscr{C}^\infty_{U(m),p}$ *with* $dP = 0$. *If* $p = 2m$, *we assume* $P(G)(T^{2m}) = 0$ *for every* G. *Then there exists* $Q \in \mathscr{C}^\infty_{\nu,p-1}$, *a* $U(m)$ *characteristic form* R, *and an error term* $S \in \mathscr{P}_{U(m),p}$ *such that* $P = dQ + R + S$. *If* P *is polynomial,* Q *is polynomial*; *if* P *is real-analytic,* Q *is real-analytic. The map* $P \to Q$ *is continuous in the* C^∞ *topology. The error term* S *has the following properties*:

(a) $S(G) = 0$ *if* G *is a Kähler metric.*

(b) $S(cG) = S(G)$ *for any constant* $c > 0$.

(c) $S(G) = 0$ *if* $G = G_0 \times 1$ *is a product metric on* $N \times T^2$. N *is a* $2m - 2$ *real-dimensional complex manifold with Hermitian metric* G_0 *and* 1 *is the flat metric on the torus* T^2.

(d) *The cohomology class represented by* $S(G)$ *is independent of* G *for any* M.

COROLLARY 2.2. *Let* $P \in \mathscr{C}^\infty_{U(m),p}$ *with* $dP = 0$. *If the cohomology class represented by* $P(G)$ *is independent of* G, *this class is a* $U(m)$ *characteristic class for any manifold which admits a Kähler metric.*

Corollary 2.2 is a weaker statement than can be proved using the methods of cobordism. E. Miller [6] proved that this theorem is true even if M does not admit a Kähler metric. We conjecture that Theorem 2.1 is true for $S = 0$; this would enable us to remove the hypothesis that M admits a Kähler metric in Corollary 2.2. This is an open problem, as is the generalization of Theorem 2.1 to embeddings of one complex manifold in another complex manifold. There are similar open problems for pseudo-Riemannian complex manifolds. The structure group in this case is $U(p, q)$.

REFERENCES

1. S. Chern, *A simple intrinsic proof of the Gauss-Bonnet formula for closed Riemannian manifolds*, Ann. of Math. (2) **45** (1944), 747–752. MR **6**, 106.

2. P. Gilkey, *Local invariants of an embedded Riemannian manifold*, Ann. of Math. (2) **102** (1975), 187–204.

3. ———, *Local invariants of a pseudo-Riemannian manifold*, Math. Scand. **36** (1975), 109–130.

4. ———, *Local invariants of a Riemannian manifold*, Advances in Math. (to appear).

5. R. Kulkarni, *Index theorems of Atiyah-Bott-Patodi and curvature invariants*, Montreal University, Montreal, Canada.

6. E. Miller, Doctoral Thesis, M. I. T., Cambridge, Mass.

7. I. M. Singer, private communication.

PRINCETON UNIVERSITY

Proceedings of Symposia in Pure Mathematics
Volume 30, 1977

FUNCTION THEORY ON COMPLETE OPEN HERMITIAN MANIFOLDS

PAUL KLEMBECK

1. Introduction. It is well know that a convex domain in C^n is a pseudoconvex domain. Therefore it is natural to ask whether or not convexity has function theoretic consequences in the more general setting of an open complex manifold. In particular, one may ask if geometric conditions which cause a complex manifold to admit a strictly geodesically convex exhaustion function imply that the manifold is a Stein manifold. Two theorems to this effect are due to Greene and Wu.

THEOREM 1.1 [**3**]. *A complete open simply connected Kähler manifold of nonpositive sectional curvatures is a Stein manifold.*

THEOREM 1.2 [**5**]. *A complete noncompact Kähler manifold of everywhere positive sectional curvatures is a Stein manifold.*

Both these theorems result from the existence of strictly geodesically convex exhaustion functions on the manifolds in question, namely distance squared from a fixed point for Theorem 1.1 and a more complex function used by Cheeger and Gromoll [**1**] in Theorem 1.2.

The purpose of this paper is to study the effect of requiring a Hermitian rather than a Kähler metric in Theorems 1.1 and 1.2. To this end, in §2 we examine the relationship between convexity and pseudoconvexity on Hermitian manifolds, in §3 we provide a counterexample to the general extension of Theorem 1.1 and in §4 we discuss some positive results towards dropping the Kähler condition in Theorems 1.1 and 1.2.

I would like to express my appreciation to Robert E. Greene for his generous help in this work. This work may be found in more detail in [**6**].

2. Geodesic convexity and plurisubharmonicity. A function on a Riemannian

AMS (*MOS*) *subject classifications* (1970). Primary 53C55; Secondary 53B35, 32F05.

manifold M is called (strictly) geodesically convex if its restriction to each geodesic in M is (strictly) convex in the one variable sense. For a function $f \in C^2(M)$ this means $(d^2/d\tau^2)f(c(\tau)) \geqq 0$ for each geodesic $c(\tau)$ in M, and if one defines the second covariant derivative by $D_f^2(v, w) = v(wf) - (D_v w)f$, this is the same as $D_f^2(T, T) \geqq 0$ for all $T \in TM_p$ at each $p \in M$. The latter observation arises from the fact that $D_T T = 0$ if T is the tangent vector field to a geodesic.

If

$$L_f = 4 \sum_{i,j} \frac{\partial^2 f}{\partial z_i \partial \bar{z}_j} dz_i \otimes d\bar{z}_j$$

is the Levi form of f, then in a Kähler manifold an easy computation based on the existence of holomorphic normal coordinates yields the formula

$$L_f(w, w) = D_f^2(\operatorname{Re} w, \operatorname{Re} w) + D_f^2(J \operatorname{Re} w, J \operatorname{Re} w), \qquad w \in TM^h. \tag{1}$$

Here TM^h denotes the holomorphic tangent space, $\operatorname{Re}(\partial/\partial z_i) = \partial/\partial x_i$, and J is the almost-complex structure. Formula (1) immediately yields the following lemma.

LEMMA 2.1 [4]. *Let M be a Kähler manifold and $f \in C^2(M)$. Then if f is a strictly geodesically convex function, f is strictly plurisubharmonic.*

If one requires M to be a Hermitian rather than a Kähler manifold then formula (1) becomes

$$\begin{aligned} L(_f w, w) = {} & D_f^2(\operatorname{Re} w, \operatorname{Re} w) + D_f^2(J \operatorname{Re} w, J \operatorname{Re} w) \\ & + (D_{\operatorname{Re} w} \operatorname{Re} w + D_{J \operatorname{Re} w} J \operatorname{Re} w) f, \qquad w \in TM^h, \end{aligned} \tag{2}$$

and combining formula (2) with some straightforward calculations yields the following:

THEOREM 2.2. *Let M be a Hermitian manifold and $p \in M$. Then every strictly geodesically convex function at p is plurisubharmonic at p if and only if $D_{\operatorname{Re} w} \operatorname{Re} w = - D_{J\operatorname{Re} w} J \operatorname{Re} w$ for all $w \in TM_p^h$.*

For complex dimension two the condition $D_{\operatorname{Re} w} \operatorname{Re} w = - D_{J\operatorname{Re} w} J \operatorname{Re} w$ turns out to be equivalent to the Kähler condition, but for higher complex dimensions it is more general.

3. A counterexample. §2 shows that if M is a Hermitian manifold rather than a Kähler manifold, then strictly geodesically convex functions need not be plurisubharmonic. In particular the geodesically convex functions constructed in the proofs of Theorems 1.1 and 1.2 need not be plurisubharmonic, so these theorems might not extend to the Hermitian case. In fact Theorem 1.1 does not extend.

EXAMPLE 3.1. Let

$$\begin{aligned} f(z) = {} & 1 + \sum_{i=1}^{n-2} z_i \bar{z}_i + \varepsilon \left(\frac{z_{n-1} + \bar{z}_{n-1}}{2} \right)^2 + \left(\frac{z_{n-1} - \bar{z}_{n-1}}{2\sqrt{-1}} \right)^2 \\ & + \delta \left(\frac{z_n - \bar{z}_n}{2\sqrt{-1}} \right)^2 + \left(\frac{z_n + \bar{z}_n}{2} \right)^3 + \alpha \left(\frac{z_n + \bar{z}_n}{2} \right)^4 \end{aligned}$$

on the set $U = \{z \in C^n | f(z) < 1\}$ where $n \geqq 2$. The metric $ds^2 = (1 - f)^{-2} \cdot (\sum_{i=1}^n dz_i d\bar{z}_i)$ on U is a complete Hermitian metric of strictly negative

Riemannian sectional curvature and U is a simply connected nonpseudoconvex domain and therefore U is not a Stein manifold, for $\alpha, \delta, \varepsilon > 0$ and sufficiently small.

The proof of this example is a messy computation so we will only remark that it is derived by placing a Poincaré-like metric on a nonpseudoconvex domain which is homeomorphic to the ball.

Whether a counterexample to extending Theorem 1.2 to the Hermitian case exists is an open question.

4. Perturbations of Kähler metrics. Example 3.1 shows that Theorem 1.1 cannot, in general, be extended to Hermitian manifolds. However, formula (2) suggests that if a function on a manifold M is sufficiently geodesically convex and if the manifold M is close to being a Kähler manifold, then the function will be plurisubharmonic and therefore tell us something of the function theory on M.

With this in mind let us look at the proof of Theorem 1.1. The basic idea is that the Rauch comparison theorem may be interpreted to mean that smaller curvature makes the distance from a fixed point a more geodesically convex function. Thus on a manifold of nonpositive curvature the geodesic convexity of distance squared from a fixed point must exceed that in Euclidean space, that is, it must exceed 2. Therefore combining this remark with Hadamard's theorem and completeness shows M admits a C^∞ strictly geodesically convex exhaustion function. By Lemma 2.1 this function is a C^∞ strictly plurisubharmonic exhaustion function, so Grauert's solution of the Levi problem [**2**] implies Theorem 1.1.

If M is a Hermitian rather than a Kähler manifold we may apply formula (2) to find out when distance squared is a strictly plurisubharmonic exhaustion function. This gives us the following:

THEOREM 4.1. *Let M be a complete open simply connected Hermitian manifold of nonpositive Riemannian sectional curvature satisfying $|d\Phi_x| < 2/3d_p(x)$ for all $x \in M$ for some $p \in M$ where $d_p(x) = \operatorname{dist}(x, p)$ and Φ is the Kähler form. Then M is a Stein manifold.*

This may be interpreted as saying that if the deviation from being a Kähler manifold dies out at infinity then Theorem 1.1 still holds.

If we require that the sectional curvatures are bounded away from zero, then rather than comparing M to Euclidean space we can compare M to the Poincaré ball and obtain a stronger result.

THEOREM 4.2. *Let M be a complete open simply connected Hermitian manifold with all Riemannian sectional curvatures $\leqq -C < 0$. Then if $|d\Phi| \leqq 2\sqrt{C}/3$, M is a Stein manifold.*

Finally, by a careful analysis of the function used by Cheeger and Gromoll on manifolds of positive curvature, we can extend Theorem 1.2.

THEOREM 4.3. *If M is a complete noncompact Hermitian manifold of everywhere positive Riemannian sectional curvature satisfying $|d\Phi| < rn_r(q)/6$ for each $q \in M$ for some $r \in (0, \infty)$, then M is a Stein manifold.*

Here we choose $n(r, q) \leqq \inf\{\text{sectional curvatures on } B_r(p)\}$ and choose $n_r(q) \leqq n(q, r)$ so that $r < \pi/\sqrt{n_r(q)}$. This last assumption is required so that the sphere of radius $1/\sqrt{n_r(q)}$ has no conjugate points with the distance between them less than r. Also this assumption implies that $\lim_{r\to\infty} rn_r(q) = 0$, so the condition in the theorem is nontrivial. Here $B_r(q)$ is the Riemannian ball about q of radius r.

We should note that as q goes to infinity, $n_r(q)$ goes to zero by Meyer's theorem, so we may interpret Theorem 4.3 as saying that if the deviation from being Kähler dies out at infinity then Theorem 1.2 holds.

REFERENCES

1. J. Cheeger and D. Gromoll, *On the structure of complete manifolds of nonnegative curvature*, Ann. of Math. (2) **96** (1972), 413–443. MR **46** #8121.

2. H. Grauert, *On Levi's problem and the imbedding of real analytic manifolds*, Ann. of Math. (2) **68** (1958), 460–472. MR **20** #5299.

3. R. E. Greene and H. Wu, *Curvature and complex analysis*, Bull. Amer. Math Soc. **77** (1971), 1045–1049. MR **44** #473.

4. ———, *On the subharmonicity and plurisubharmonicity of geodesically convex functions*, Indiana Univ. Math. J. **22** (1973), 641–653.

5. ———, *A theorem in complex geometric function theory*, Value Distribution Theory, Part A (Proc. Tulane Univ. Program on Value Distribution Theory in Complex Analysis and Related Topics in Differential Geometry, 1972/73), Dekker, New York, 1974, pp. 145–167. MR **50** #5021.

6. P. Klembeck, *Function theory on complete open Hermitian manifolds*, Ph.D. Thesis, UCLA, 1975.

PRINCETON UNIVERSITY

Proceedings of Symposia in Pure Mathematics
Volume 30, 1977

THE QUESTION OF HOLOMORPHIC CARRIERS

H. BLAINE LAWSON, JR.

One of the fundamental problems in complex geometry is that of determining which homology classes on a complex manifold are carried by holomorphic cycles. Specifically, suppose X is an n-dimensional complex manifold and let S be a reasonably nice compact subset of X. By a *holomorphic p-cycle in* (X, S) we mean an integral $2p$-current T with compact support on X such that the boundary of T is supported in S and, in $X - S$, T can be expressed as a locally finite sum,

$$T = \sum n_i[V_i] \tag{1}$$

where for each i, $n_i \in \boldsymbol{Z}$ and V_i is an irreducible, p-dimensional complex subvariety of $X - S$. If S is a Lipschitz neighborhood retract in X (for example, if S is a C^1 submanifold or a real analytic subvariety), then the homology groups $H_*(X, S; \boldsymbol{Z})$ can be computed using integral currents (cf. **[5]**). Our problem then is to find necessary and sufficient conditions for a given class $\alpha \in H_{2p}(X, S; \boldsymbol{Z})$ to contain a holomorphic p-cycle.

The object of this article is to discuss some recent progress that has been made on this problem by the methods of geometric measure theory. Throughout the discussion I shall use the notation and terminology from the excellent article of Reese Harvey **[12]** which appears in these PROCEEDINGS. All homology groups with integer coefficients will be defined using integral currents, and homology with real coefficients will be defined using de Rham currents (cf. **[19]**, **[6]**).

The most basic condition[1] necessary for a class $\alpha \in H_{2p}(X; \boldsymbol{Z})$ to contain a holomorphic cycle is that it be of *bidimension* (p, p). This means that the image of α under the homomorphism

AMS (MOS) subject classifications (1970). Primary 32C10, 49F20.

[1] In the general case this condition follows from an unpublished theorem of Harvey and Lawson concerning Aeppli cohomology.

$$H_{2p}(X, S; \mathbf{Z}) \to H^{2n-2p}(X - S; \mathbf{R})$$

is represented by a differential form ϕ of type $(n-p, n-p)$. If $S = \emptyset$, we may further assume that ϕ has compact support.

EXAMPLE 1. Suppose X is a compact Kähler manifold and $S = \emptyset$. Let $\alpha_{\mathbf{R}}$ denote the image of α in $H_{2p}(X; \mathbf{R}) \cong H_{2p}(X; \mathbf{Z}) \otimes \mathbf{R}$. Then α is of bidimension (p, p) if and only if the Poincaré dual of $\alpha_{\mathbf{R}}$ is represented by a harmonic form of type $(n-p, n-p)$.

EXAMPLE 2. Suppose X is a projective algebraic manifold and S is a smooth hyperplane section. Then $X - S$ is Stein and every class in $H_{2p}(X, S; \mathbf{Z})$, $p \geqq 0$, is of bidimension (p, p). Moreover, by the work of Grauert every cohomology class in $H^{2n-2p}(X - S; \mathbf{Q})$ is represented (via intersection) by chains of complex subvarieties in $X - S$. Our question now reduces to a question of the order of growth of these subvarieties in $X - S$. In particular, a class $\alpha \in H_{2p}(X, S; \mathbf{Z})$ will contain a holomorphic cycle if and only if its corresponding dual class in $H^{2n-2p}(X - S; \mathbf{Z})$ is represented by a chain of subvarieties of *finite mass* in the projective metric on $X - S$. We shall discuss this below in greater detail.

Note that any torsion class α is trivially of bidimension (p, p) since its image in $H^{2n-2p}(X - S; \mathbf{R})$ is zero.

The problem we are considering falls naturally into two stages. We begin by considering classes which are in the image of the inclusion map

$$j : H_{2p}(X; \mathbf{Z}) \to H_{2p}(X, S; \mathbf{Z}). \tag{2}$$

A class $\alpha = j(\hat{\alpha})$ will clearly contain a holomorphic cycle if $\hat{\alpha}$ does. In fact, if S is a complex subvariety, Bishop's theorem [2] states that α contains a holomorphic cycle if and only if $\hat{\alpha}$ does. (In particular, Bishop shows that any holomorphic cycle in (X, S) is a holomorphic cycle in X.)

The first case we shall consider then is where $S = \emptyset$. For a class $\alpha \in H_{2p}(X; \mathbf{Z})$ to contain a holomorphic cycle there is a second set of necessary conditions due to Atiyah and Hirzebruch [1]. These are homotopy conditions on the class α and do not involve the choice of complex structure on X. They arise from the Atiyah-Hirzebruch spectral sequence

$$E_2^* = H^*(X; \mathbf{Z}) \Rightarrow K^*(X).$$

Roughly, the idea is as follows. Suppose $\alpha \in H_{2p}(X; \mathbf{Z})$ contains a holomorphic cycle which we may assume, without loss of generality, to be an irreducible, p-dimensional subvariety V. The group $H^{2n-2p}(X, X - V; \mathbf{Z}) \cong \mathbf{Z}$ has a canonical generator u which lifts to the class $\alpha^* \in H^{2n-2p}(X; \mathbf{Z})$ given by intersection with V (the Poincaré dual of $\hat{\alpha}$ if X is compact). Similarly, the group $K^*(X, X - V)$ has a canonical element ξ given by a locally free resolution of the structure sheaf of V. The classes u and ξ are intimately related. To be precise, we have that $\operatorname{ch}(\xi) = u_{\mathbf{Q}} +$ higher degree terms, where ch denotes the Chern character and $u_{\mathbf{Q}}$ is the image of u in $H^{2n-2p}(X, X - V; \mathbf{Q})$. This implies that the dual class α^* survives the spectral sequence, and so each of the differentials d_r, $r \geqq 2$, must be zero on α^*. This implies, in particular, that

$$\delta_q \mathscr{P}_q^1(\alpha^*) = 0 \quad \text{for all primes } q \tag{3}$$

where $\mathscr{P}_q^1 : H^{2(n-p)}(X; \boldsymbol{Z}) \to H^{2(n-p+q-1)}(X; \boldsymbol{Z}/q)$ is the first q-primary operation and δ_q is the Bockstein homomorphism for the sequence $0 \to \boldsymbol{Z} \to \boldsymbol{Z} \to \boldsymbol{Z}/q \to 0$. For $q = 2$, $\mathscr{P}_2^1 = Sq^2$ and $\delta_2 = Sq^1$, and so the condition (3) can be written as

$$Sq^3(\alpha^*) = 0.$$

If $H^*(X; \boldsymbol{Z})$ is torsion free, the spectral sequence collapses and all these conditions are trivially satisfied. Moreover, given any class α, there is some multiple $N\alpha$ for which all the conditions are satisfied. However, there are examples of torsion classes $\alpha \in H_{2p}(X; \boldsymbol{Z})$ for X Stein and for X projective algebraic, such that $\delta_q \mathscr{P}_q^1(\alpha^*) \neq 0$ for some q. In general, therefore, we must consider arbitrary integer multiples of α when looking for holomorphic chains.

Suppose now that X is a compact submanifold of $\boldsymbol{P}^N$. Then by Kodaira and Spencer **[14]** every class $\alpha \in H_{2n-2}(X; \boldsymbol{Z})$ of bidimension $(n-1, n-1)$ contains a holomorphic cycle, and therefore, by the hard Lefschetz theorem, so does an appropriate multiple of any class of bidimension (1, 1). For the intervening dimensions, not so much is known.

However, assuming X is projective, we can reduce our question to an equivalent and somewhat more tractable one, namely: Under what conditions does α contain a *positive* holomorphic cycle, i.e., one where each integer n_i in the expression (1) is positive? The relationship between these problems is given as follows. Let h^p denote the class corresponding to a p-dimensional linear section of X. Then α contains a holomorphic cycle iff $\alpha + mh^p$ contains a positive holomorphic cycle for all sufficiently large positive integers m. The nontrivial half of this statement follows from the fact that any purely p-dimensional subvariety V of X can be expressed as the union of components of a variety Y which is the intersection on X of $n - p$ hypersurfaces from $\boldsymbol{P}^N$. (Consider $n - p$ generic projections onto $\boldsymbol{P}^{p+1}$.) Thus there is a positive p-dimensional subvariety W such that $[V] + [W] \sim mh^p$ (i.e., $-[V] \sim [W] - mh^p$) for some $m \in \boldsymbol{Z}^+$.

For the narrower question of finding positive holomorphic cycles in α there are further necessary conditions. The first is that the *corresponding real class*

$$\alpha_{\boldsymbol{R}} \in H_{2p}(X; \boldsymbol{R}) = H_c^{2n-2p}(X; \boldsymbol{R})$$

(where H_c^* denotes cohomology with compact supports) *be represented by a strongly positive* $(n-p, n-p)$*-current.*

Note. It is not true in general that $\alpha_{\boldsymbol{R}}$ will be represented by a *smooth* positive $(n-p, n-p)$-form. To see this, consider the class α carried by an exceptional curve C on an algebraic surface X. If $\alpha_{\boldsymbol{R}}$ contained a positive (1, 1)-form ϕ, then we would have

$$\alpha_{\boldsymbol{R}} \cup \alpha_{\boldsymbol{R}}([X]) = \int_X \phi \wedge \phi > 0,$$

in contradiction to the fact that $C \cdot C < 0$. Note, however, that from the arguments two paragraphs above, it suffices for the general question to consider only classes which *are* dual to smooth positive forms.

Suppose now that $\alpha_{\boldsymbol{R}}$ contains a strongly positive, $(n-p, n-p)$-current $\mathscr{T}$. Then this assumption has interesting consequences in terms of the generalized

Plateau problem on X. We know from the Harvey-Knapp-Wirtinger inequality [12, Theorem 1.23] that for any Kähler metric on X, *$\mathscr{T}$ is a current of least mass among all de Rham currents in $\alpha_{\boldsymbol{R}}$*. (This minimum mass depends only on the cohomology class of the Kähler form.) Furthermore, *any other current $\mathscr{T}' \in \alpha_{\boldsymbol{R}}$ is of least mass in $\alpha_{\boldsymbol{R}}$ if and only if $\mathscr{T}'$ is also positive and of type $(n-p, n-p)$*. It then follows from the structure theorem of King [13] that any *integral* current $T \in \alpha \subset \alpha_{\boldsymbol{R}}$, which is of least mass in $\alpha_{\boldsymbol{R}}$ must be a positive holomorphic chain

This leads us to an interesting and subtle geometric question concerning the following norms:

$$\|\alpha\| = \inf\{M(T) : T \in \alpha\}, \qquad \|\alpha_{\boldsymbol{R}}\| = \inf\{M(\mathscr{T}) : \mathscr{T} \in \alpha_{\boldsymbol{R}}\}.$$

We recall from the compactness theorems of Federer and Fleming [5] that there are always currents $T_0 \in \alpha$ and $\mathscr{T}_0 \in \alpha_{\boldsymbol{R}}$ such that $M(T_0) = \|\alpha\|$ and $M(\mathscr{T}_0) = \|\alpha_{\boldsymbol{R}}\|$. Hence, from our discussion above we have the following result.

THEOREM 1. *Let X be a compact Kähler manifold and suppose $\alpha \in H_{2p}(X; \boldsymbol{R})$ satisfies the condition that $\alpha_{\boldsymbol{R}}$ contains a strongly positive $(n-p, n-p)$-current. Then $\|\alpha_{\boldsymbol{R}}\| = \|\alpha\|$ if and only if α contains a positive holomorphic cycle.*

Since $\alpha_{\boldsymbol{R}} \supset \alpha$, we always have $\|\alpha_{\boldsymbol{R}}\| \leqq \|\alpha\|$. In fact, $\|\alpha_{\boldsymbol{R}}\| \leqq (1/m)\|m\alpha\|$ for all $m \in \boldsymbol{Z}^+$ and Federer [7] has shown that $\|\alpha_{\boldsymbol{R}}\| = \lim_{m\to\infty}(1/m)\|m\alpha\|$. An interesting set of examples where $\|\alpha_{\boldsymbol{R}}\| < \|\alpha\|$ is provided by the following result of Matsusaka. Let A be a principally polarized abelian variety of dimension n. This is a complex n-dimensional torus together with a class $\phi \in H^2(A; \boldsymbol{Z})$ of bidegree $(1, 1)$, which, considered as an integer-valued, skew-symmetric form on $H_1(A; \boldsymbol{Z})$, has determinant 1. By the Frobenius theorem (cf. [15]), there exists a basis $c_1, \cdots, c_{2n}$ of $H^1(A; \boldsymbol{Z})$ such that $\phi = \sum_{i=1}^{n} c_{2i-1} \cup c_{2i}$. Let $\alpha \in H_2(A; \boldsymbol{Z})$ be the Poincaré dual of the class $(1/(n-1)!)\,\phi^{n-1}$. Then Matsusaka's theorem [18] states that *α contains a positive holomorphic chain if and only if A is, up to finite coverings, a product of Jacobian varieties*. Most principally polarized abelian varieties are not of this form. Hence, generically, the class α satisfies $\|\alpha_{\boldsymbol{R}}\| < \|\alpha\|$. However, in each case we have $(n-1)!\,\|\alpha_{\boldsymbol{R}}\| = \|(n-1)!\,\alpha\|$, since $(n-1)!\,\alpha$ always contains the intersection of $(n-1)$ translates of the divisor corresponding to ϕ.

This leads to the following

DEFINITION. A class $\alpha \in H_k(X; \boldsymbol{Z})$ is said to be *stable* if there is a positive integer m such that $m\|\alpha_{\boldsymbol{R}}\| = \|m\alpha\|$.

Theorem 1 can now be rephrased in terms more consistent with the Matsusaka examples and with the results of Atiyah and Hirzebruch.

THEOREM 1′. *Let X and α be as in Theorem 1. Then some integer multiple of α contains a positive holomorphic cycle if and only if α is stable.*

Unfortunately, it is not true that every such class α on a projective manifold X is stable. In [17] it is shown that *for each $n \geqq 2$ there exists an abelian variety A of complex dimension $2n$ and a class $\alpha \in H_{2n}(A; \boldsymbol{Z})$ such that $\alpha_{\boldsymbol{R}}$ contains a strongly positive (n, n)-form but $\|\alpha_{\boldsymbol{R}}\| < (1/m)\|m\alpha\|$ for all $m \in \boldsymbol{Z}^+$*. The positive forms in these examples, however, lie on the boundary of the cone of all positive forms; and it still seems reasonable to ask the following

Question. Is every $\alpha \in H_{2p}(X; \boldsymbol{Z})$, such that $\alpha_{\boldsymbol{R}}$ contains a form in the interior of the cone of strongly positive $(n-p, n-p)$-forms on X, a stable class?

An affirmative answer to this question would be very strong since from the discussion above we have the following result.

THEOREM 2. *Let X be a projective algebraic manifold and let $h^p \in H_{2p}(X; \boldsymbol{Z})$ denote the class corresponding to a p-dimensional linear section of X. Suppose $\alpha \in H_{2p}(X; \boldsymbol{Z})$ is any class of bidimension (p, p). Then some integer multiple of α contains a holomorphic chain if and only if $\alpha + mh^p$ is stable for all sufficiently large integers m.*

Note that for m large, $(\alpha + mh^p)_{\boldsymbol{R}}$ contains a form in the interior of the cone of positive $(n-p, n-p)$-forms, since $(h^p)_{\boldsymbol{R}}$ contains the pth power of the Kähler form on X (cf. [**12**, Corollary 1.17]).

We shall now formulate a quite different sufficient condition for a class $\alpha \in H_{2p}(X; \boldsymbol{Z})$ to contain a positive holomorphic cycle. This condition is a consequence of a measure-theoretic characterization of the holomorphic chains on $\boldsymbol{P}^N$. Let J denote the almost complex structure on $\boldsymbol{P}^N$ and extend J to $\bigwedge^* T(\boldsymbol{P}^N)$ as a derivation. (That is, for a simple k-vector $\xi = v_1 \wedge \cdots \wedge v_k$, we set

$$J\xi = \sum_{i=1}^{k} v_1 \wedge \cdots \wedge Jv_i \wedge \cdots \wedge v_k.)$$

The kernel of J on $\bigwedge^{2p} T_x(\boldsymbol{P}^N)$ is precisely the set of real (p, p)-vectors at x. We recall that any $2p$-current T of finite mass on $\boldsymbol{P}^N$ determines a Radon measure $\|T\|$ and a field of unit $2p$-vectors $\boldsymbol{T}$, defined $\|T\|$-almost everywhere, such that

$$T(\omega) = \int \omega(\boldsymbol{T}_x)\, d\|T\|(x)$$

for any exterior $2p$-form ω. It is easy to see that T has bidimension (p, p) iff $\boldsymbol{T}_x$ is a (p, p)-vector (i.e., $J\boldsymbol{T}_x = 0$) for $\|T\|$-almost all x.

We now consider the Lie algebra $\mathscr{K}$ of Killing vector fields (infinitesimal isometries) on $\boldsymbol{P}^N$. To each k-current T of finite mass on $\boldsymbol{P}^N$ we associate a quadratic form Q_T on $\mathscr{K}$ as follows. For $V \in \mathscr{K}$, let ϕ_t be the flow generated by the vector field JV, and set

$$Q_T(V) = \frac{d^2}{dt^2} M(\phi_{t*}T)\Big|_{t=0}.$$

Then it is shown in [**16**] that, with respect to the natural inner product on $\mathscr{K}$,

$$\text{trace}(Q_T) = -\int \|J\boldsymbol{T}_x\|^2\, d\|T\|. \tag{4}$$

This formula has several immediate consequences. We say that a current T of finite mass on a compact Riemannian manifold X is *stable* if

$$\frac{d}{dt} M(\phi_{t*}T)\Big|_{t=0} = 0, \qquad \frac{d^2}{dt^2} M(\phi_{t*}T)\Big|_{t=0} \geqq 0$$

for all 1-parameter groups of diffeomorphisms ϕ_t on X. Clearly, any current of least mass in its homology class is stable. However, it is easy to find stable currents which are not homologically mass-minimizing. Formula (4) together with the structure theorem of Harvey and Shiffman ([**9**] or [**12**, Theorem 2.1]) gives the following result.

THEOREM 3. *Let T be a d-closed integral current on $\boldsymbol{P}^N$. Then T is stable if and only if T is a holomorphic cycle.*

In fact, formula (4) says much more. We recall that the Lie algebra $\mathscr{H}$ of holomorphic vector fields (infinitesimal holomorphic transformations) on $\boldsymbol{P}^N$ has a natural decomposition $\mathscr{H} = \mathscr{K} \oplus J(\mathscr{K})$. $\mathscr{H}$ is just the algebra of vector fields generated by the natural action of $PGL_{N+1}(\boldsymbol{C})$ on $\boldsymbol{P}^N$.

THEOREM 3′. *Let T be a d-closed integral k-current on $\boldsymbol{P}^N$ such that the Hausdorff $(k+2)$-measure of* supp T *is zero. (This is automatically true if, for example, T is homologically mass-minimizing in some subvariety of $\boldsymbol{P}^N$.) Then T is a holomorphic cycle if and only if the average second derivative of the mass of T with respect to holomorphic deformations of $\boldsymbol{P}^N$ is $\geqq 0$.*

Suppose now that X is a complex subvariety of $\boldsymbol{P}^N$ and $\alpha \in H_{2p}(X; \boldsymbol{Z})$. Each $g \in PGL_{N+1}(\boldsymbol{C})$ gives an embedding $X \cong g(X) \subset \boldsymbol{P}^N$ and a corresponding length $\|\alpha\|_g$ for α. The function $\phi_\alpha: PGL_{N+1}(\boldsymbol{C}) \to \boldsymbol{R}^+$ defined by $\phi_\alpha(g) = \|\alpha\|_g$ is clearly fixed under left multiplication by elements in PU_{N+1}, and therefore descends to a function

$$\phi_\alpha: \mathscr{D}_{N+1} \to \boldsymbol{R}^+$$

on the domain $\mathscr{D}_{N+1} = PGL_{N+1}(\boldsymbol{C})/PU_{N+1}$ of positive definite hermitian matrices mod $\boldsymbol{R}^+$.

COROLLARY 4. *Let X be a complex subvariety of $\boldsymbol{P}^N$ and let $\alpha \in H_{2p}(X; \boldsymbol{Z})$. Then the function ϕ_α is superharmonic on $\mathscr{D}_{N+1}$. (In particular, $\Delta\phi_\alpha$ is a negative measure.) Furthermore, if at any point $x \in \mathscr{D}_{N+1}$, $\Delta\phi_\alpha = 0$, in the sense that for any $\varepsilon > 0$, $\Delta\phi_\alpha + \varepsilon$ is a positive measure on some neighborhood of x, then ϕ_α is constant and α contains a holomorphic cycle.*

PROOF. Let T be any integral k-current on $\boldsymbol{P}^N$ and consider the C^∞ function ϕ_T: $\mathscr{D}_N \to \boldsymbol{R}^+$ given by $\phi_T(g) = M(g_* T)$. It follows from formula (4) that $\Delta\phi_T \leqq 0$. (In fact, if k is odd, $\Delta\phi_T \leqq -\phi_T$.) Hence, ϕ_T is superharmonic. It follows that $\phi_\alpha(x) = \inf\{\phi_T(x): T \in \alpha\}$ is also superharmonic.

Suppose now that $\Delta\phi_\alpha(x) = 0$. By the compactness of integral currents (cf. [5]) there is a current $T \in \alpha$ such that $\phi_T(x) = \phi_\alpha(x)$. Then $\Delta\phi_T(x) = 0$ for otherwise $\phi_T - \phi_\alpha$ would be a nonnegative, strictly superharmonic function in a neighborhood of x, which vanished at x and thereby contradicted the minimum principle. Since $\Delta\phi_T(x) = 0$, it follows from Theorem 3′ that T is a holomorphic cycle.

Corollary 4 has the following immediate consequence.

COROLLARY 5. *Suppose X is a complex submanifold of $\boldsymbol{P}^N$ and let $\alpha \in H_{2p}(X; \boldsymbol{Z})$. Suppose there is a Kähler metric h_0 on X which is cohomologous to the one induced from $\boldsymbol{P}^N$ and suppose that the length $\|\alpha\|_{h_0}$ of α in this metric satisfies $\|\alpha\|_{h_0} \leqq \|\alpha\|_h$ for all such metrics on some C^∞ neighborhood of h_0. Then α contains a holomorphic cycle.*

REMARK. Formula (4) has another interesting consequence. Let T be any integral current of odd dimension on $\boldsymbol{P}^N$, and let $\varepsilon > 0$. Then there exists $g \in PGL_{N+1}(\boldsymbol{C})$ such that $M(T) < \varepsilon$. Moreover, if $\alpha_1, \cdots, \alpha_r$ is a basis for the odd-dimensional homology of $X \subset \boldsymbol{P}^N$ and if $\varepsilon > 0$, then there exists $g \in PGL_{N+1}(\boldsymbol{C})$ such that

$$\sum_{i=1}^{r} \|\alpha_i\|_g < \varepsilon .$$

At the same time the norm of each (even-dimensional) class α_R which contains a positive (k, k)-current remains constant.

Let us return now to the general question of a class $\alpha \in H_{2p}(X, S; Z)$. If $\alpha = j(\hat{\alpha})$ for some $\hat{\alpha} \in H_{2p}(X; Z)$ (cf. [2]) and if $\hat{\alpha}$ contains a holomorphic cycle, then, of course, so does α. On the other hand, for appropriate choices of S, the existence of a holomorphic chain in α will imply the existence of one in $\hat{\alpha}$.

Consider, for example, the case where S is a complex subvariety of X. Then by Bishop's theorem [2] the closure in X of any complex subvariety *of finite volume* in $X - S$ is a subvariety of X. This means precisely that any holomorphic cycle in (X, S) extends to a holomorphic cycle in X.

As a consequence we have a "filtration" of the problem of finding holomorphic cycles on X. If we can solve the problem for a subvariety S, then from the sequence

$$H_*(S; Z) \xrightarrow{i} H_*(X; Z) \xrightarrow{j} H_*(X, S; Z)$$

and the observation above, it remains only to solve the problem for $H_*(X, S; Z)$. The group $j(H_*(X; Z))$ will be called the *S-primitive homology*[2] of X.

Note that an S-primitive class will contain a holomorphic cycle only if it is the image of a class of bidimension (p, p) in $H_{2p}(X; Z)$.

Let us consider the case where X is projective and S is a smooth hyperplane section on X (cf. Example 2). In this case, $X - S$ is Stein, and by Grauert's proof of the Oka principle we have that some integer multiple of every class $\alpha \in H^{2n-2p}(X - S; Z)$ is represented by a p-dimensional analytic subvariety of $X - S$. (See [4], for example.) Suppose now that $\alpha \in H_{2p}(X, S; Z) \cong H^{2n-2p}(X - S; Z)$ is the image of a (p, p)-class in $H_{2p}(X; Z)$. Then our problem reduces to showing that we can find an analytic subvariety representing (some multiple of) α in $X - S$, and having *finite volume* in the projective metric. If we consider $X - S$ embedded in C^N (an affine coordinate chart on P^N), this amounts to representing $m\alpha$ (some $m \in Z$) by a variety V such that the "growth function"

$$N(V, r) \equiv [V \cap B(r)](\omega^p) = O(1),$$

where $B(r) = \{z \in C^N : |z| \leqq r\}$ and

$$\omega = dd^c \log(1 + |z|^2).$$

Note that ω is just the Kähler form on P^N and so $N(T, r)$ is just ($p!$ times) the volume in P^N of $V \cap B(r)$. A beautiful study of the problem from this point of view has been made by Cornalba and Griffiths [4], [8], using methods of value distribution theory and L^2-estimates for the $\bar{\partial}$-operator.

One could rephrase this question of growth in terms of a relative Plateau problem on X. For $\varepsilon > 0$, let S_ε be the set of points in X a distance $\leqq \varepsilon$ from S in the projective metric on X. Then there are natural isomorphisms

$$H_{2p}(X, S) \xrightarrow{\approx} H_{2p}(X, S_\varepsilon) \xrightarrow{\cong} H_{2p}(X_\varepsilon, dX_\varepsilon)$$

[2] If X is compact and S is a divisor dual to the Kähler form on X, then the S-primitive homology is naturally the dual of the classical primitive cohomology.

where $X_\varepsilon = X - \mathrm{int}(S_\varepsilon)$ for ε sufficiently small; and from the Oka principle referred to above we know that $H_{2p}(X, S_\varepsilon; \boldsymbol{Q})$ is generated by holomorphic cycles. Hence, given $\alpha \in H_{2p}(X, S; \boldsymbol{Z})$ and $\varepsilon > 0$, we can define a length $\|\alpha\|_\varepsilon^{\mathscr{H}} = \inf M(T)$ over holomorphic cycles T in (X, S_ε), whose homology class corresponds to α. By the compactness theorem [**12**, Theorem 4.2] we have the following.

PROPOSITION 6. *Let S be a smooth hyperplane section on a projective manifold X and consider a class $\alpha \in H_{2p}(X, S; \boldsymbol{Z})$. Then α contains a holomorphic cycle if and only if $\gamma(\varepsilon) \equiv \|\alpha\|_\varepsilon^{\mathscr{H}}$ is bounded as $\varepsilon \to 0$.*

Recall that for α to contain a holomorphic cycle, $\alpha_{\boldsymbol{C}}$ must lie in the (p, p)-component of image$(j) \cong H_{2p}(X; \boldsymbol{C})/H_{2p}(S; \boldsymbol{C})$. To date no one has succeeded in using this condition to prove that $\|\alpha\|_\varepsilon^{\mathscr{H}}$ is bounded for $p < n - 1$. However, Cornalba and Griffiths have established a beautiful related result.

THEOREM 7 [**4**]. *Let X, S and α be as in Proposition 6, and suppose $\alpha_{\boldsymbol{C}} \in$ image(j) $\cong H_{2p}(X; \boldsymbol{C})/H_{2p}(S; \boldsymbol{C})$. If $\alpha_{\boldsymbol{C}}$ has a nonzero Dolbeault component of bidimension $(p - k, p + k)$, then $\|\alpha\|_\varepsilon^{\mathscr{H}} \geqq c\varepsilon^{-k}$ for some $c > 0$.*

Let us now turn to the question of more general subsets of X. To begin suppose S is any compact subset with $\mathscr{H}^{2p-1}(S) = 0$. ($\mathscr{H}^k$ denotes k-dimensional Hausdorff measure.) Then by the Remmert-Stein-Shiffman theorem [**20**], any holomorphic p-cycle in $X - S$, with or without compact support, extends to a holomorphic cycle in X. Hence, the groups of holomorphic p-cycles on X and on $X - S$ are essentially the same.

Suppose now that S has Hausdorff dimension $2p - 1$. In fact, let us suppose that S is a *compact C^1 submanifold with singularities in X*, i.e., outside a compact subset of $\mathscr{H}^{2p-1}$-measure zero, S is an oriented, $(2p - 1)$-dimensional, real submanifold of class C^1 and of finite volume in X. Again by arguments of Shiffman, every holomorphic p-chain T (with or without compact support) in $X - S$ has finite mass in a neighborhood of S (cf. [**10**, Theorem 4.7]). Therefore, T defines a rectifiable current in X, whose boundary dT is supported in S. It then follows from Federer [**6**, 4.1.7 and 4.1.15] that $dT = \sum n_i[S_i]$ where the n_i are integers and the S_i are connected components of the regular part of S.

Observe now that since T has bidimension (p, p), $dT = \partial T + \bar{\partial} T$ has Dolbeault components only in bidimensions $(p - 1, p)$ and $(p, p - 1)$. It is then an easy consequence that at each point $x \in S_i$ where $n_i \neq 0$, the tangent space contains a complex subspace of maximal dimension. That is,

$$\dim_{\boldsymbol{R}}[T_x S \cap J(T_x S)] = 2p - 2 \tag{5}$$

where J is the almost complex structure on X. Any $(2p - 1)$-dimensional submanifold with singularities which satisfies condition (5) at each regular point x is called *maximally complex*.

Interestingly, in spaces which are sufficiently small holomorphically, the condition of maximal complexity is enough to guarantee the existence of a cobounding holomorphic p-cycle.

THEOREM 8 (CF. [**10**], [**11**]). *Let X be a complex subvariety of $\boldsymbol{P}^N - \boldsymbol{P}^{N-q}$, and suppose S is a compact, $(2p - 1)$-dimensional, C^1 submanifold with singularities in X, where $p > q$. If S is maximally complex and $d[S] = 0$, then there is a unique holomorphic p-cycle T in (X, S) with $dT = [S]$.*

In the limiting case $p = q$,there is also a theorem of this type, although there are further necessary conditions.

THEOREM 9 ([**10**], [**11**]). *Let X and S be as in Theorem* 8, *but suppose now that $p = q$. Then there is a (unique) holomorphic p-cycle T in (X, S) with $dT = [S]$ if and only if S satisfies the following "moment condition":*

$$[S](\phi) = 0$$

for all $(p, p-1)$-forms ϕ on X with $\bar{\partial}\phi = 0$.

Note. There are strong boundary regularity results accompanying Theorems 8 and 9. See [**10**] or [**12**] for details.

The moment condition in Theorem 9 is necessary even in the cases $p > 1$. For example, let S be a real hypersurface in a p-dimensional abelian variety $A \subset \boldsymbol{P}^N$ and suppose $[S] \not\sim 0$ in $H_{2p-1}(A; \boldsymbol{Z})$. Then it is possible to find a linear subspace $\boldsymbol{P}^{N-p} \subset \boldsymbol{P}^N$ such that $S \subset \boldsymbol{P}^N - \boldsymbol{P}^{N-p}$. However, by the strong uniqueness [**10**, Proposition 8.5], $[S]$ is not the boundary of any holomorphic chain in $\boldsymbol{P}^N$.

When $p < q$, the group $H^{p,p-1}(\boldsymbol{P}^N - \boldsymbol{P}^{N-q}) = 0$ and the moment condition is trivial. Consequently, it is easy to find manifolds in $\boldsymbol{P}^N - \boldsymbol{P}^{N-q}$ which are both maximally complex and satisfy the moment condition but which fail to bound holomorphic chains. For example, choose S as in the paragraph above and choose any linear subspace $\boldsymbol{P}^{N-q}$ with $S \subset \boldsymbol{P}^N - \boldsymbol{P}^{N-q}$.

These last results can be interpreted in terms of our original problem. Let X be a subvariety of $\boldsymbol{P}^N - \boldsymbol{P}^{N-q}$ and let $S = \prod_{i=1}^n S_i$ be a compact subset of X where each S_i is a compact, oriented submanifold of dimension $2p - 1$ and where $p > q$. Consider the sequence

$$0 \to H_{2p}(X; \boldsymbol{Z}) \xrightarrow{j} H_{2p}(X, S; \boldsymbol{Z}) \xrightarrow{\delta} H_{2p-1}(S) \cong \bigoplus_{i=1}^{k} \boldsymbol{Z} \cdot [S_i].$$

A class $\alpha \in H_{2p}(X, S; \boldsymbol{Z})$ can contain a holomorphic cycle only if $\delta\alpha$ lies in the subgroup generated by the maximally complex components S_i. Let us therefore assume that each S_i is maximally complex. Then by Theorem 8 there is a homomorphism

$$\sigma\colon H_{2p-1}(S; \boldsymbol{Z}) \to H_{2p}(X, S; \boldsymbol{Z})$$

such that $\delta \circ \sigma = \mathrm{Id}$, and for each $\beta \in H_{2p-1}(S; \boldsymbol{Z})$ there is a holomorphic p-cycle in $\sigma(\beta)$.

We can conclude, in particular, that $H_{2p}(X, S; \boldsymbol{Z})/H_{2p}(X; \boldsymbol{Z}) \approx H_{2p-1}(S; \boldsymbol{Z})$, and that, if $H_{2p}(X; \boldsymbol{Z})$ is generated by holomorphic cycles, so is $H_{2p}(X, S; \boldsymbol{Z})$.

One interesting application of Theorem 8 is the following generalization of Bochner's extension theorem.

Let Ω be a domain in $\boldsymbol{C}^n$ with a compact connected boundary $d\Omega$ of class C^1. By a *meromorphic* CR *map* of $d\Omega$ into a complex manifold Y we mean a C^1 mapping $f\colon (d\Omega - \Sigma) \to Y$ defined outside a compact set Σ with $\mathscr{H}^{2n-1}(\Sigma) = 0$, such that

(i) f satisfies the tangential Cauchy-Riemann equations in $d\Omega - \Sigma$, i.e., f_* is $\boldsymbol{C}$-linear on the complex subspaces of $T(d\Omega - \Sigma)$.

(ii) The graph Γ of f in $\boldsymbol{C}^n \times Y$ satisfies the conditions

(a) $\bar{\Gamma}$ is compact and $\mathscr{H}^{2n-1}(\bar{\Gamma}) = \mathscr{H}^{2n-1}(\Gamma) < \infty$.

(b) $d[\Gamma] = 0$.

If $\Sigma = \emptyset$, condition (ii) is trivial and f is called a *regular* CR map. If, furthermore, $Y = C^n$, f is a classical *CR function*.

The restriction of any meromorphic function $F : \bar{\Omega} \to Y$ to $d\Omega$ is a meromorphic CR map. The following converse is a version of Bochner's theorem for meromorphic maps.

COROLLARY 10 [**11**]. *Let $\Omega \subset C^n$ be a bounded domain with connected C^1 boundary $d\Omega$, and suppose f is a meromorphic CR map of $d\Omega$ into a projective variety Y. Then, if $n > 2$, f extends to a unique meromorphic map $F : \Omega \to Y$.*

If we replace f with a CR function in the above corollary, we obtain Bochner's theorem [**3**], which holds for $n \geqq 2$. Both Corollary 10 and Bochner's theorem can be generalized, by means of Theorem 8, to CR maps (respectively, functions) on the boundary of a subvariety of $P^N - P^{N-k}$. (See [**10**], [**11**], [**12**] for details.)

REFERENCES

1. M. F. Atiyah and F. Hirzebruch, *Analytic cycles on complex manifolds*, Topology **1** (1962), 25–45. MR **26** #3091.

2. Errett Bishop, *Conditions for the analyticity of certain sets*, Michigan Math. J. **11** (1964), 289–304. MR **29** #6057.

3. S. Bochner, *Analytic and meromorphic continuation by means of Green's formula*, Ann. of Math. (2) **44** (1943), 652–673. MR **5**, 116.

4. M. Cornalba and P. Griffiths, *Analytic cycles and vector bundles on non-compact algebraic varieties*, Invent. Math. **28** (1975), 1–106.

5. H. Federer and W. Fleming, *Normal and integral currents*, Ann. of Math. (2) **72** (1960), 458–520. MR **23** #A588.

6. H. Federer, *Geometric measure theory*, Springer-Verlag, New York, 1969. MR **41** #1976.

7. ———, *Real flat chains, cochains and variational problems* (to appear).

8. P. A. Griffiths, *Function theory of finite order on algebraic varieties*. I(A), (B), J. Differential Geometry **6** (1971/72), 285–306; ibid. **7** (1972), 45–66. MR **46** #3829; **48** #4345.

9. R. Harvey and B. Shiffman, *A characterization of holomorphic chains*, Ann. of Math. (2) **99** (1974), 553–587. MR **50** #7572.

10. R. Harvey and H. B. Lawson, Jr., *On boundaries of complex analytic varieties*. I, Ann. of Math. (2) **102** (1975), 223–290.

11. ———, *On boundaries of complex analytic varieties*. II (to appear).

12. R. Harvey, *Holomorphic chains and their boundaries*, Proc. Sympos Pure Math., Vol. 30, part 1, Amer. Math. Soc., Providence, R.I., 1977, pp. 309–382.

13. J. King, *The currents defined by analytic varieties*, Acta. Math. **127** (1971), 185–220.

14. K. Kodaira and D. C. Spencer, *Divisor class groups on algebraic varieties*, Proc. Nat. Acad. Sci. U.S.A. **39** (1953), 872–877. MR **16**, 75.

15. S. Lang, *Introduction to algebraic and abelian functions*, Addison-Wesley, Reading, Mass., 1972. MR **48** #6122.

16. H. B. Lawson, Jr. and J. Simons, *On stable currents and their application to global problems in real and complex geometry*, Ann. of Math. (2) **98** (1973), 427–450. MR **48** #2881.

17. H. B. Lawson, Jr., *The stable homology of a flat torus*, Math. Scand. **36** (1975), 49–73.

18. T. Matsusaka, *On a characterization of a Jacobian variety*, Mem. Coll. Sci. Univ. Kyoto. Ser. A. Math. **32** (1959), 1–19. MR **21** #7213.

19. G. de Rham, *Variétés différentiables*, Actualités Sci. Indust., no. 1222, Hermann, Paris, 1955. MR **16**, 957.

20. B. Shiffman, *On the removal of singularities of analytic sets*, Michigan Math. J. **15** (1968), 111–120. MR **37** #464.

UNIVERSITY OF CALIFORNIA, BERKELEY

Proceedings of Symposia in Pure Mathematics
Volume 30, 1977

DEFORMATIONS OF STRONGLY PSEUDOCONVEX DOMAINS IN C^2

R. O. WELLS, JR.*

1. Deformations of complex structures on strongly pseudoconvex domains. If D is a simply connected domain in the complex plane C with a smooth boundary Γ then it follows that any smooth perturbation $\tilde{\Gamma}$ of Γ will bound a domain $\tilde{D}$ which is still biholomorphic to the original domain D (Riemann mapping theorem). In C^n, $n > 1$, this is no longer the case, and the purpose of this note is to announce some theorems which describe the variation in complex structure of a given domain in C^2 by simply perturbing its boundary appropriately. Let D be a strongly pseudoconvex domain in C^2, i.e., D is a relatively compact open set in C^2 with a smooth (C^∞) boundary, and there is a neighborhood U of ∂D and a real-valued function φ: $U \to R$ with the property that $d\varphi \neq 0$ in U, $i\partial\bar{\partial}\varphi$ is a positive-definite Hermitian quadratic form in U, and $D \cap U = \{z \in U: \varphi(z) < 0\}$. The level sets $M_\varepsilon = \{z \in U: \varphi(z) = \varepsilon\}$, for $0 \leqq |\varepsilon| < \varepsilon_0$, bound domains D_ε which are perturbations of the original domain $D = D_0$. If, for instance, $\varphi = |z|^2 - 1$, then the domains D_ε are simply balls of radius ε and are biholomorphically equivalent to D. In general, this is not the case, as we see in the following theorem. To formulate the theorem consider a sufficiently small neighborhood N of φ in the function space $C^\infty(U, R)$, so that any $\tilde{\varphi} \in N$ has the property that $\tilde{\varphi}^{-1}(\varepsilon)$ is diffeomorphic to $M = M_0 = \partial D_0$, and bounds a domain $\tilde{D}_\varepsilon$ in C^2. These are simply all smooth perturbations of the boundary of D. For each $\tilde{\varphi} \in N$, there is an $\tilde{\varepsilon}_0$ so that $\tilde{\varphi}^{-1}(\varepsilon)$ is a well-defined perturbation of $\varphi^{-1}(0)$, for $0 \leqq |\varepsilon| < \tilde{\varepsilon}_0$.

THEOREM 1. *There exists a set N_0 of second category in N such that for all $\tilde{\varphi} \in N_0$ one has*:

(a) $\tilde{D}_{\varepsilon_1} \not\cong \text{bih } \tilde{D}_{\varepsilon_2}$, $0 \leqq |\varepsilon_1|, |\varepsilon_2| < \tilde{\varepsilon}_0$;

AMS (MOS) subject classifications (1970). Primary 32F15, 32G05.

*Research supported partially by NSF grant MPS 75-05270; Alexander von Humboldt awardee.

(b) $\operatorname{Aut}(\tilde{D}_\varepsilon) = \{\mathrm{id}\}$, $0 \leqq |\varepsilon| < \varepsilon_0$.

Here $\operatorname{Aut}(\tilde{D}_\varepsilon) = \{$biholomorphic self-mappings of $\tilde{D}_\varepsilon\}$. Thus this theorem is in stark contrast to the situation in one complex variable, which we discussed above. In particular, Theorem 1 implies that almost all perturbations of the boundary of the unit ball have the property that different sublevel sets $(\tilde{D}_\varepsilon)$ are biholomorphically inequivalent and have no automorphisms. This result for the unit ball has been proven by Fefferman (unpublished) and by Burns and Shnider [2], and moreover, is true for the unit ball in $\boldsymbol{C}^n$, $n \geqq 2$.

We note that in Theorem 1, all of the domains $\tilde{D}_\varepsilon$, $\tilde{\varphi} \in N_0$, $0 \leqq |\varepsilon| < \tilde{\varepsilon}_0$, are diffeomorphic to the original domain D, assuming the original neighborhood N of $\varphi \in C^\infty(U, \boldsymbol{R})$ was chosen sufficiently small. Thus we can envision each of the domains $\tilde{D}_\varepsilon$ as giving possibly different complex structures on an original domain D. Theorem 1 asserts that for a fixed $\tilde{\varphi}$, the complex structures induced on D in this manner vary in a nontrivial manner with the parameter ε.

THEOREM 2. *The complex structures on D induced by the domains $\tilde{D}_\varepsilon$ depend on an infinite number of independent parameters.*

In other words, the "number of moduli" of an open strongly pseudoconvex domain in $\boldsymbol{C}^2$ is infinite. This contrasts to the finite-dimensional parameter space for the complex structures on compact complex manifolds (Kuranishi [12]), compact complex spaces (Grauert [8]), or isolated singularities (Grauert [7]).

Theorems 1 and 2 are both existential in nature, and there remain many difficult problems in analyzing the deformations of complex structures on open complex manifolds, in particular, in finding specific examples of such deformations, or in determining the variation of complex structure for a given geometric situation. Here we are deforming the boundary geometrically to deform the complex structure. This problem was first studied in this manner by Poincaré [13] who, in particular, studied perturbations of the unit ball in $\boldsymbol{C}^2$ of a specific nature, and found necessary and sufficient conditions that the perturbed domain be equivalent to the original unit ball. A different approach to this same general problem is to look at the abstract Kuranishi-type deformation of the compact CR manifold which is the boundary of the given domain D. This is the point of view taken by Hamilton [9] and Kiremidjian [10], [11]. The relation between our work and theirs is not evident at the present time.

In the next two sections we outline the elements of the proofs of Theorems 1 and 2. The details will appear in a joint paper with Daniel Burns and S. Shnider in which the deformation of the ball in $\boldsymbol{C}^n$ is also carried out in detail [3].

2. Reduction of the problem to the boundary. To show that D_1 and D_2, two strongly pseudoconvex domains in $\boldsymbol{C}^n$, are biholomorphically equivalent, it suffices to know that their boundaries are CR-equivalent, i.e., that there exists a diffeomorphism $f: \partial D_1 \to \partial D_2$ which is a CR mapping, and whose inverse is a CR mapping. Briefly, this means that $df|_{H_x(D_1)}$ is a $\boldsymbol{C}$-linear mapping into $H_{f(x)}(\partial D_2)$, where $H_x(\partial D_1)$, $H_{f(x)}(\partial D_2)$ are the maximal complex subspaces of $T_x(\partial D_1)$ and $T_{f(x)}(\partial D_2)$, respectively (cf. Wells [16] for a discussion of CR function theory). That the condition above is necessary and sufficient follows from Fefferman's deep theorem that any biholomorphic mapping F: $D_1 \to D_2$ extends to the boundary in a C^∞ manner [6], and from Bochner's generalization of Hartogs' theorem [1] which asserts that any CR function in the boundary of say D_1 extends to a holomorphic function on the interior of D (cf. [16]).

E. Cartan [4], following up on the seminal work of Poincaré [13] on the equiv l-ence problem of domains in C^2, gave a complete set of differential-geometric invariants which classify equivalence of CR hypersurfaces in C^2. Here we mean by a CR hypersurface in C^2 a real C^∞ hypersurface equipped with the structure of $H(M) \subset T(M)$, where $H_x(M)$ is the maximal complex subspace of $T_x(M)$ considered as a subspace of $T_x(C^2)$. The CR structure is the tangent bundle and its distinguished subbundle $H(M)$ inherited from the complex structure of C^2. This is a generalization of an almost-complex manifold, and Cartan's invariant theory has been generalized to C^n by Tanaka [14], [15], Chern-Moser [5], and Burns-Shnider [2].

In C^2, under appropriate nondegeneracy conditions which are generic in nature, Cartan associates to a CR hypersurface nine scalar functions $\{\alpha_1, \cdots, \alpha_9\}$ which have the property that:

(a) If $f: M \to \tilde{M}$ is a CR equivalence, then $f^*(\tilde{\alpha}_j) = \alpha_j$, $j = 1, \cdots, 9$, pointwise on M.

(b) If $\{\alpha_1, \cdots, \alpha_9\}$ are functionally independent on M and if $f: M \to \tilde{M}$ satisfies $f^*(\tilde{\alpha}_j) = \alpha_j$, then f is a CR equivalence.

These scalar functions are analogues, for the geometry under consideration, of Gaussian curvature and scalar functions associated with covariant derivatives of the Gaussian curvatures for the equivalence problem of isometric 2-dimensional surfaces. They are derived from curvature forms on an associated principal bundle of frames associated with the CR structure. The generalization of these scalar invariants to C^n, $n > 2$, is a difficult open problem, and that is why Theorems 1 and 2 are only known for domains in C^2 at present.

3. Transversality theory. In this section we indicate briefly how Cartan's invariants $\{\alpha_1, \cdots, \alpha_9\}$ are used to prove Theorems 1 and 2. Let φ be defined in U as in §1. Consider the jet bundle $J^8(U, R)$ consisting of eighth-order jets of C^∞ mappings of $U \to R$. We consider only those jets in $J^8(U, R)$ corresponding to the open neighborhood N of $\varphi \in C^\infty(U, R)$. Call this set (open) N^8. Inside N^8 there is a submanifold S of real codimension 2 outside of which the invariants $\{\alpha_1, \cdots, \alpha_9\}$ of Cartan are well-defined algebraic functions of the 8-jet of a given $\tilde{\varphi} \in N$ for the level set of $\tilde{\varphi}$ passing through a given $p \in U$. The functions $\{\alpha_1, \cdots, \alpha_9\}$ then define a mapping K: $N^8 - S \to R^9$ and have the property that, for a given $\tilde{\varphi} \in N$, $K \circ j^8(\tilde{\varphi})$: $U - \pi(S \cap j^8(\tilde{\varphi})) \to R^9$ has the values $\{\alpha_1(p), \cdots, \alpha_9(p)\}$, for the surface $\tilde{\varphi}(z) = \tilde{\varphi}(p)$. Here π is the projection $J^8(U, R) \to U$, and $j^8(\tilde{\varphi})$ is the 8-jet extension of $\tilde{\varphi}$ to $J^8(U, R)$, i.e., the local 8-jet of $\tilde{\varphi}$ at a point p. Extending this situation to the product jet bundles we perturb φ to $\tilde{\varphi}$ so that $j^8(\tilde{\varphi})$ in the product jet bundle is transversal to the pull-back of the diagonal in $R^9 \times R^9$ by the mapping $K \times K$. By a long computation, using Cartan's formulas [4] and Moser's normal coordinates (Chern-Moser [5]), one can show that K has maximal rank outside of S, and thus the pull-back of the diagonal has codimension 9 in the product bundle. It then follows, as in the proof of injectivity for Whitney's embedding theorem, that $K_{\tilde{\varphi}} = K \circ j^8(\tilde{\varphi})$ is one-to-one on $U - \Sigma$, where $\Sigma = \pi(S \cap j^8(\tilde{\varphi}))$ and Σ has real codimension 2 in U. Theorem 1 now follows from this construction by observing that if f:

$\tilde{\varphi}^{-1}(\varepsilon_1) \to \tilde{\varphi}^{-1}(\varepsilon_2)$, $\varepsilon_1 \neq \varepsilon_2$, were a CR equivalence mapping, then we would have on the complement of Σ

$$(*) \qquad f^*(K_{\tilde{\varphi}})(p) = K_{\tilde{\varphi}}(p)$$

(by the CR invariance of the Cartan functions α_j, $j = 1, \cdots, 9$). But $K_{\tilde{\varphi}}$ is one-to-one on the complement of Σ and $(*)$ means that for some p we would have

$$K_{\tilde{\varphi}}(f(p)) = K_{\tilde{\varphi}}(p), \qquad f(p) \neq p, p, f(p) \notin \Sigma,$$

and this is a contradiction. The fact that $\tilde{D}_\varepsilon$ has no automorphisms is proved similarly.

For Theorem 2 we will simply mention that after a deformation $\tilde{\varphi}$ of φ is found where there are no automorphisms of $\tilde{D}_\varepsilon$ as above, then one can introduce Moser normal coordinates near a single point, and let the (infinitely many) higher order Taylor coefficients vary in an independent manner. The details of this process are carried out in [**3**].

REFERENCES

1. S. Bochner, *Analytic and meromorphic continuation by means of Green's formula*, Ann. of Math (2) **44** (1943), 652–673. MR **5**, 116.

2. D. Burns, Jr. and S. Shnider, *Pseudoconformed geometry of hypersurfaces in* C^{n+1}, Proc. Nat. Acad. Sci. U.S.A. **72** (1975), 2433–2436.

3. D. Burns, Jr., S. Shnider and R. O. Wells, Jr., *Deformations of strongly pseudoconvex domains* (to appear).

4. E. Cartan, *Sur la géométrie pseudo-conforme des hypersurfaces de deux variables complexes*. I, II, Ann. Math. Pura Appl. (4) **11** (1932), 17–90; Ann. Scuola Norm. Sup. Pisa (2) **1** (1932), 333–354 (or Oeuvres II, 2, 1231–1304; ibid. III, 2, 1217–1238).

5. S. S. Chern and J. K. Moser, *Real hypersurfaces in complex manifolds*, Acta Math. **133** (1974), 219–271.

6. C. Fefferman, *The Bergman kernel and biholomorphic mappings of pseudoconvex domains*, Invent. Math. **26** (1974), 1–65. MR **50** #2562.

7. H. Grauert, *Über die Deformation isolierter Singularitäten analytischer Mengen*, Invent. Math. **15** (1972), 171–198. MR **45** #2206.

8. ———, *Der Satz von Kuranishi für kompakte komplexe Räume*, Invent. Math. **25** (1974), 107–142. MR **49** #10920.

9. R. Hamilton, *Deformation of complex structures on manifolds with boundary*. I, II, III (preprints).

10. Garo K. Kiremidjian, *Deformations of complex structures on certain non-compact complex manifolds*, Ann. of Math. (2) **98** (1973), 411–426. MR **48** #8853.

11. ———, *On deformations of complex compact manifolds with boundary*, Pacific J. Math. **54** (1974), 177–190.

12. M. Kuranishi, *On the locally complete families of complex analytic structures*, Ann. of Math. (2) **75** (1962), 536–577. MR **25** #4550.

13. H. Poincaré, *Les functions analytique de deux variables et la représentation conforme*, Rend. Circ. Mat. Palermo **23** (1907), 185–220 (or Oeuvres IV, 224–289).

14. N. Tanaka, *On the pseudo-conformal geometry of hypersurfaces of the space of n complex variables*, J. Math. Soc. Japan **14** (1962), 397–429. MR **26** #3086.

15. ———, *On generalized graded Lie algebras and geometric structures*. I, J. Math. Soc. Japan **19** (1967), 215–254. MR **36** #4470; erratum, **36**, p. 1568.

16. R. O. Wells, Jr., *Function theory on differentiable submanifolds*, Contributions to Analysis, Academic Press, New York, 1974, pp. 407–441. MR **50** #10322.

RICE UNIVERSITY

UNIVERSITÄT GÖTTINGEN

Proceedings of Symposia in Pure Mathematics
Volume 30, 1977

THE BEHAVIOR OF THE INFINITESIMAL KOBAYASHI PSEUDOMETRIC IN DEFORMATIONS AND ON ALGEBRAIC MANIFOLDS OF GENERAL TYPE*

MARCUS W. WRIGHT

1. Introduction. In this paper we discuss the behavior of the infinitesimal form of the Kobayashi pseudometric in two situations. Namely, we are interested in the case of a fixed compact complex-analytic manifold and in the case where a complex manifold is deformed. In the former it is shown that the form is continuous if the manifold is in a class which contains algebraic manifolds of general type and in the latter we discuss the continuity of the form in the parameters of the deformation; much stronger results hold true when the original manifold is hyperbolic, and in fact, we show that hyperbolic compact manifolds have Hausdorff moduli spaces.

Here are some definitions and remarks about hyperbolicity. Let M be a complex-analytic manifold, Δ the unit disc in C, and TM the bundle of holomorphic tangent vectors to M. The Kobayashi form F_M on TM is defined as (see [11]): If $\langle x, \xi\rangle \in TM$, with $x \in M$ and ξ a vector at x, then $F_M(x, \xi) = \inf \gamma = \inf(1/R)$ where the first infimum is taken over all complex numbers γ such that there exists a holomorphic mapping $f: \Delta \to M$ with $f_*(\langle 0, \gamma\partial/\partial z\rangle) = \langle x, \xi\rangle$, and the second is over all real $R > 0$ for which there is a holomorphic mapping $f: \Delta_R \to M$ (Δ_R is the disc of radius R in C) with $f_*(\langle 0, \partial/\partial z\rangle) = \langle x, \xi\rangle$. F_M is easily seen to be homogeneous in ξ; as proved by H. L. Royden [11], it is also upper semicontinuous on TM. So given p and q in M, we may define a pseudodistance $d_M(p, q)$ by

$$d_M(p, q) = \inf_\sigma \int F_M(\sigma(t), \dot{\sigma}(t))\, dt,$$

where the infimum is over all piecewise smooth curves from p to q. Royden has

AMS (MOS) subject classifications (1970). Primary 32G05, 32G13, 32H15, 32H20.

*This work was supported in part by NSF.

shown [11] that this pseudodistance coincides with the Kobayashi pseudodistance $\bar{d}_M$ defined as

$$\bar{d}_M(p, q) = \inf \frac{1}{2} \sum_{i=1}^{n} \log \frac{1 + |a_i|}{1 - |a_i|},$$

where the infimum is over all finite sets $\{a_i\} \subset \Delta$ for which there exist holomorphic mappings $f_i : \Delta \to M$, $i = 1, \cdots, n$, with $f_1(0) = p$, $f_n(a_n) = q$, and $f_j(a_j) = f_{j+1}(0)$, $j = 1, \cdots, n - 1$. It is immediate that $\bar{d}_M$ has the property that $\bar{d}_M(f(p), f(q)) \leqq \bar{d}_N(p, q)$ whenever $f : N \to M$ is holomorphic; also, $\bar{d}_M$ is the largest pseudodistance on M with the same property for all such f when $N = \Delta$. From this it follows that $\bar{d}_M$ is inner, i.e.,

$$\bar{d}_M(p, q) = \inf_{\sigma} L_{\bar{d}_M}(\sigma),$$

where the infimum is over all piecewise smooth curves from p to q and $L_{\bar{d}_M}$ is the length function associated to $\bar{d}_M$, i.e.,

$$L_{\bar{d}_M}(\sigma) = \sup_{\{x_i\} \subset \sigma} \sum_i d_M(x_i, x_{i+1}).$$

Somewhat more difficult to prove is that, also, $\int F_M(\sigma(t), \dot{\sigma}(t))\, dt = L_{\bar{d}_M}(\sigma)$. See [15].

DEFINITION. A complex-analytic manifold M is *hyperbolic* if and only if $d_M(p, q) \neq 0$ whenever $p \neq q$.

REMARKS. (1) It can be shown [11] that M is hyperbolic if and only if F_M satisfies the following condition: For every $p \in M$ there is a coordinate neighborhood U of p and a positive constant C_U such that $F_M(y, \eta) > C_U \|\eta\|$ for all $\langle y, \eta \rangle \in TM|U$. Here $\|\eta\|$ can be defined using the coordinate system or a hermitian metric.

(2) If M is compact, then (1) implies that M is hyperbolic if and only if $\sup_{H(\Delta,M)} \|f'(0)\| < \infty$, where $\| \ \|$ is defined using a hermitian metric and $H(\Delta, M)$ denotes the set of all holomorphic mappings from Δ to M.

2. Behavior of F_M in deformations. Suppose that M is a (not necessarily compact) complex manifold. We are interested in "small" deformations of M represented by integrable almost-complex structures; such a structure can be represented by a C^∞ TM-valued (0, 1)-form φ on M with $\bar{\partial}\varphi - \frac{1}{2}[\varphi, \varphi] = 0$ and all the coefficients of φ small in Sobolev norms. A good reference for this situation is [7]. By a (smooth) deformation of M we shall mean a family $F = \{\varphi(s) \mid s \in S\}$ of such integrable almost-complex structures parametrized by an analytic set S, with smooth dependence on S. The Newlander-Nirenberg theorem [9] asserts that M can be given the structure of a complex manifold M_s so that $TM_s = \{X - \varphi(s)X \mid X \in TM\}$. If M is compact, then there is a complex space V and a mapping $w : V \to S$ such that $M_s = w^{-1}(s)$ for all $s \in S$. In any case, there is a C^∞ bundle isomorphism $\Phi(s) : TM \to TM_s$ (both domain and range are thought of as subbundles of $CT\,|M|$, the complexification of the tangent bundle of $|M|$, the underlying differentiable manifold) defined by $\Phi(s)\xi = \xi - \overline{\varphi(s)}\xi$. Let $F_s = F_{M_s}$ denote the infinitesimal Kobayashi pseudometric on TM_s. The basic theorem on the behavior of F_s as s varies is

THEOREM 1. *Let $\langle x, \xi \rangle \in TM$ and $\varepsilon > 0$ be given. Then there is a $\delta > 0$ such that if $|s| < \delta$ then $F_s(x, \Phi(s)\xi) \leqq F_0(x, \xi) + \varepsilon$. And, in fact, this inequality holds for all $\langle y, \eta \rangle$ in a neighborhood of $\langle x, \Phi(s)\xi \rangle$ in TM_s.*

REMARKS. (1) The proof of this theorem can be found in the author's thesis [13] or in [14]. If $f : \Delta_R \to M$ is such that $f_*(0, \partial/\partial z) = \langle x, \xi \rangle$ and $F_0(x, \xi) + \varepsilon > 1/R$, then what must be done is show that f can be deformed to be analytic with respect to the structures on M_s, provided only that $|s|$ is small enough. This is done by first extending f to be an equidimensional embedding of a polydisc into $M \times \Delta$ using a theorem of H. L. Royden [12] and then deforming this mapping using a modified Newlander-Nirenberg theorem.

(2) Moreover, the technique of proof does not utilize the existence of a smooth family of deformations; the conclusion of Theorem 1 could read: There is a $\delta > 0$ such that if φ is an integrable almost-complex structure with $\|\varphi\| < \delta$ ($\| \ \|$ referring to a suitable Sobolev norm) then $F_\varphi(x, \Phi\xi) \leqq F_0(x, \xi) + \varepsilon$, and there is a neighborhood of $\langle x, \Phi\xi \rangle$ in TM_φ on which the inequality remains true.

(3) Because F_M is homogeneous in vectors, we immediately obtain the following

COROLLARY 2. *With the hypotheses as in Theorem* 1, *there is a δ such that if $|s| < \delta$, then $F_s(x, \Phi(s)\xi) \leqq F_0(x, \xi) + \varepsilon\|\xi\|$.*

(4) Concerning deformations of hyperbolic manifolds we can conclude

COROLLARY 3. *If $\{\varphi(s) | \ s \in S\}$ is a deformation of a complex manifold M with M_{s_i} biholomorphic to M' and M' hyperbolic for $s_i \to 0$, then M is hyperbolic.*

Recently M. Green and R. Brody [2] have given an example of a deformation $\{M_s | s \in \Delta\}$ of a nonhyperbolic simply connected algebraic manifold $M = M_0$ such that M_s is hyperbolic for $s \neq 0$.

The δ in Corollary 2 depends on x and the direction of ξ. The main difficulty in removing dependence on $\langle x, \xi \rangle$ is that F_M is in general only known to be upper semicontinuous on TM. There is an unpublished example of Royden of a domain D in $\boldsymbol{C}^2$ such that $F_D(x, \xi) \geqq C_x\|\xi\|$ for all $\langle x, \xi \rangle \in TD_x$, but D is also nonhyperbolic. When F_0 is continuous, we can obtain improvements of Theorem 1; see [13] for details. We mention here the case of a compact hyperbolic manifold.

THEOREM 4. *If M_0 is a compact hyperbolic manifold and $\varepsilon > 0$ is given, together with a deformation of M_0, then there is a $\delta > 0$ such that if $|s| < \delta$, then all for $\langle x, \xi \rangle \in TM_0$, $F_s(y, \eta) \leqq F_0(x, \xi) + \varepsilon\|\xi\|$ for all $\langle y, \eta \rangle$ in a neighborhood of $\langle x, \Phi(s)\xi \rangle$ in TM_s.*

A proof can be found in [13]. The main point is that $H(\Delta, M_0)$ is compact, so the lower semicontinuity of F_0 follows. The continuity of F_0 then implies that the conclusion of Theorem 1 for a finite number of points of TM_0 suffices to obtain the same inequality on all of TM_0.

If M is an arbitrary compact complex manifold, then it is unknown whether F_M is lower semicontinuous on TM or whether F_{M_s} is lower semicontinuous in s in the context of a deformation of M. However, if M is compact hyperbolic, a recent result of R. Brody [1] together with Theorem 4 imply that F_s is continuous on $\bigcup_{s \in U} TM_s$, if U is any sufficiently small neighborhood of 0 in S. Brody actually proves that M_s is hyperbolic for s in a neighborhood of 0 in S. Let $C_s = \sup_{f \in H(\Delta, M_s)} \|f'(0)\|$. Then, as mentioned in Remark (2) of §1, M_s is hyperbolic if and only if $C_s < \infty$. Brody proves that C_s is lower semicontinuous, while Theorem 1 shows that it is upper semicontinuous, both results holding when M is an arbitrary compact complex manifold. If $C_0 < \infty$, then the lower semicontinuity of F_{M}, follows for small s, since then $\bigcup_{s\in U} H(\Delta, M_s)$ is a compact family of mappings for U a sufficiently small neighborhood of 0 in S. Another, more quantitative proof that a small deformation of a compact hyperbolic manifold is hyperbolic can be found in **[15]**, where we modify the proof of Theorem 1 to establish uniform upper semicontinuity of F_s.

Whenever F_s is continuous on $\bigcup_{s\in S} TM_s$, we have the following

THEOREM 5. *Let $\{\varphi(s) | s \in S\}$ be a deformation of a complex manifold M. If F_s is continuous on $\bigcup_{s\in S} TM_s$, then d_s is continuous on $M \times M \times S$.*

A proof of this theorem can be found in R. Brody's thesis **[1]**.

COROLLARY 6. *If $\{\varphi(s) | s \in S\}$ is a deformation of a compact hyperbolic manifold M, then d_s is continuous on $M \times M \times U$ for U a sufficiently small neighborhood of 0 in S.*

Corollary 6 leads to the following result about biholomorphic mappings between small deformations of a compact hyperbolic manifold. See **[14]** for proof.

THEOREM 7. *If $\{\varphi(s) | s \in S\}$ is a deformation of a compact hyperbolic manifold M, then for a sufficiently small neighborhood U of zero in S, the family $\mathscr{A} = \bigcup_{s,t\in \bar{U}} \mathrm{Bih}(M_s, M_t)$ is compact.*

Using this result and the Kuranishi theory of versal deformations of a compact complex manifold (see, e.g., **[7]**) we can obtain the following, just as the corresponding result for manifolds with ample canonical bundle is proved by M. S. Narasimhan and R. R. Simha in **[8]**.

THEOREM 8. *Let M be a compact complex manifold and let $\mathscr{M}$ denote the collection of isomorphism classes of hyperbolic complex structures on M. Then $\mathscr{M}$ has the structure of a Hausdorff complex space such that if $\{M_s\}_{s\in S}$ is any family of hyperbolic complex structures on M then the map sending s to the isomorphism class of M_s is a morphism from S to $\mathscr{M}$.*

3. The continuity of F_M. If M is a compact complex manifold and f_i is a holomorphic mapping form Δ_{r_i}, the disc of radius r_i in C, into M with $r_i \to r$ and $f_{i*}(\langle 0, \partial/\partial z\rangle) \to \langle x, \xi\rangle$ as $i \to \infty$, then it is natural to ask whether there is a "limit" mapping, i.e., an $f : \Delta_r \to M$ with $f_*(0, \partial/\partial z) = \langle x, \xi\rangle$. If the answer is always yes, F_M will be lower semicontinuous on TM as well as upper semicontinuous. The work of R. Brody in **[1]** shows that there is always a nonconstant mapping $f : \Delta_r \to M$ with $f(0) = x$, but does not allow us to conclude that $f'(0) \neq 0$. If M is not hyperbolic then we can find such a situation with $r_i \to \infty$, and thus Brody proves that M is compact hyperbolic if and only if there does not exist a nonconstant mapping form C to M.

There is one class of not necessarily hyperbolic manifolds for which we can prove continuity for F_M. It contains compact algebraic manifolds of general type.

DEFINITIONS. If M is a complex-analytic manifold of dimension n and w is a coordinate system at $p \in M$, let $k_w(p) = \inf\{|\det J_f^w(0)|^{-2}\}$ where the infimum is taken overall holomorphic mappings from the n-dimensional unit ball in C^n into M with $f(0) = p$; $J_f^w(0)$ is the Jacobian. The *hyperbolic volume form* $\eta(p)$ is defined by

$$\eta(p) = n!\,(1/2i)^n\, k_w(p) d\bar{w}^1 \wedge dw^1 \wedge \cdots \wedge dw^n \wedge d\bar{w}^n.$$

D. Pelles shows in **[10]** that η is upper semicontinuous on M and defines a measure on M which differs from the Kobayashi hyperbolic measure (defined in **[4]**) by a constant factor depending only on n.

THEOREM 9. *Suppose that M is a compact definite measure hyperbolic manifold with canonical bundle K. Assume that for some m there exist global sections $s_1, \cdots, s_k$ of K^m such that $z \mapsto [s_1(2), \cdots, s_k(2)] \in P^{k-1}$ defines a meromorphic projective embedding of M and such that $(\overline{s_i s_i})^{1/m}/\eta$ is bounded on M for all i. Then F_M is continuous on TM.*

The proof of this theorem, which can be found in **[14]**, uses the Royden extension theorem **[12]** and techniques of Yau **[16]** which go back to Griffiths **[3]**.

DEFINITION. An n-dimensional compact algebraic manifold M is *of general type* if and only if

$$\limsup_{m \to +\infty} m^{-n} \dim H^0(M, \mathcal{O}(K^m)) > 0.$$

COROLLARY 10. *If M is compact algebraic of general type, then F_M is continuous on TM.*

PROOF. See Kodaira **[6]** and Kobayashi and Ochiai **[5]**.

We remark that, because of Corollary 10, Theorem 4 applies to algebraic manifolds of general type.

REFERENCES

1. Robert Brody, *Intrinsic metrics and measures on compact complex manifolds*, Thesis, Harvard Univ., 1975.

2. Robert Brody and Mark Green, *A family of smooth hyperbolic hypersurfaces in P_3* (unpublished).

3. P. A. Griffiths, *Holomorphic mapping into canonical algebraic varieties*, Ann. of Math. (2) **93** (1971), 439–458. MR **43** #7668.

4. S. Kobayashi, *Hyperbolic manifolds and holomorphic mappings*, Pure and Appl. Math. 2, Dekker, New York, 1970. MR **43** #3503.

5. S. Kobayashi and T. Ochiai, *Mappings into compact complex manifolds with negative first Chern class*, J. Math. Soc. Japan **23** (1971), 137–148. MR **44** #5514.

6. K. Kodaira, *Holomorphic mappings of polydiscs into compact complex manifolds*, J. Differential Geometry **6** (1971/72), 33–46. MR **46** #386.

7. M. Kuranishi, *Deformations of compact complex manifolds*, Séminaire Mathématiques Supérieures, no. 39 (Été 1969), Press Univ. Montréal, Montréal, Qué., 1971. MR **50** #7588.

8. M. S. Narasimhan and R. R. Simha, *Manifolds with ample canonical class*, Invent. Math. **5** (1968), 120–128. MR **38** #5253.

9. A. Newlander and L. Nirenberg, *Complex analytic coordinates in almost complex manifolds*, Ann of Math. (2) **65** (1957), 391–404. MR **19**, 577.

10. D. A. Pelles (Formerly D. A. Eisenman), *Holomorphic maps which preserve intrinsic measure*, Amer. J. Math. **97** (1975), 1–15.

11. H. L. Royden, *Remarks on the Kobayashi metric*, Several Complex Variables, II (Proc. In-

ternat. Conf. Univ. of Maryland, College Park, Md., 1970), Lecture Notes in Math., vol. 185, Springer-Verlag, Berlin and New York, 1971, pp. 125–137. MR **46** #3826.

12. H. L. Royden, *The extension of regular holomorphic maps*, Proc. Amer. Math. Soc. **43** (1974), 306–310. MR **49** #629.

13. M. W. Wright, *The behavior of the differential Kobayashi pseudo-metric in deformations of complex manifolds*, Thesis, Stanford Univ., 1974.

14. ———, *The Kobayashi pseudo-metric on algebraic manifolds of general type and in deformations of complex manifolds*, Trans. Amer. Math. Soc. (to appear).

15. ———, *The Kobayashi pseudo-metric and deformations of hyperbolic manifolds* (to appear).

16. S.-T. Yau, *Intrinsic measures of compact complex manifolds*, Math. Ann. **212** (1975), 317–329.

RICE UNIVERSITY

Proceedings of Symposia in Pure Mathematics
Volume 30, 1977

CURVATURE OF COMPLEX SUBMANIFOLDS OF C^n

PAUL YANG

By a complex submanifold of C^n we will mean a holomorphic immersion $\varphi: M^k \to C^n$. The standard flat Kähler metric of C^n restricts to a Kähler metric on M^k. It is a well-known fact in Kähler geometry that M is a minimal submanifold in the sense of riemannian geometry and all the holomorphic sectional curvatures of M are nonpositive. In [1] Bochner showed that the unit disk equipped with the Poincaré metric of constant negative curvature cannot be holomorphically isometrically imbedded in C^n, even locally. It is therefore of interest to study:

Question I. Does there exist a complete complex submanifold M^k of C^n with holomorphic sectional curvature bounded above by a negative constant?

A riemannian manifold M is said to be complete if every path tending to the ideal boundary of M has infinite length. By a *complete complex submanifold of* C^N we will mean a holomorphic immersion $\varphi: M^k \to C^N$ such that the induced Kähler metric on M^k is a complete riemannian metric on M^k. Thus the image of φ is not assumed to be closed. It is a natural condition to impose on M when we are interested in its global properties. It follows from the maximum principle for holomorphic functions that compact complex manifolds cannot be found in C^n, thus one is led to ask:

Question II. Does there exist a complete complex submanifold $\varphi: M^k \to C^n$ which has a bounded image?

As complex submanifolds are minimal submanifolds of C^n, this is a particular case of a question raised by Chern [3].

I will discuss some partial answers to these questions. Details will appear in [6].

I would like to thank Professors S. S. Chern and H. Wu for their generous support and advice.

AMS (MOS) subject classifications (1970). Primary 53C40; Secondary 32H20.

Relationship between I and II. Let F: $B^n(\{z \in C^n \mid |z| < 1\}) \to C^{2n}$ be given by $F(z_1, \cdots, z_n) = (z_1, \cdots, z_n, e^{z_1}, \cdots, e^{z_n})$; then

(a) $|dF(v)| \geqq |v|$ for each $v \in T_p(B^n)$, $p \in B^n$;

(b) the holomorphic sectional curvatures of $F(B^n)$ are strongly negative: $K(v) \leqq -c < 0$ for all $v \in T_p(B^n)$, $p \in B^n$.

The properties (a) and (b) show that if $\varphi : M^k \to B^n$ is a complete bounded complex submanifold of C^n, then $F \circ \varphi$: $M^k \to C^{2n}$ is a complete complex submanifold of C^{2n} with strongly negative holomorphic sectional curvature. Thus to answer Question II in the negative it suffices to do likewise for Question I.

The case $n = k + 1$. For a hypersurface $\varphi : M^k \to C^{k+1}$ we can answer Question I in the negative due to the equidimensional character of the Gauss map which we define: Let v_p be a unit normal to M^k at p. v_p is determined up to multiplication by $e^{i\theta}$, so v_p determines a point $[v_p]$ in $P_n(C)$. This will be the Gauss map $G(p) = [v_p]$. The relevant properties of G are (1) G is conjugate holomorphic; (2) $G^*\omega$, the pull-back of the Kähler form ω of the Fubini-Study metric on $P_n(C)$, is equal to the negative of the Ricci form of M, i.e., the (1, 1)-form corresponding to the Ricci tensor: $G^*\omega = -S$. These facts lead immediately to

THEOREM 1. *If* φ : $M^k \to C^{k+1}$ *is a complete complex hypersurface, then* $\sup_{v \in T_pM; p \in M} K(v) = 0$.

Holomorphic curves. For the general case, it is easy to see that it suffices to consider the case of $k = 1$. For if M^k is a complete complex submanifold of C^n with strongly negative holomorphic sectional curvature, then so is a slice of M^k by a generic affine subspace L^{M-k+1} of C^n. Furthermore, in view of the uniformization theorem for Riemann surfaces, it suffices to consider the case when M is the unit disk Δ, since the relevant properties are preserved when we pass to the universal covering of M.

Calabi [2] has introduced in this situation φ: $M^1 \to C^n$ a sequence of curvature functions which measure the higher order osculations of M, $K_1, \cdots, K_{n-1} \geqq 0$ where $K_1 = -K$ (K being the Gauss curvature). These are precisely the complex analogues of the curvature functions of a real curve in R^n. In terms of these curvature functions we have

THEOREM 2. *If* φ: $M \to C^n$ *is a complete holomorphic curve, then* inf $K_1 = 0$ *or* inf $K_1 \cdots K_{n-1} = 0$.

We observe that in the case of holomorphic curves φ: $M^1 \to C^2$, this recovers our theorem for hypersurfaces; in fact, we can sharpen the result in this case:

THEOREM 3. *If* φ: $\Delta \to C^2$ *is a complete holomorphic curve, then there cannot exist a point* $p_0 \in M$ *such that, for some constant* $c > 0$ *and compact set* $G \subset M$ $(p_0 \in G)$,

$$K(p) \leqq -c/d(p_0, p)^2 \quad \text{for } p \notin G,$$

where $d(p_0, p) =$ *geodesic distance between* p_0 *and* p.

The proof depends on a criterion for hyperbolicity which is of independent interest:

LEMMA. *Let* M *be a simply connected Riemann surface with a complete hermitian metric on* M *with nonpositive curvature satisfying*: $\exists\, p_0 \in M$, $c > 0$ *and a compact*

$G \subset M$ such that $K(p) \leqq -c/d(p_0, p)^2$ for $p \notin G$; then M is conformally equivalent to the disk.

REMARKS. (1) This result complements the following theorem of Greene and Wu [**4**]: If M is a simply connected riemann surface with complete hermitian metric, suppose $\exists\, p_0 \in M$, $c > 0$ such that

$$-c/d(p_0, p)^{2+\varepsilon} \leqq K(p) \leqq 0;$$

then M is conformally equivalent to $\boldsymbol{C}$.

(2) This criterion has been extended by Greene and Wu [**5**] to a condition for a complete Kähler manifold of arbitrary dimension to be complete hyperbolic in the sense of Kobayashi.

REFERENCES

1. S. Bochner, *Curvature in Hermitian metric*, Bull. Amer. Math. Soc. **53** (1947), 179–195. MR **8**, 490.

2. E. Calabi, *Metric Riemann surfaces*, Contributions to the Theory of Riemann Surfaces, Ann. of Math. Studies, no. 30, Princeton Univ. Press, Princeton, N. J., 1953, pp. 77–85. MR **15**, 863.

3. S. S. Chern, *The geometry of G-structures*, Bull. Amer. Math. Soc. **72** (1966), 167–219. MR **33** #661.

4. R. E. Greene and H. Wu, *Curvature and complex analysis*, Bull. Amer. Math. Soc. **77** (1971), 1045–1049. MR **44** #473.

5. ———, (in preparation).

6. P. Yang, *Curvature of complex submanifolds of* C^n, J. Differential Geometry (to appear).

RICE UNIVERSITY

PRINCIPAL LECTURE

D. BURNS, JR. AND S. SHNIDER

REAL HYPERSURFACES IN COMPLEX MANIFOLDS

Proceedings of Symposia in Pure Mathematics
Volume 30, 1977

REAL HYPERSURFACES IN COMPLEX MANIFOLDS

D. BURNS, JR.* AND S. SHNIDER*

Introduction. We present here a survey of recent work on the "equivalence problem" for real hypersurfaces in complex manifolds and related areas. The problem is, roughly speaking, to find local differential invariants of a hypersurface M in a complex manifold D such that two such hypersurfaces M_1 and M_2 are locally equivalent if and only if the corresponding invariants agree. Obviously, several things have to be made precise here, and we have more than the usual number of choices to make because such an M may be viewed *extrinsically*, as a hypersurface contained in D, or *intrinsically*, as an integrable CR manifold. (The definitions are recalled in §1.)

From the extrinsic point of view, the natural notion of equivalence is that $M_1 \subset D_1$ and $M_2 \subset D_2$ are (locally) equivalent if there is a (local) biholomorphic map $\phi: D_1 \to D_2$ with $\phi(M_1) = M_2$, and a natural notion of differential invariants is combinations of Taylor series coefficients for local defining functions for the M_i. In practice, one assumes the hypersurfaces are *nondegenerate*, i.e., their Levi forms are nonsingular. This point of view was initiated by Poincaré around the turn of the century, but the equivalence problem in this framework was only recently solved explicitly by Moser in the nondegenerate case [**11**]. As is typical in such problems, the existence of the map ϕ can be demonstrated only when M_1 and M_2 are real analytic and have the same invariants. Of course, when M_1 and M_2 are not real analytic, the invariants will be useful as obstructions to equivalence. In practice, one can rarely compute the infinite number of invariants involved here, and so their importance is more often as obstructions or additional basic structures on the hypersurfaces.

From the intrinsic point of view, the natural invariants are the usual auxiliary

AMS (MOS) subject classifications (1970). Primary 32C05, 32F15, 53A55, 53B25.

*Partially supported by the National Science Foundation.

objects of differential geometry: frame bundles, connections, curvature, etc. E. Cartan **[9]** solved the local equivalence problem for nondegenerate hypersurfaces in C^2 in this setting. His work has been considerably extended to cover hypersurfaces in C^{n+1} ($n > 1$) by Tanaka and Chern **[22]**, **[11]**, independently. (Tanaka's work even extends to some real submanifolds of codimension > 1, but with a condition on the Levi form which is not generic in most cases.) The solution is stated in terms of a principal bundle with connection which is constructed functorially for CR manifolds. Again, these invariants are probably more important as additional structures on M rather than for actual existence statements, even in the real analytic case. After the machinery has been developed, it turns out that both notions of equivalence agree in the real analytic case.

Aside from the intrinsic geometric and analytic interest the circle of local geometric ideas above has, it has been given greatly added significance by C. Fefferman's recent theorem **[13]** on the boundary regularity of biholomorphic maps of strictly pseudoconvex domains. This theorem gives global significance on such a domain D to the local geometric invariants on ∂D, justifying Poincaré's original interest in the equivalence problem for hypersurfaces in relation to the problem of "uniformization" in several variables (i.e., the biholomorphic equivalence problem for domains in C^n).

For more comments on the history of these questions, see Wells's article **[25]** in these PROCEEDINGS.

In spite of its age, the subject presented here has only recently received considerable renewed interest, and this applies most especially to its role in local and global analysis problems in domains which our hypersurfaces bound. This has the drawback that several aspects of the subject presented here are tentative, not only in the sense that some formulations of results or concepts are not "definitive" yet, but also in that many of the problems scattered throughout the text (explicitly or implicitly!) and in §5 are obviously groping towards a definition of which aspects are most significant and useful and which are superfluous. On the other hand, this semiempirical stage is novel and refreshing, and still seems to contain some surprises. Almost all serious questions here are still open, though it is clear, for example, that the strength of Fefferman's regularity theorem and the invariants used in tandem is far from fully exploited. At any rate, this word should warn the wise sufficiently!

This survey covers many of the results in the field now, if only by passing reference at times (several were found in the time between the Summer Institute and this writing). §§1 and 3 give outlines of the intrinsic and extrinsic approaches to the equivalence problem, respectively. §2 gives several applications of the differential geometric machinery, to problems motivated either by differential geometry itself or by related problems in complex analysis. (Of course, we are still learning which problems are "related".) §4 is more speculative, and treats the circle of ideas pertaining to a family of distinguished, biholomorphically invariant curves in nondegenerate hypersurfaces called *chains*. It especially discusses a new viewpoint in the geometry of hypersurfaces introduced by Fefferman **[14]**, motivated by the problem of interpreting chains, and suggested by his work on the Bergman kernel. As already mentioned, §5 is a list of problems.

It is very hard for us to thank fully all the colleagues who have so generously shared with us their ideas on this subject: K. Diederich, C. Fefferman, J. Moser, S. Webster and R. O. Wells, Jr. We would also like to thank J. J. Kohn for first arousing our interest in the subject.

1. E. Cartan's method. E. Cartan's method of studying the equivalence problem for a geometric structure can be described briefly as follows. Associate to any manifold M with the given structure a second manifold P, fibered over M, such that any structure-preserving map f lifts to an $\tilde{f}$ for which the diagram commutes:

$$\begin{array}{ccc} P_1 & \xrightarrow{\tilde{f}} & P_2 \\ {\scriptstyle\pi_1}\downarrow & & \downarrow{\scriptstyle\pi_2} \\ M_1 & \xrightarrow{f} & M_2 \end{array} \tag{1.1}$$

Then define on P a parallelism, a trivialization of the tangent bundle $T(P)$, such that the maps $g : P_1 \to P_2$ of the form $\tilde{f}$ are precisely those for which dg carries the parallelism on P_1 onto that of P_2. This leads one to the study of mappings preserving a parallelism for which Cartan has developed a general method.

Suppose that on $P_j, j = 1, 2$, there is a parallelism determined by a vector-valued one-form $\omega_j : T(P_j) \to \boldsymbol{R}^k$ which is an isomorphism at each point. Under certain nondegeneracy conditions, one can give necessary and sufficient conditions for there to exist an $f : U \to V$, $U \subset P_1$, $V \subset P_2$ with $f^*\omega_2 = \omega_1$.

For the moment we drop the subscripts. Let ω^α be the components of ω. Since ω defines a parallelism, ω_p^α form a basis for $T_p^*(P)$ and we have

$$d\omega^\alpha = c^\alpha_{\beta\gamma}\omega^\beta \wedge \omega^\gamma \quad \text{(summation convention)} \tag{1.2}$$

for some smooth functions $c^\alpha_{\beta\gamma}$, uniquely determined if we set $c^\alpha_{\beta\gamma} = -c^\alpha_{\gamma\beta}$. Similarly,

$$dc^\alpha_{\beta\gamma} = c^\alpha_{\beta\gamma,\delta}\,\omega^\delta. \tag{1.3}$$

Continuing to express differentials in this way, we get a set of functions $\mathscr{F} = \{c^\alpha_{\beta\gamma}, \cdots, c^\alpha_{\beta\gamma,\delta_1\cdots\delta_n}, \cdots\} = \{c^\alpha_A\}$.

Let V_p be the subspace of $T_p^*(P)$ spanned by $\{dc^\alpha_A\}$. Call a point p *regular* if the dimension of V_q is constant for q in some open set U containing p. Let $r(p)$ be the minimum number such that a basis for V_p can be found with $|A| \leqq r(p)$. Let $\{y^1, \cdots, y^n\}$ be functions chosen from among the c^α_A with $|A| \leqq r(p)$ whose differentials give a basis for V_q when $q \in U$. Complete this to a chart on U, $\{y^1, \cdots, y^n, x^1, \cdots, x^m\}$, which will be called an *adapted chart*.

THEOREM 1.4 (CARTAN [10], [20]). *Given (P_1, ω_1), (P_2, ω_2) manifolds with complete parallelism, suppose $p_1 \in P_1$ is a regular point and that $\{y_1^1, \cdots, y_1^n, x_1^1, \cdots, x_1^m\}$ is an adapted chart on U. A necessary and sufficient condition for the existence of a local equivalence $f : U \to V$, $f(p_1) = p_2$ is that p_2 be regular with adapted chart $\{y_2^1, \cdots, y_2^{n'}, x_2^1, \cdots, x_2^{m'}\}$ on V with $n = n'$ and $r(p_1) = r(p_2)$; further, that y_1^i and y_2^i correspond to the same multi-indices and that C^α_{1A} and C^α_{2A} are the same functions in terms of the adapted charts for $|A| \leqq r(p_1) + 1$.*

What follows is a description of the way this method works in the case of CR structures on real hypersurfaces. First we will outline the construction of the bundle

P with parallelism, and then show how it can be used to prove theorems of independent interest.

Let M be a real hypersurface in $\boldsymbol{C}^{n+1}$. On $T_x(\boldsymbol{C}^{n+1})$ there is the complex structure transformation J_x. Define a complex vector bundle $H(M)$ over M by

$$H_x(M) = T_x(M) \cap JT_x(M) \subset T_x(\boldsymbol{C}^{n+1}). \tag{1.5}$$

$H(M)$ defines a codimension 1 distribution in $T(M)$ with a complex structure depending smoothly on $x \in M$. Let E_x be the annihilator of $H_x(M)$ in $T_x^*(M)$. If $\mathscr{D}$ is the sheaf of germs of $\boldsymbol{R}$-valued smooth functions on M, then the sheaf of germs of smooth sections of E, denoted $\mathscr{I}$, is a sheaf of $\mathscr{D}$-modules. By $\mathscr{D}_{\boldsymbol{C}}$ denote the sheaf $\mathscr{D} \otimes \boldsymbol{C}$, and by $\mathscr{C}$, the sheaf of germs of $\boldsymbol{C}$-valued 1-forms on M, which are $\boldsymbol{C}$-linear when restricted to $H(M)$. $\mathscr{C}$ is a sheaf of $\mathscr{D}_{\boldsymbol{C}}$-modules.

Define the CR structure on M to be the pair $(\mathscr{I}, \mathscr{C})$. A smooth map $f : (M, \mathscr{I}, \mathscr{C}) \to (M'; \mathscr{I}', \mathscr{C}')$ is called a CR map if

$$f^*\mathscr{I}' \subset \mathscr{I}, \qquad f^*\mathscr{C}' \subset \mathscr{C}. \tag{1.6}$$

A complex-valued smooth function f is a CR function iff

$$df \wedge \theta \wedge \theta' \wedge \cdots \wedge \theta^n = 0 \tag{1.7}$$

for $\theta \in \mathscr{I}$, $\theta^\alpha \in \mathscr{C}$, $\alpha = 1, \cdots, n$.

In other words, a CR manifold is a contact manifold with a complex structure on the real codimension 1 distribution; CR maps are contact transformations whose tangent maps are complex linear.

The Levi form at $x \in M$, defined only up to a multiple, is the quadratic form on H_xM

$$(X, Y) = d\theta(X, JY). \tag{1.8}$$

If the Levi form is definite, then M is called strictly pseudoconvex (s. ψ. c.).

The integrability condition which is a necessary condition for an abstract CR manifold to be CR equivalent to an imbedded hypersurface in $\boldsymbol{C}^{n+1}$ can be stated

$$d(\mathscr{I} + \mathscr{C}) \subset \mathscr{I} + \mathscr{C}. \tag{1.9}$$

Recently Boutet de Monvel **[3]** has shown that this condition is also sufficient for compact strictly pseudoconvex CR manifolds of real dimension $2n + 1$ to imbed locally in $\boldsymbol{C}^{n+1}$ ($n \geqq 2$).

Following E. Cartan, one would like to construct a space P with a trivialization of $T(P)$ that is preserved by lifts of CR maps to P, i.e., such that

$$\begin{array}{ccc} P \times \boldsymbol{R}^k & \xrightarrow{\tilde{f} \times \mathrm{id}} & P' \times \boldsymbol{R}^k \\ \uparrow & & \uparrow \\ TP & \xrightarrow{d\tilde{f}} & TP' \end{array} \tag{1.10}$$

commutes. It would be natural to use the extension and reduction procedure known as *prolongation*, which was developed by Cartan to attack this kind of problem in relation to pseudogroups. (See Cartan **[10]**, Kobayashi **[15]**, Sternberg **[20]**.) Unfortunately, for technical reasons, this will not work if one starts with $\mathscr{I} + \mathscr{C}$ and considers the naturally associated real frames

(1.11) $\{\theta, \operatorname{Re}(\theta^\alpha), \operatorname{Im}(\theta^\alpha)\}, \qquad \theta^\alpha_x | H_x M$ complex linear.

Nonetheless, using other techniques, Cartan [9] succeeded for $\dim M = 3$ in defining the bundle P and a parallelism invariant under CR maps, thus solving the equivalence problem.

The first generalization of Cartan's new techniques was due to Tanaka [**21**]. He originally made the restrictive assumption of some homogeneity in M, specifically, the existence at each $p \in M$ of an infinitesimal CR automorphism transverse to $H_x M$. In a later paper [**22**] he removed the restriction by defining a new method of prolongation.[1] The essential idea (which is also used in the study of PDE problems arising from CR structures) is to grade the structure inhomogeneously by giving different weights to different tangent directions.

In his approach, although it is completely independent of Tanaka, Chern implicitly uses the same idea, first prolonging the Pfaffian system $\mathscr{I}$ and getting the line bundle E, then carrying out a similar procedure for $\pi^*\mathscr{C}$ on E.

By orienting M we can reduce E to a ray bundle which we continue to denote E. The desired principal bundle is a subbundle of the frame bundle on E. The final picture is

(1.12)

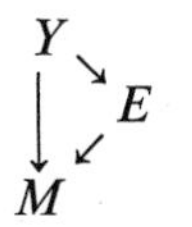

where Y is principal over both E and M.

At this point we will give a quick description of Y; more details can be found in the appendix to this section.

From now on we will restrict to strictly pseudoconvex M for ease of notation. The construction applies as well to any CR manifold whose Levi form is nondegenerate.

As a subbundle of the cotangent bundle, E has a tautological one-form ν (at $\theta \in E$, $\langle \nu, X\rangle = \langle \theta, \pi_* X\rangle$). Y is the bundle of coframes on E, $\{\nu, \operatorname{Re}(\omega^\alpha), \operatorname{Im}(\omega^\alpha), \phi\}$, $\alpha = 1, \cdots, n$, such that

$$(1.13) \qquad \begin{aligned} &\text{(a)} \quad d\nu = -\overline{i\omega^\alpha} \wedge \omega^\alpha + \nu \wedge \phi, \\ &\text{(b)} \quad \omega^\alpha_p \in \pi^*(\mathscr{C}_{\pi(p)}). \end{aligned}$$

On Y there are tautological forms which will be written $\{\nu, \omega^\alpha, \phi\}$ for economy. The integrability condition implies that

$$(1.14) \qquad d\omega^\alpha = \phi^\alpha_\beta \wedge \omega^\beta + \phi^\alpha \wedge \nu$$

for some choice of $n^2 + n$ complex one-forms on Y. One can show that ϕ^α_β can be chosen such that $\phi^\alpha_\beta + \frac{1}{2}\phi\delta^\alpha_\beta \in \mathfrak{U}(n)$ (the Lie algebra of $U(n)$). Further, one can find a one-form ψ such that

$$(1.15) \qquad d\phi = -\overline{i\phi^\alpha} \wedge \omega^\alpha + i\phi^\alpha \wedge \overline{\omega^\alpha} + \psi \wedge \nu.$$

Chern [**11**] proves that it is possible to specify $\{\phi^\alpha_\beta, \phi^\alpha, \psi\}$ uniquely such that they are CR invariants.

[1] This paper seems to have been overlooked by Chern and Moser in their description of Tanaka's work.

Let $G = SU(n+1, 1)/K$, where K is the center, i.e., the $n+2$ roots of 1. Let $H = (CU(n) \ltimes N)/K$, where $CU(n)$ is the conformal unitary group $U(n) \times \boldsymbol{R}^+$, N is the Heisenberg group, and "$\ltimes$" means semidirect product.

THEOREM 1.16 (E. CARTAN, TANAKA, CHERN). *There exist a principal H-bundle Y over M and an* $\mathfrak{su}(n+1, 1)$*-valued one-form* ω *on Y such that*

$$(1.17) \qquad \begin{array}{ll} \text{(a)} & \omega_p(X) = 0 \Leftrightarrow X_p = 0, \\ \text{(b)} & R_a^* \omega = \mathrm{Ad}(a^{-1})\omega \quad \text{for } a \in H. \end{array}$$

A smooth map $g: Y_1 \to Y_2$ *is the lift of a CR map* $f: M_1 \to M_2$ *if and only if* $g^*\omega_2 = \omega_1$.[2]

The homogeneity used in the proof of this theorem arises from a natural grading on the Lie algebra $\mathfrak{g} = \mathfrak{su}(n+1, 1)$. That is, $\mathfrak{g}$ can be expressed as the direct sum of vector subspaces $\mathfrak{g}^j$:

$$(1.18) \qquad \mathfrak{g} = \mathfrak{g}^{-2} + \mathfrak{g}^{-1} + \mathfrak{g}^0 + \mathfrak{g}^1 + \mathfrak{g}^2$$

with $[\mathfrak{g}^j, \mathfrak{g}^k] \subset \mathfrak{g}^{j+k}$. In fact

$$(1.19) \qquad \mathfrak{g}^{-2} \cong \mathfrak{g}^2 \cong \boldsymbol{R}, \quad \mathfrak{g}^{-1} \cong \mathfrak{g}^1 \cong \boldsymbol{C}^n, \quad \mathfrak{g}^0 \cong \mathfrak{cu}(n).$$

Thus the form ω can be written

$$(1.20) \qquad \omega = \omega^{(-2)} + \omega^{(-1)} + \omega^{(0)} + \omega^{(1)} + \omega^{(2)},$$

and if we define the *curvature* by

$$(1.21) \qquad \Omega = d\omega + \tfrac{1}{2}[\omega, \omega],$$

it, too, can be expressed in homogeneous components. The normalizations in the proof of 1.16 guarantee:

$$(1.22) \qquad \Omega^{-2} = 0, \qquad \Omega^{-1} = 0.$$

The uniqueness of ω is achieved by placing further auxiliary conditions on Ω^0, Ω^1 and Ω^2.

Appendix to §1. For $x, y \in \boldsymbol{C}^n$ let $F(x, y)$ be a nondegenerate Hermitian form of type $(k, n-k)$. With respect to the standard complex basis for $\boldsymbol{C}^n$, we can represent F by a Hermitian matrix $(g_{\alpha\beta})$. If $(x) = (x', \cdots, x^n)$ and $(y) = (y', \cdots, y^n)$ then $F(x, y) = (x)(g_{\alpha\beta})(y)^*$ where * indicates transpose conjugate. Let $\boldsymbol{P}^{n+1}$ be $(n+1)$-dimensional complex projective space and let $(s, z^1, \cdots, z^n, \omega)$ be homogeneous complex coordinates. Define a real quadric Q by the equation:

$$(1.23) \qquad \sum_{\alpha,\beta=1}^{n} g_{\alpha\beta} z^\alpha \overline{z^\beta} + i/2(w\bar{s} - \bar{w}s) = 0, \qquad i = \sqrt{-1},$$

or in matrix notation

$$(1.24) \qquad (s, z^\alpha, w) \begin{pmatrix} 0 & 0 & -i/2 \\ 0 & g_{\alpha\beta} & 0 \\ i/2 & 0 & 0 \end{pmatrix} \begin{pmatrix} \bar{s} \\ \overline{z^\alpha} \\ \bar{w} \end{pmatrix} = 0.$$

[2] Tanaka's formulation is slightly different.

If F' is the Hermitian bilinear form defined by the matrix in (1.24), then F' is of type $(k+1, n-k+1)$, and the group preserving F' is isomorphic to $U(k+1, n-k+1)$. Let G be the factor group of $SU(k+1, n-k+1)$ modulo the center, a finite group of diagonal matrices ζI with $\zeta^{n+2} = 1$. The group G is a subgroup of the group of holomorphic transformations on $\boldsymbol{P}^{n+1}$, and it acts transitively and effectively on the quadric Q. The definition of the left action λ of G on Q is

$$G \ni A \mapsto \lambda(A) \in \mathrm{Aut}(Q), \qquad \lambda(A)\,(s, z^\alpha, w) = (s, z^\alpha, w)\,A^{-1}. \tag{1.25}$$

If we define H to be the subgroup of G which leaves fixed the point in Q with homogeneous coordinates $(s, 0, 0)$, then G/H is real analytically isomorphic to Q. The mapping of G onto $G/H = Q$ defines a real analytic fiber bundle over Q. The Maurer-Cartan forms on G define, at every point $g \in G$, an isomorphism of the tangent space $T_g(G)$ to G at g with the Lie algebra $\mathfrak{g} = \mathfrak{su}(k+1, n-k+1)$.

In the special case $g_{\alpha\beta} = \delta_{\alpha\beta}$ the hyperquadric Q is given by

$$\sum_{\alpha=1}^{n} |z^\alpha|^2 + i/2(w\bar{s} - \bar{w}s) = 0. \tag{1.26}$$

Under the mapping (inverse Cayley transformation)

$$\eta = -i\,\frac{w - is}{w + is}, \qquad \zeta^\alpha = -2\,\frac{z^\alpha}{w + is}, \tag{1.27}$$

an open neighborhood of Q is mapped biholomorphically onto a neighborhood of the sphere in $\boldsymbol{C}^{n+1}$

$$\sum_{\alpha=1}^{n} |\zeta^\alpha|^2 + |\eta|^2 = 1.$$

G becomes the group of linear fractional transformations of the unit ball in $\boldsymbol{C}^{n+1}$. H is the subgroup leaving $(\zeta, \eta) = (0, i)$ fixed.

Before we proceed with the study of the Cartan connection on a general M, we must study the Lie algebra $\mathfrak{g}$. Let $\mathfrak{sl}(n+2, \boldsymbol{C})$ be the algebra of $n+2$ by $n+2$ complex matrices of trace 0. If B is the matrix defining F' in (1.24), then $A \in \mathfrak{g}$ if and only if $A \in \mathfrak{sl}(n+2, \boldsymbol{C})$ and

$$AB + B\overline{A^t} = 0. \tag{1.28}$$

Setting

$$A = \begin{pmatrix} a_0^0 & a_0^\alpha & a_0^{n+1} \\ a_\beta^0 & a_\beta^\alpha & a_\beta^{n+1} \\ a_{n+1}^0 & a_{n+1}^\alpha & a_{n+1}^{n+1} \end{pmatrix}, \qquad B = \begin{pmatrix} 0 & 0 & -i/2 \\ 0 & g_{\alpha\beta} & 0 \\ i/2 & 0 & 0 \end{pmatrix}$$

we have

$$\begin{aligned} &a_0^0 + \overline{a_{n+1}^{n+1}} = 0, \qquad && a_\beta^\gamma g_{\gamma\alpha} + g_{\beta\gamma}\overline{a_\alpha^\gamma} = 0, \\ &a_0^0 + \mathrm{tr}\, a_\beta^\alpha + a_{n+1}^{n+1} = 0, && \\ &a_0^{n+1} = \overline{a_0^{n+1}}, && a_{n+1}^0 = \overline{a_{n+1}^0}, \\ &a_\beta^{n+1} = 2i g_{\alpha\gamma}\overline{a_0^\gamma}, && a_\beta^0 = -2i g_{\beta\gamma}\overline{a_{n+1}^\gamma}. \end{aligned} \tag{1.29}$$

Notation will be much easier if we assume $g_{\alpha\beta} = \delta_{\alpha\beta}$; however, all our calculations will apply equally well to the case of an arbitrary Hermitian matrix. In the case $g_{\alpha\beta} = \delta_{\alpha\beta}$, the group G is $SU(n+1, 1)/\text{center}$ and $\mathfrak{g} = \mathfrak{su}(n+1, 1)$, and if $A \in \mathfrak{g}$

$$A = \begin{pmatrix} a_0^0 & b^\alpha & b \\ -2ia^{*\alpha} & a_\beta^\alpha & 2ib^{*\alpha} \\ a & a^\alpha & -\overline{a_0^0} \end{pmatrix}, \tag{1.30}$$

with $a_\beta^\alpha \in \mathfrak{U}(n)$, $a_0^0 - \overline{a_0^0} = -\operatorname{tr} a_\beta^\alpha$, a, b real.

The algebra $\mathfrak{g}$ is a graded Lie algebra with five-step grading $\mathfrak{g} = \mathfrak{g}^{-2} \oplus \mathfrak{g}^{-1} \oplus \mathfrak{g}^0 \oplus \mathfrak{g}^1 \oplus \mathfrak{g}^2$, with:

$$\begin{aligned} A \in \mathfrak{g}^{-2} &\Leftrightarrow A = \begin{pmatrix} 0 & 0 & t \\ 0 & 0 & 0 \\ 0 & 0 & 0 \end{pmatrix}, \\ A \in \mathfrak{g}^{-1} &\Leftrightarrow A = \begin{pmatrix} 0 & b^\alpha & 0 \\ 0 & 0 & 2ib^{*\alpha} \\ 0 & 0 & 0 \end{pmatrix}, \\ A \in \mathfrak{g}^{0} &\Leftrightarrow A = \begin{pmatrix} a_0^0 & 0 & 0 \\ 0 & a_\beta^\alpha & 0 \\ 0 & 0 & -\overline{a_0^0} \end{pmatrix}, \\ A \in \mathfrak{g}^{1} &\Leftrightarrow A = \begin{pmatrix} 0 & 0 & 0 \\ -2ia^{*\alpha} & 0 & 0 \\ 0 & a^\alpha & 0 \end{pmatrix}, \\ A \in \mathfrak{g}^{2} &\Leftrightarrow A = \begin{pmatrix} 0 & 0 & 0 \\ 0 & 0 & 0 \\ r & 0 & 0 \end{pmatrix}. \end{aligned} \tag{1.31}$$

If $\mathfrak{h}$ is the Lie algebra of the subgroup H, then $\mathfrak{h} = \mathfrak{g}^0 \oplus \mathfrak{g}^1 \oplus \mathfrak{g}^2$. $\mathfrak{g}^{-2} \oplus \mathfrak{g}^{-1}$ and $\mathfrak{g}^1 \oplus \mathfrak{g}^2$ are subalgebras, isomorphic to the Heisenberg algebra. For if $x, y \in \boldsymbol{C}^n$, $b \in \boldsymbol{R}$, and $\langle x, y\rangle = F(x, y) = (x)(y)^*$, let

$$\sigma(x) = \begin{pmatrix} 0 & x & 0 \\ 0 & 0 & 2ix^* \\ 0 & 0 & 0 \end{pmatrix}, \qquad \tau(b) = \begin{pmatrix} 0 & 0 & b \\ 0 & 0 & 0 \\ 0 & 0 & 0 \end{pmatrix}.$$

Then $[\sigma(x), \sigma(y)] = \tau(a)$, with $a = -4\operatorname{Im}(F(x, y))$, and $[\sigma(x), \tau(b)] = 0$.

Let N^- be the subgroup of G corresponding to $\mathfrak{g}^{-2} \oplus \mathfrak{g}^{-1}$, let N^+ be the subgroup corresponding to $\mathfrak{g}^1 \oplus \mathfrak{g}^2$, and let G^0 be the subgroup corresponding to $\mathfrak{g}^0$. Then, if $[A]$ means the coset of A and ζ some $(n+2)$nd root of 1

$$[A_1] \in N^- \Leftrightarrow A = \zeta \begin{pmatrix} 1 & b^\alpha & r + i\,|b|^2 \\ 0 & I & 2ib^{*\alpha} \\ 0 & 0 & 1 \end{pmatrix}, \tag{1.32}$$

where $|b|^2 = \sum |b^\alpha|^2$, r real;

$$[A_2] \in G^0 \Leftrightarrow A = \zeta \begin{pmatrix} a & 0 & 0 \\ 0 & \tilde{A} & 0 \\ 0 & 0 & \bar{a}^{-1} \end{pmatrix}$$

with $(a/\bar{a})\det \tilde{A} = 1$, and $\tilde{A} \in U(n)$;

$$[A_3] \in N^+ \Leftrightarrow A = \zeta \begin{pmatrix} 1 & 0 & 0 \\ -2ia^{*\alpha} & I & 0 \\ s - i\,|a|^2 & a^\alpha & 1 \end{pmatrix},$$

$|a|^2 = \sum_{\alpha=1}^{n} |a^\alpha|^2$, s real. The subgroup $H = G^0 \ltimes N^+$ leaves $p = (s, 0, 0)$ fixed, so we can define the linear isotropy representation

$$\tau : H \to \mathrm{Aut}(T_p(Q)), \qquad \tau : a \mapsto d(\lambda(a))_p. \tag{1.33}$$

The differential of this representation defines a representation of Lie algebras

$$d\tau : \mathfrak{h} \to \mathrm{End}(T_p(Q)) \tag{1.34}$$

which allows one to interpret the grading of the space of infinitesimal automorphisms. If one identifies $T_p(Q)$ with the vector space $\mathfrak{g}/\mathfrak{h}$, then $d\tau$ is induced by the adjoint representation of $\mathfrak{h}$ on $\mathfrak{g}$. Then it is immediate that $\mathfrak{g}^{-2}$ is a real line transversal to $H_p(Q)$, the distribution of complex subspaces, and $\mathfrak{g}^{-1} = H_p(Q)$. One checks that $d\tau(\mathfrak{g}^0)$ consists of the skew Hermitian matrices on $\mathfrak{g}^{-1}$ extended by all matrices which act as a real scalar ρ on $\mathfrak{g}^{-1}$ and 2ρ on $\mathfrak{g}^{-2}$. Further, $d\tau(\mathfrak{g}^1)$ consists of transformations which map $\mathfrak{g}^{-2}$ into $\mathfrak{g}^{-1}$ and are 0 on $\mathfrak{g}^{-1}$; these correspond to an infinitesimal change in the direction of the real transversal. Finally, $d\tau(\mathfrak{g}^2) = 0$. This says that the one-parameter group of transformations $(s, z^\alpha, w) \mapsto (s + rw, z^\alpha, w)$ looks like the identity to second order at p (in terms of affine coordinates). This fact explains the need for a bundle which contains more information than the bundle of frames on M to study CR structures, just as in conformal geometry, as opposed to Riemannian geometry. What is surprising is that, also as in conformal geometry, one only needs second-order jet information.

Now we will give a precise definition of the CR structure bundle Y. Let Y_p be the fiber of Y over $p \in E$:

$$Y_p = \text{Set of } (2n+2)\text{-tuples } \{\phi, \mathrm{Re}(\omega^\alpha), \mathrm{Im}(\omega^\alpha), \nu\}$$

such that

(1.35)
(1) they form a basis for $T_p^* E$,
(2) ν is the canonical one-form on E and $d\nu = -i\omega^\alpha \wedge \overline{\omega^\alpha} + \nu \wedge \phi$,
(3) $\omega^\alpha = \pi^*\theta^\alpha$, $\alpha = 1, \cdots, n$, for $\theta^\alpha \in T_x^*(M) \otimes \mathbf{C}$ $(\pi(p) = x$, where θ^α restricted to $H_x M$ is complex linear).

To determine the structure group of Y we look for all changes of frame which preserve the structure equation (1.35(2)). Using the summation convention on repeated indices:

$$\nu = \nu, \quad \omega'^\alpha = a^\alpha_\beta \omega^\beta + a^\alpha \nu, \quad \phi' = \phi + \mathrm{Re}(b_\beta \omega^\beta) + c\nu, \tag{1.36}$$

where $A = (a^\alpha_\beta)$ is an $n \times n$ complex matrix, $a = (a^\alpha)$ and $b = (b_\beta)$ are complex n-vectors, and c is a real number. Then equation (1.35(2)):

$$d\nu = -i\omega'^\alpha \wedge \overline{\omega'^\alpha} + \nu \wedge \phi'$$

implies

$$\begin{aligned} a^\alpha_\beta \overline{a^\alpha_\gamma} &= \delta_{\beta\gamma}, & AA^* &= I, \\ 2a^\alpha_\beta \overline{a^\alpha} + \tfrac{1}{2} b_\beta &= 0, & -2iAa^* &= b. \end{aligned} \tag{1.37}$$

Thus the transformation is given by

$$(\phi', \omega'^\alpha, \nu) = (\phi, \omega^\alpha, \nu) \begin{pmatrix} 1 & 0 & 0 \\ -2iAa^* & A & 0 \\ c & a & 1 \end{pmatrix} \tag{1.38}$$

where ϕ' is interpreted as the real part of the first entry in the product on the right side. In fact, with this convention, we can replace the real number c by $c - i\,|a|^2$. Then the matrix on the right side is in $SU(n+1, 1)$ provided $\det A = 1$. However, the only way to restrict admissible changes of frame so that $\det A = 1$ is by choosing a volume element on $H(M)$. Such a choice globally would amount to a distinguished choice of Levi form, which we do not want to assume. To avoid this difficulty, we notice that if $B \in U(n)$ and λ is a unit complex number such that $\det B = \lambda^2$, then

$$\tilde{B} = \begin{pmatrix} \lambda^{-1} & 0 & 0 \\ 2iBb^* & B & 0 \\ c'i\,|b|^2 & b & \lambda^{-1} \end{pmatrix} \in SU(n+1, 1) \tag{1.39}$$

and

$$B^{-1}\begin{pmatrix} \phi & \omega^\alpha & \nu \\ 0 & 0 & 2i\omega^{*\alpha} \\ 0 & 0 & -\phi \end{pmatrix} B \equiv \begin{pmatrix} \phi' & \omega'^\alpha & \nu \\ 0 & 0 & 2i\omega'^{*\alpha} \\ 0 & 0 & -\phi \end{pmatrix} \quad \mathrm{mod}\,(\mathfrak{g}^0 + \mathfrak{g}^1 + \mathfrak{g}^2) \tag{1.40}$$

where

$$\begin{gathered} \nu' = \nu, \qquad \omega'^\alpha = \lambda B\omega^\alpha + \lambda b^\alpha \nu, \\ \mathrm{Re}\,(\phi') = \phi - \mathrm{Re}\,(\lambda(2iBb^*)_\alpha \omega^\alpha + \lambda(c' - i\,|b|^2)\nu). \end{gathered} \tag{1.41}$$

Set

$$\lambda B = A, \quad \lambda b^\alpha = a^\alpha, \quad \mathrm{Re}\,(\lambda(c' - i\,|b|^2)) = c. \tag{1.42}$$

$\det A = \det \lambda B = \lambda^{n+2}$ solves for λ up to an $(n+2)$nd root of unity. For $\zeta^{n+2} = 1$, the action of $\tilde{B} = \zeta I$ is trivial, so the adjoint representation of $U(n) \ltimes N^+$ on frames $\{\phi, \omega^\alpha, \nu\}$ considered as forms with values in $\mathfrak{g}/\mathfrak{h}$ makes Y into a principal bundle over E with structure group the image of $U(n) \ltimes N^+$ under the adjoint representation.

To see that Y is a principal bundle over M, we note that multiplication μ_{t^2} on E defines an action of $\boldsymbol{R}^+$ on Y as follows.

$$R_t\{\phi, \omega^\alpha, \nu\} = \{\mu_{t^2}^*\phi, \mu_{t^2}^*\omega^\alpha/t, \mu_{t^2}^*\nu/t^2\}. \tag{1.43}$$

If $A \in U(n)$, $a \in \boldsymbol{C}^n$ and $C \in \boldsymbol{R}$, and $R_{(A,a,c)}$ is the change of frame in (1.36), then

$$R_t^{-1} R_{(A,a,c)} R_t = R_{(A,a/t,c/t^2)} \tag{1.44}$$

which agrees with the representation of H as a semidirect product of $\boldsymbol{R}^+$ with $(U(n)/K) \ltimes N$. Thus, Y is a principal H-bundle over M.

Let $\pi\colon Y \to E$ be the projection. On Y we have $n+2$ differential one-forms defined at $(\phi, \omega^\alpha, \nu) \in Y$ by

$$\begin{aligned} \tilde{\phi}_{(\phi,\omega^\alpha,\nu)}(X) &= \langle \phi, \pi_* X\rangle, \\ \tilde{\omega}^\alpha{}_{(\phi,\omega^\alpha,\nu)}(X) &= \langle \omega^\alpha, \pi_* X\rangle, \\ \tilde{\nu}_{(\phi,\omega^\alpha,\nu)}(X) &= \langle \nu, \pi_* X\rangle. \end{aligned} \tag{1.45}$$

As in the text of §1, we abuse notation and denote $\tilde{\nu}$ by ν, etc. Define one-forms with values in $\mathfrak{g}$:

$$\omega^{(-2)} = \begin{pmatrix} 0 & 0 & 2\nu \\ 0 & 0 & 0 \\ 0 & 0 & 0 \end{pmatrix},$$

(1.46)
$$\omega^{(-1)} = \begin{pmatrix} 0 & \omega^\alpha & 0 \\ 0 & 0 & 2i\omega^{*\alpha} \\ 0 & 0 & 0 \end{pmatrix},$$

$$\mathrm{Re}(\omega_0^0) = \begin{pmatrix} \frac{1}{2}\phi & 0 & 0 \\ 0 & 0 & 0 \\ 0 & 0 & -\frac{1}{2}\phi \end{pmatrix}.$$

$\mathrm{Im}(\omega_0^0)$ will be defined later. To complete a Cartan connection we must extend ω_0^0 to a complex-valued form and define a matrix-valued one-form $\omega^{(0)}$ with values in $\mathfrak{cu}(n)$, a vector-valued one-form $\omega^{(1)}$ with values in $\boldsymbol{C}^n$, and a real one-form $\omega^{(2)}$ such that the $\mathfrak{g}$-valued one-form $\omega = \omega^{(-2)} + \omega^{(-1)} + \omega^{(0)} + \omega^{(1)} + \omega^{(2)}$ satisfies (a) $\omega(X) = 0 \Leftrightarrow X = 0$, and (b) $R_a^* \omega = \mathrm{Ad}(a^{-1})\, \omega$ for $a \in H$. We will define $\omega^{(0)}, \omega^{(1)}, \omega^{(2)}$ rather arbitrarily and then show that there are natural normalizations which make the choice unique.

First we note that if H' is the structure group of $Y \to E$, then the maximal compact subgroup of H' is $L = U(n)/(\zeta I)$, where ζ is a primitive $(n + 2)$nd root of 1. We can reduce the structure group of $Y \to E$ to L, and choose a representative L-subbundle Y_0 of Y. Then for $p \in Y_0$, we have a splitting of $T_p(Y)$ into $T_p(Y_0) \oplus V_p$ where V_p is a real, $(2n + 1)$-dimensional subspace of the tangents to the fibers of $Y \to E$. Since the vertical vector fields generated by the action of N^+ on Y are transverse to Y_0, we have an isomorphism at each $p \in Y$, $V_p \cong \mathfrak{g}^1 \oplus \mathfrak{g}^2$. Let $(\phi^\alpha)_{\alpha=1,\cdots,n}$ and ϕ be, respectively, the $\boldsymbol{C}^n$-valued and $\boldsymbol{R}$-valued forms defined by this isomorphism. Choose an L-connection on Y_0 with the associated $\mathfrak{U}(n)$-valued one-form $(\tilde{\phi}_\beta^\alpha)$. Then we define the forms $\omega^{(0)}, \omega^{(1)}, \omega^{(2)}$ as follows.[3]

$$\omega^{(0)} = \begin{pmatrix} \frac{1}{2}\phi - \frac{\tilde{\phi}_\lambda^\lambda}{n+2} & 0 & 0 \\ 0 & \tilde{\phi}_\beta^\alpha - \frac{\tilde{\phi}_\lambda^\lambda}{n+2} & 0 \\ 0 & 0 & -\frac{1}{2}\phi - \frac{\tilde{\phi}_\lambda^\lambda}{n+2} \end{pmatrix},$$

(1.47) $\qquad \omega_0^0 = $ entry in first column, first row of $\omega^{(0)}$,

$$\omega^{(1)} = \begin{pmatrix} 0 & 0 & 0 \\ -i\phi^{*\alpha} & 0 & 0 \\ 0 & \frac{1}{2}\phi^\alpha & 0 \end{pmatrix},$$

$$\omega^{(2)} = \begin{pmatrix} 0 & 0 & 0 \\ 0 & 0 & 0 \\ -\frac{1}{4}\phi & 0 & 0 \end{pmatrix}.$$

For each $p \in Y_0$, the form ω is a $\mathfrak{g}$-valued one-form on $T_p(Y)$ which is equivariant with respect to L. Extend the definition to all of Y by equivariance with respect to H. In this way we have defined a Cartan connection on Y.

Define a $\mathfrak{g}$-valued two-form Ω which will be called Cartan *curvature*:

[3] The seemingly strange definitions are adopted to be consistent with **[11]**, where $\tilde{\phi}_\beta^\alpha = \phi_\beta^\alpha + \frac{1}{2}\phi$.

$$(1.48) \qquad d\omega + \tfrac{1}{2}[\omega, \omega] = \Omega, \qquad \Omega = \Omega^{-2} + \Omega^{-1} + \Omega^{0} + \Omega^{1} + \Omega^{2}.$$

Note that (1.35) shows that $\Omega^{-2} = 0$. Using a technique implicit in Tanaka and developed by Ochiai [**18**] of interpreting curvature in terms of functions with values in Lie algebra cohomology groups for certain graded Lie algebras, one can show that ω is unique if its curvature is a harmonic representative of the cohomology classes intrinsically associated to Y [**5**].

2. Some applications. What is the advantage of looking at the bundle Y, the form ω, and its associated curvature? First we note that the structure functions c_{ijk} described in §1 consist of the constants which occur in the structure equations of $\mathfrak{su}(n+1, 1)$ and the components of curvature expressed in terms of $\{\nu, \omega^\alpha, \bar{\omega}^\alpha\}$. In the notation of Chern-Moser [**11**, 4.53, A.8′, A.9′].

$$(2.1) \quad \begin{aligned} (\Omega^0)^\alpha_\beta &= S^\alpha_{\beta\rho\cdot\bar\sigma}\omega^\rho \wedge \overline{\omega^\sigma} + V^\alpha_{\beta\cdot\sigma}\omega^\sigma \wedge \omega - V^{\alpha}_{\cdot\,\beta\bar\sigma}\overline{\omega^\sigma} \wedge \nu, \qquad (\Omega^0)^0_0 = 0, \\ (\Omega^1)^\alpha &= V^{\alpha}_{\cdot\,\beta\bar\sigma}\omega^\beta \wedge \overline{\omega^\sigma} + P^\alpha_{\beta\cdot}\omega^\beta \wedge \nu + Q^\alpha_{\bar\beta\cdot}\overline{\omega^\beta} \wedge \nu, \\ \Omega^2 &= 2iP^\alpha_{\beta\cdot}\,\overline{\omega^\alpha} \wedge \omega^\beta + i(R^\alpha\overline{\omega^\alpha} - \overline{R^\alpha}\omega^\alpha) \wedge \nu. \end{aligned}$$

If one can define a section of $Y \to M$ which is invariant under lifts of CR maps, then the functions $\{c^\alpha_A\}$ on Y which solve the equivalence problem can be pulled back down to M by the section and one defines a map $\phi\colon M \to N$. If ϕ is an imbedding, then M_1 is equivalent to M_2 iff $\phi_1(M_1) = \phi_2(M_2)$. This is the way Cartan constructed the nine scalar invariants referred to by Wells in his article in these PROCEEDINGS on deformations of s. ψ. c. domains in $\boldsymbol{C}^2$ [**25**]. Let D be a smooth, bounded s. ψ. c. domain in $\boldsymbol{C}^2$, and let U be an open neighborhood of ∂D. A defining function ϕ for D means $\phi = 0$, $d\phi \neq 0$ along ∂D. Wells proves:

THEOREM 2.2. *There is a neighborhood N of any defining function $\phi \in C^\infty(U, \boldsymbol{R})$ of a strictly pseudoconvex domain, and a set of second category $N_0 \subset N$ such that if $\tilde\phi \in N_0$, $\tilde D_\varepsilon = \{z \in \boldsymbol{C}^2 \mid \tilde\phi(z) < \varepsilon\}$, then*
(a) $\tilde D_{\varepsilon_1} \not\cong_{\text{bihol}} \tilde D_{\varepsilon_2}$, $0 \leqq |\varepsilon_1|, |\varepsilon_2| < \varepsilon_0$.
(b) $\operatorname{Aut} \tilde D_\varepsilon = \{\mathrm{id}\}$, $0 \leqq |\varepsilon| < \varepsilon_0$.

The curvature functions are related to the coefficients of Moser's normal form for the defining function of M (cf. §3 below). A rather long calculation shows that if one picks the initial normalizations for the normal form at $x \in M$ compatible with the frame $p \in Y_x$, then

$$(2.3) \qquad S(p) = c_1 N_{22}, \quad V(p) = c_2 \operatorname{tr} N_{32}, \quad P(p) = c_3 \operatorname{tr}{}^2 N_{33} + c_4 \|S\|^2,$$

where c_1, c_2, c_3, c_4 are constants depending only on dim M. The remaining relations are more complicated. Thus, differential equations among S, V, P, Q, R imply corresponding differential equations among the coefficients of Moser's normal form. One important such set of equations is given by the Bianchi identities

$$(2.4) \qquad d\Omega^j = \sum_{k=-2}^{2} [\Omega^{j+k}, \omega^{(-k)}],$$

which one obtains simply by taking the exterior derivative of (1.21), keeping track of the homogeneity of terms. The grading and bracket structure of $\mathfrak{g}$ show that from (2.4) it follows easily that if $S^\alpha_{\beta\rho\cdot\bar\sigma} \equiv 0$ for all indices identically on some open set, then $\Omega \equiv 0$, provided $n \geqq 3$. (For $M \subset \boldsymbol{C}^2$ one has $S^\alpha_{\beta\rho\cdot\bar\sigma} \equiv 0$, $V^\alpha_{\beta\cdot\sigma} \equiv 0$, $P^\alpha_{\beta\cdot} \equiv 0$ for dimension reasons. In this case $Q^\alpha_{\bar\beta\cdot} \equiv 0$ is sufficient to conclude

$\Omega \equiv 0$.) If $\Omega \equiv 0$, then locally Y looks like the Lie group $SU(n+1,1)$, and M is locally CR equivalent to S^{2n+1}.

Define an *umbilic point* $x \in M$ to be any point such that for $p \in Y_x$, $S^{\alpha}_{\beta\rho\cdot\bar{\sigma}}(p) = 0$ if $\dim M > 3$, and $Q^{\alpha}_{\bar{\beta}}.(p) = 0$ if $\dim M = 3$. These conditions are equivalent to saying that the osculation of a CR image of S^{2n+1} to M achieved byMoser's normal coordinates is one order better than generic. Hence

COROLLARY 2.5 (to the BIANCHI identities). *If M is umbilic at each point, then it is locally CR equivalent to S^{2n+1}. Call such a hypersurface ($\Omega \equiv 0$) spherical.*

Returning to the general case, we have the following

THEOREM 2.6. *The sheaf of infinitesimal automorphisms of M has a finite-dimensional stalk at each point. The dimension is at most $(n+2)^2-1$.*

PROOF. Let X be an infinitesimal CR automorphism of M defined on some connected open set U. Let $\tilde{X}$ be the lift of X to Y and suppose $\tilde{X}_p = 0$. We will show that this implies $X_x = 0$ for $x \in U$. This will prove both parts of the theorem, since the linear transformation of lifting a vector field on M to Y and evaluating at p has finite-dimensional range and the assertion is that it is injective. Let $Z_1, \cdots, Z_N$ be a basis of vector fields on Y with $\omega(Z_i) \in \mathfrak{g}$ constant for each i. Since $\mathscr{L}_{\tilde{X}}\omega = 0$, one has $[\tilde{X}, Z_j] = 0$. Then, if exp is the exponential of the parallelism on Y,

$$(2.7)\quad \exp(t\tilde{X})\exp(x_1Z_1 + \cdots + x_NZ_N)p = \exp(x_1Z_1 + \cdots + x_NZ_N)\exp(t\tilde{X})p.$$

Thus $\tilde{X}_{p'} = 0$ for $p' \in$ domain of exponential coordinates. A connectedness argument concludes the proof.

COROLLARY 2.8. *The group $\mathrm{Aut}_{CR}(M)$ of global automorphism is a Lie group of dimension at most* $\dim Y = (n+2)^2 - 1$.

PROOF. Apply the following theorem of Kobayashi [**15**, p. 13].

THEOREM. *Let G be a group of differentiable transformations of a manifold M. Let S be the set of all vector fields which generate global one-parameter groups of transformations $\phi_t \in G$. If the set S generates a finite dimensional Lie algebra, then G is a Lie transformation group and S is the Lie algebra.*

The next theorem can be used in completing the classification of compact homogeneous CR manifolds begun by Morimoto and Nagano [**16**] and Rossi [**19**].

THEOREM 2.9 (WEBSTER). *If $\mathrm{Aut}^0_{CR}(M)$ is noncompact for a compact CR manifold, then M is locally CR equivalent to S^{2n+1}.*

PROOF. Assume $\mathrm{Aut}^0_{CR}(M)$ is noncompact. Then there exists a one-parameter group with noncompact closure. Suppose X is the infinitesimal generator; let $\tilde{X}$ be the lift to Y. $\mathscr{L}_{\tilde{X}}\omega = 0$ implies $\mathscr{L}_{\tilde{X}}\Omega = 0$, which implies $\tilde{X}S^{\alpha}_{\beta\rho\cdot\bar{\sigma}} = 0$ for any component of the curvature term S. (In dimension 3 substitute Q.) Let M_1 be the open subset of M consisting of those x such that $S(p) \neq 0$, for $p \in Y_x$. If M_1 is not empty, then one can show that over M_1 there is a CR invariant reduction of Y to a $U(n)$ bundle. The inclusion $U(n) \subset O(2n) \subset O(2n+1)$ defines a CR invariant Riemannian structure. Let g be the associated metric. One shows the orbits of X stay away from ∂M_1 by considering the function $f(x) = g_x(X_x, X_x)$, which cannot vanish on any open subset of M_1, by Theorem 2.6, and which is constant on the orbits of X. But then $M_\varepsilon \subset M_1 = \{x \mid f(x) \geqq \varepsilon\}$ is a compact Riemannian manifold with boundary and the group of isometries preserving the boundary imbeds into a compact space, namely, the $O(2n+1)$ frame bundle over M_ε.

To apply this result to the classification of homogeneous compact CR manifolds, one needs the following classification theorem.

THEOREM 2.10. *The homogeneous, simply connected CR manifolds with $\Omega \equiv 0$ are the following or their simply connected coverings:*

(1) S^{2n+1},

(2) $\partial \mathscr{U}_{n+1} - \partial \mathscr{U}_{n+1} \cap V$,

where $\mathscr{U}_{n+1} = \{(z, w) \in \mathbf{C}^n \times \mathbf{C} \mid |z|^2 - (w - \bar{w})/2i < 0\}$, *and V is a subspace of* $\mathbf{C}^{n+1}$ *which is either complex (possibly* 0) *or such that* $V + iV = \mathbf{C}^{n+1}$.

THEOREM 2.11. *The compact homogeneous CR manifolds in dimension $\geqq 5$ are covered by*

(1) *the sphere,*

(2) *the unit tangent bundle to a compact rank one Riemannian symmetric space,*

(3) *the unit circle bundle of a homogeneous negative line bundle over a homogeneous algebraic variety.*

PROOF. If $\mathrm{Aut}^0_{\mathrm{CR}}(M)$ is noncompact, then the space is locally equivalent to the sphere and one can apply the classification theorem above, and verify directly that the sphere is the only example which covers a compact *homogeneous* CR manifold.

If $\mathrm{Aut}^0_{\mathrm{CR}}(M)$ is compact, then one can show that the center is at most of dimension 1. If the center $\mathscr{Z}$ is of dimension 1, then the infinitesimal automorphism generating the center is always transverse to $H(M)$. Thus, $V = M/\mathscr{Z}$ is a homogeneous complex manifold and $M \to V$ is the unit circle bundle of a homogeneous negative line bundle, and V is projective algebraic.

If $\mathrm{Aut}^0_{\mathrm{CR}}(M)$ is compact semisimple, then $\pi_1(M)$ is finite. Using Rossi's classification, one knows from the structure of $\mathrm{Aut}_{\mathrm{CR}}(M)$ that the Stein analytic space "filling in" M is nonsingular, hence a Stein manifold, and the classification of Morimoto and Nagano shows that M is of type (2).

We remark that Cartan's complete classification of all homogeneous simply connected (not necessarily spherical) CR manifolds shows that Rossi's exceptional structures on S^3 complete the list in the case $\dim M = 3$.

We mention an application of the existence of Y and ω to the problem of extending automorphisms across a boundary.

THEOREM 2.12. *Let D_1 and D_2 be strongly pseudoconvex domains with real analytic boundaries M_1 and M_2, respectively. Then any biholomorphic $f: D_1 \to D_2$ extends holomorphically to a neighborhood of $\bar{D}_1$.*

PROOF. Fefferman's theorem proves that f extends to a smooth CR map, still denoted $f: M_1 \to M_2$. Then $\tilde{f}^*\omega_2 = \omega_1$. This is an elliptic quasilinear PDE with real analytic coefficients; thus the solution is real analytic. But real analytic CR functions extend to holomorphic functions on a neighborhood.

A similar method is described in Naruki **[17]**.

One may pursue the ideas in the proof of Theorem 2.9 a bit further to obtain other boundary regularity theorems for mappings in the s.ψ.c. case. Webster's CR invariant metric is constructed from a CR invariant 1-form θ_0 annihilating $H(M)$. (The construction of θ_0 is based on an earlier construction of Obata in conformal geometry.) Denote by Σ the locus of umbilics on M, and set $U = M - \Sigma$. The smooth, positive $(2n + 1)$-form on U,

$$d\mu = \theta_0 \wedge (d\theta_0)^n \tag{2.13}$$

extends to a nonnegative, continuous $(2n + 1)$-form on all of M, which gives a CR invariant measure (still denoted $d\mu$) on M. This measure is positive on U, and vanishes along Σ. Using $d\mu$ and the techniques discussed by H. Alexander in these PROCEEDINGS, it is surprisingly easy to prove a generalization of his theorem that proper self-maps of B_{n+1} are biholomorphic.

THEOREM 2.14. *Let D be a compact, complex manifold with smooth, s.ψ.c. boundary, and f: $D \to D$ a proper holomorphic map. Then f is smooth at the boundary.*

COROLLARY 2.15. *If D in Theorem* 2.14 *is Stein, f is biholomorphic.*

This follows from a lemma of Pinčuk.

COROLLARY 2.16. *If the curvature $\Omega \not\equiv 0$ on ∂D, then f is biholomorphic.*

This follows from the method of proof of Theorem 2.14, which distinguishes the case $\Omega \equiv 0$ from $\Omega \not\equiv 0$. This distinction is not illusory, since there are examples of proper f: $D \to D$ with $\Omega \equiv 0$ on ∂D, such that f is a branched cover, branching along the maximal compact subvariety of D.

Geometric techniques arising from the canonical metric in the proof of Theorem 2.9 also give partial results on boundary regularity for proper maps f: $D_1 \to D_2$ ($D_1 \neq D_2$), if ∂D_i are real analytic. For example, if cod $\Sigma_1 \geqq 3$, then f extends across the boundary ∂D_1 (here Σ_1 = umbilic locus of ∂D_1). For more details and related results cf. **[6]**.

Finally, we add that recent work of Webster on s.ψ.c. hypersurfaces with a distinguished, nonvanishing 1-form $\theta \in H(M)^\perp$ (what he calls a *pseudo-Hermitian manifold*) has led to an interesting class of hypersurfaces, analogous to Riemannian symmetric spaces. Webster classifies these and finds that they are, essentially, circle bundles of negative line bundles over Hermitian symmetric spaces.

3. Normal forms for hypersurfaces. We now describe Moser's extrinsic approach to the equivalence problem. The idea is simply to apply local biholomorphic maps fixing a point until we find a simple *normal form* for an equation defining the transformed hypersurface in $\boldsymbol{C}^{n+1}$. Formal *uniqueness* results are obtained at first, working with Taylor expansions of defining functions, and are applicable to $\mathscr{C}^\infty$ hypersurfaces. For analytic hypersurfaces, the *existence* of a convergent solution is proved.

Fix reference coordinates (z, w) for $\boldsymbol{C}^{n+1}$, $z \in \boldsymbol{C}^n$, $\omega = u + iv \in \boldsymbol{C}$, and let M be a real hypersurface through 0 which we will assume to be $\mathscr{C}^\infty$ and strictly pseudoconvex for convenience (nonsingular Levi form is sufficient). By a complex linear transformation we may arrange that $T_0(M) =$ kernel of $dv \subset T_0(C^{n+1})$, $H_0(M) =$ kernel of $dw \subset T_0(C^{n+1})$. M is then defined uniquely by an equation

$$v = F(z, \bar{z}, u), \qquad F \text{ real}, \ dF(0, 0, 0) = 0.$$

A standard biholomorphic transformation shows we may further assume M is defined as

$$v = |z|^2 + F(z, \bar{z}, u), \qquad F \text{ real}, \ dF(0) = 0, \tag{3.1}$$

with F free of terms quadratic in z.

Consider again the one-parameter family A of dilations ρ: $(z, w) \to (\rho z, \rho^2 w)$, $\rho > 0$, which preserve the quadric Q: $v - |z|^2 = 0$. The function F is of *weight* ν if $\rho^* F = \rho^\nu F$. Let $\mathscr{F}$ denote the space of real formal power series in $z, \bar{z}, u$ (i.e., $\overline{F(z, \bar{z}, u)} = F(\bar{z}, z, u)$, and set $\mathscr{F}_\nu = \{F \in \mathscr{F} \mid F = \sum_{\mu \geqq \nu} F_\mu, F_\mu \text{ of weight } \mu\}$. Then (3.1) may be written

$$v = |z|^2 + F(z, \bar{z}, u), \qquad F \in \mathscr{F}_3. \tag{3.2}$$

Denote by $M(F)$ the formal hypersurface determined by (3.2). We want to consider first the relation of formal biholomorphic equivalence among the hypersurfaces so parametrized by $F \in \mathscr{F}_3$, so we consider the group $\mathscr{G}$ of all formal biholomorphic transformations fixing the origin and preserving the family of hypersurfaces given as in (3.2). This gives us a right action of $\mathscr{G}$ on $\mathscr{F}_3$, defined by $\tilde{F} \circ \phi = F$ if and only if

$$\phi^*(v - |z|^2 - \tilde{F}) = h \cdot (v - |z|^2 - F) \tag{3.3}$$

where h is a (real) formal power series in $(z, \bar{z}, w, \bar{w})$ with nonvanishing constant term. Our formal problem is to find a set of representatives in $\mathscr{F}_3$ for the action of $\mathscr{G}$.

The group $\mathscr{G}$ consists of formal transformations $\phi = (f, g)$, with $f = (f_1, \cdots, f_n)$, such that

$$f(z, w) = C \cdot z + \sum_{\nu \geqq 2} f_\nu(z, w), \qquad g(z, w) = \rho \cdot w + \sum_{\nu \geqq 3} g_\nu(z,w) \tag{3.4}$$

where $|Cz|^2 = \rho |z|^2$, f_ν and g_ν of weight ν. The $\mathscr{G}$-isotropy group at $F \in \mathscr{F}_3$ is the subgroup of $\mathscr{G}$ transforming $M(F)$ into itself. $\mathscr{G}$ contains the group H (as in §1), which we know fixes $0 \in \mathscr{F}_3$ (since $M(0)$ is the standard quadric). In fact, $\mathscr{G}$ is a semidirect product, $\mathscr{G} = H \ltimes \mathscr{G}_1'$, where $\mathscr{G}_1'$ is the subgroup of all $\phi = (f, g)$ in $\mathscr{G}$ satisfying

$$\begin{aligned} &\text{(i)} \ \ d\phi(0) = I, \\ &\text{(ii)} \ \operatorname{Re}(\partial^2 g(0)/\partial w^2) = 0. \end{aligned} \tag{3.5}$$

(I = identity matrix.) Since isotropy groups usually give trouble when trying to compute orbit classes, it will be more convenient to find representatives for the action of $\mathscr{G}_1'$. The data $d\phi(0)$, $\operatorname{Re}(\partial^2 g(0)/\partial w^2)$ specify uniquely an element of H and correspond geometrically to the choice of a frame (in the sense of Y) at 0 for M. Thus, a section of the orbits of $\mathscr{G}_1'$ would give the formal biholomorphic equivalence classes of hypersurfaces with such additional data.

Let $\mathfrak{X}_1'$ denote the Lie algebra of $\mathscr{G}_1'$, that is, the set of all formal holomorphic vector fields X at $0 \in C^{n+1}$ such that $\exp X$ lies in $\mathscr{G}_1'$ (for more detail here, see the appendix to this section). As a first approximation to sectioning the orbits of $\mathscr{G}_1'$, we would like to find the tangent space to the orbit of $\mathscr{G}_1'$ through $0 \in \mathscr{F}_3$, and choose a linear complement to this space. Letting $L: \mathfrak{X}_1' \to \mathscr{F}_3 = T_0(\mathscr{F}_3)$ denote the differential of the action of $\mathscr{G}_1'$, we have $\mathscr{R} = L(\mathfrak{X}_1')$ is the tangent space to $0 \circ \mathscr{G}_1'$. The space $\mathscr{R}$ is easily seen to be A-stable: $\mathfrak{X}_1'$ is graded by weights, and L is a graded map (cf. the formula in the appendix to this section). Hence, we may choose a subspace $\mathscr{N} \subset \mathscr{F}_3$, complementary to $\mathscr{R}$, stable by A. An important lemma is

LEMMA 3.6. (i) *$L: \mathfrak{X}_1' \to \mathscr{F}_3$ is injective.*

(ii) *For any $\mathscr{N}$ as above, the map $(N, \phi) \to N \circ \phi$ gives a bijection $\mathscr{N} \times \mathscr{G}_1' \simeq \mathscr{F}_3$.*

Lemma 3.6(i) is proved in **[11]**, but follows from earlier work of Tanaka **[23**, Theorem 4.1]. (ii) says that an infinitesimal section is a global section. (ii) is a formal induction argument, the induction being on the weights of homogeneous terms. (This is carried out in the appendix to this section.)

The preceding has been abstract nonsense. The crux of the matter is to compute $\mathscr{R}$ and a suitable $\mathscr{N}$ explicitly. Say that $F = F(z, \bar{z}, u)$ is of *type* (k, l) if $F(tz, s\bar{z}, u) = t^k s^l F(z, \bar{z}, u)$. Let $\tilde{\mathscr{R}}$ be the space of $F \in \mathscr{F}_3$ such that

$$F = \sum_{\min(k,l) \leq 1} G_{k,l} + G_{1,1}|z|^2 + (G_{1,0} + G_{0,1})|z|^4 + G_{0,0}|z|^6,$$

where $G_{i,j}$ is of type (i, j). Take $\mathscr{N}$ to be the space of $N \in \mathscr{F}_3$ such that $N = \sum N_{k,l}$, with $N_{k,l}$ of type (k, l) and subject to the conditions:

$$\text{(3.7)}\qquad \begin{aligned} &\text{(i)} && N_{k,l} = 0, \quad \text{if } \min(k, l) \leq 1, \\ &\text{(ii)} && \operatorname{tr}(N_{2,2}) = 0, \\ &\text{(iii)} && \operatorname{tr}^2(N_{3,2}) = 0, \\ &\text{(iv)} && \operatorname{tr}^3(N_{3,3}) = 0. \end{aligned}$$

Here tr is the usual Hermitian contraction, i.e., if

$$F_{k,l}(z, \bar{z}, u) = \sum a_{\alpha_1,\cdots,\alpha_k,\bar{\beta}_1,\cdots,\bar{\beta}_l}(u) z_{\alpha_1}\cdots z_{\alpha_k} \cdot \bar{z}_{\beta_1}\cdots \bar{z}_{\beta_l},$$

with the a's symmetric with respect to permutations of $\alpha_1, \cdots, \alpha_k$ and $\bar{\beta}_1, \cdots, \bar{\beta}_l$, then

$$\operatorname{tr} F_{k,l} = \sum b_{\alpha_1,\cdots,\alpha_{k-1},\bar{\beta}_1,\cdots,\bar{\beta}_{l-1}}(u) z_{\alpha_1}\cdots z_{\alpha_{k-1}} \bar{z}_{\beta_1}\cdots \bar{z}_{\beta_{l-1}},$$

where

$$b_{\alpha_1,\cdots,\alpha_{k-1},\bar{\beta}_1,\cdots,\bar{\beta}_{l-1}} = \sum_{\alpha_k,\beta_l} \delta_{\alpha_k,\bar{\beta}_l} a_{\alpha_1,\cdots,\alpha_k,\bar{\beta}_1,\cdots,\bar{\beta}_l}.$$

The significance of this choice of $\mathscr{N}$ is the following

THEOREM 3.8 (MOSER). *With $\mathscr{N}$ as in* (3.7) *above, the map $(N, \phi) \to N \circ \phi$ gives bijections*

(i) $\mathscr{N} \times \mathscr{G}_1' \simeq \mathscr{F}_3$,

(ii) $\mathscr{N}^\omega \times (\mathscr{G}_1')^\omega \simeq \mathscr{F}_3^\omega$,

where $(\ )^\omega$ *denotes convergent* (*real analytic*) *elements.*

The proof of (i) follows from the lemma, after it is shown that $\mathscr{R} = \tilde{\mathscr{R}} \bmod \mathscr{N}$.

The determination of $\mathscr{R}$ mod $\mathscr{N}$ amounts to the formal solution of linear systems of ODE in the variable u.

We make two closely related remarks about the proof of (ii): First, the statement (ii) appears to depend on the choice of $\mathscr{N}$, unlike (i);[4] second, the usual argument for proving convergence of a formal power series seems difficult to apply here. These are both because the algorithm of Lemma 3.6 solving for N and ϕ, given $F \in \mathscr{F}_3$, does so by constructing N and ϕ by terms of pure *weight*, and one has little or no control over how terms are entering the power series construction by *degree*. What has been achieved in (i) is the *uniqueness* of N and ϕ. Hence, given $F \in \mathscr{F}_3^\omega$, to prove the convergence of N and ϕ, it suffices to find a convergent $\psi \in (\mathscr{G}_1')^\omega$ such that, formally, $F \circ \psi \in \mathscr{N}$. This, in turn, requires two steps.

First, one fixes a parametrized real analytic curve $\gamma = \gamma(u)$ through 0 in $M = M(F)$, transverse to $H_0(M)$, together with a real analytic frame $\{e_\alpha(u)\}$ along $\gamma(n)$ which is orthonormal for the Levi form on $H_{\gamma(u)}$. (Note that $\dot{\gamma}(u)$ determines a $\theta \in E_{\gamma(u)}$ by $\theta(\dot{\gamma}(u)) = 1$; this θ determines the specific Levi form in question at $\gamma(u)$.) Elementary arguments show that there is a (unique) biholomorphic ϕ such that $\phi(M(F))$ is defined by $v = |z|^2 + F^*(z, \bar{z}, u)$, where

$$
(3.9)\qquad
\begin{aligned}
&\text{(a)}\quad F^* = \sum_{k,l \geq 2} F^*_{k,l} \qquad (F^*_{k,l} \text{ of type } k, l),\\
&\text{(b)}\quad \phi(\gamma(u)) = (0, u),\\
&\text{(c)}\quad d\phi_*(e_\alpha(u)) = \partial/\partial x_\alpha \qquad (x_\alpha = \operatorname{Re}(z_\alpha)).
\end{aligned}
$$

Second, one examines closely the dependence of the resulting F^* on the choices of γ, the framing, and the parametrization, in that order. One arrives at nonlinear systems of ODE for γ, e_α, and the parametrization of γ which guarantee that conditions (iii), (ii) and (iv), respectively, of the definition (3.7) of $\mathscr{N}$ are satisfied by F^*. The solvability of these equations is shown, concluding the proof.

The key step in the convergence proof is the proper selection of the curve γ. These curves, each uniquely determined by a direction transverse to $H_0(M)$, are biholomorphic invariants of M. E. Cartan first discovered them ($n = 1$), and called them *chains*. In the normal form chosen here, a hypersurface has a chain in common with the standard quadric Q. On Q itself, the chains are exactly the transverse intersections of complex lines in $\boldsymbol{C}^{n+1}$ with Q. Much effort has recently gone into the study of these curves and their analytic meaning on boundaries of domains; we will discuss what is known in §4 below.

In practice, normal forms are hard to compute explicitly. But the most significant and most computable term in the power series N is that of lowest order (order 6, if $n = 1$; order 4, if $n > 1$). The origin is called an *umbilic* point if this term vanishes. (This vanishing is independent of the initial normalizations.) The relation of this lowest order term with the curvature of the Cartan connection is described in (2.3) above.

The normal forms show that the inequivalent nondegenerate hypersurfaces M through 0 depend on infinitely many parameters. Using Fefferman's theorem and

[4] What appears important about the form of $\mathscr{N}$ in Theorem 3.8 is that the three significant linear conditions defining $\mathscr{N}$ (3.7) (ii)—(iv) are independent of u, and represent, therefore, relations among functions of u: The convergence proof rests on satisfying these conditions by imposing differential equations in u.

the normal form, this should globalize to a statement about the "infinitude of moduli" of any smooth, bounded, strongly pseudoconvex domain D. So far, however, this is proved explicitly for just the ball and examples very similar to it ([8]; see also Wells's article in these PROCEEDINGS referred to earlier).

Appendix to §3. In this appendix we indicate a proof of the Lemma 3.6(ii). We consider $F \in \mathscr{F}_3$, and wish to solve uniquely for $N \in \mathscr{N}, \phi \in \mathscr{G}_1'$ such that $F = N \circ \phi$. The argument is simply an induction argument by weights, but we emphasize the nilpotence of the group $\mathscr{G}_1'$, mainly as a matter of taste.

We consider the space $\mathfrak{X}$ of formal holomorphic vector fields at the origin in C^{n+1}. For $X \in \mathfrak{X}$,

$$x = \sum_{i=1}^{n} f_i \frac{\partial}{\partial z_i} + g \frac{\partial}{\partial w} \tag{3.10}$$

with $f_i, g \in C[[z, w]]$, the ring of formal power series in z, w. $\mathfrak{X}$ is the algebra of (continuous) derivations of $C[[z, w]]$. The group A of dilations grades $C[[z, w]]$. An element $X \in \mathfrak{X}$ is homogeneous of weight μ if $(d\rho^{-1})_* X = \rho^\mu X$ (equivalently, X is a graded derivation of $C[[z, w]]$, increasing weight grading by μ). This is the same grading as in §1. Set

$$\mathfrak{X}_\nu = \left\{ X \in \mathfrak{X} \mid X = \sum_{\mu \geqq \nu} X_\mu,\ X_\mu \text{ homogeneous of weight } \mu \right\}. \tag{3.11}$$

For $\nu \geqq 1$, set

$$\mathscr{G}_\nu = \{ g \in \mathscr{G} \mid g = \exp X,\ X \in \mathfrak{X}_\nu \}. \tag{3.12}$$

Note that since $X \in \mathfrak{X}_\nu$, $\nu \geqq 1$, strictly increases the filtration by weight, the exponential series applied to X as an endomorphism of $C[[z, w]]$ makes good sense, and $\exp X$ is an automorphism of $C[[z, w]]$, i.e., a formal biholomorphic transformation. We have

(3.13) $\mathfrak{X}_\nu$ is an ideal in $\mathfrak{X}_1$, $\nu \geqq 1$, and $\mathfrak{X}_\nu / \mathfrak{X}_{\nu+1}$ is abelian.
Similarly for $\mathscr{G}_1$ and $\mathscr{G}_\nu$.

Given $\phi_\nu \in \mathscr{G}_\nu$, $\nu = 1, 2, \cdots$, it makes sense to form the infinite product $\tilde{\phi} = \phi_1 \circ \phi_2 \circ \cdots$.

Now $\mathscr{G}_1' \subset \mathscr{G}_1$. To compute the map L from $\mathfrak{X}_1' \to \mathscr{F}_3$, it suffices to compute L for a homogeneous X, and this is quite simple. One gets:

$$L(X) = \operatorname{Re}\{2 \langle z, f \rangle + ig\}_{v=|z|^2} \tag{3.14}$$

where X is as in (3.10) and $f = (f_1, \cdots, f_n)$. (X is of weight μ iff the f_i are of weight $\mu + 1$ and g of weight $\mu + 2$.) Thus, L is graded, shifting weights by $+2$. If $\mathscr{R} = L(\mathfrak{X}_1')$, then $\mathscr{R}$ is A-stable, and we take $\mathscr{N}$ complementary to $\mathscr{R}$ and A-stable. Then the associated projection P: $\mathscr{F}_3 \to \mathscr{R}$ is graded, and for $N \in \mathscr{F}_3$, $N = \sum_{\mu \geqq 3} N_\mu$, with N_μ of weight μ, is in $\mathscr{N}$ iff $N_\mu \in \mathscr{N}$, for every μ.

We have the rules: For $\phi_\nu = \exp X$ in $\mathscr{G}_\nu$, $F \in \mathscr{F}_3$,

(3.15) (a) $F \circ \phi_\nu \equiv F \bmod \mathscr{F}_{\nu+2}$,
(b) $F \circ \phi_\nu \equiv F + L(X) \bmod \mathscr{F}_{\nu+3}$.

Set $F = F^{(1)}$, and inductively, $F^{(\nu+1)} = F^{(\nu)} \circ \phi_\nu$, where $\phi_\nu \in \mathscr{G}_\nu$ are to be determined so that there are uniquely determined $N_\mu \in \mathscr{N}_\mu$ with

$$F^{(\nu)} \equiv \sum_{\mu=3}^{\nu+1} N_\mu \bmod \mathscr{F}_{\nu+2}. \tag{3.16}$$

Once this is done, set $\tilde{\phi} = \phi_1 \circ \phi_2 \circ \cdots$, and

$$F \circ \tilde{\phi} = N = \sum_{\mu=3}^{\infty} N_\mu \in \mathscr{N}. \tag{3.17}$$

If ϕ_ν is determined uniquely mod $\mathscr{G}_{\nu+1}$, given $F^{(\nu)}$, then $\tilde{\phi}$ is uniquely determined. But (3.15)(b) and Lemma 3.6(i) say that

$$\phi_\nu = -\exp(-L^{-1}(P(F^{(\nu)}_{\nu+2}))) \tag{3.18}$$

works, and is unique mod $\mathscr{G}_{\nu+1}$. (Here, $F^{(\nu)}_{\nu+2}$ is the component of weight $\nu + 2$ in $F^{(\nu)}$.) Setting $\phi = \tilde{\phi}^{-1}$, we have $F = N \circ \phi$, with $N \in \mathscr{N}, \phi \in \mathscr{G}'_1$ uniquely determined.

4. Chains and Lorentz structures. In analogy with the geodesics of classical Riemannian geometry, one expects the chains mentioned in §3 to play a basic role in studying the geometric and analytic properties of real hypersurfaces, and domains which they bound. For example, qualitative properties of chains are biholomorphic invariants, although little is known about them at this point. Before discussing the general situation, let us first examine some simple examples arising from the ball.

EXAMPLES. (1) $Q = \partial\mathscr{U}_{n+1}$. As noted earlier, the chains here are intersections of Q with complex lines in $\boldsymbol{C}^{n+1}$.

(2) $S^{2n+1} = \partial B_{n+1}$. By inverse Cayley transformation of (1), the chains here are intersections of S^{2n+1} with complex lines in $\boldsymbol{C}^{n+1}$. Hence, all chains are closed, and any two points are joined by a (unique) chain.

(3) Next, consider $\partial\mathscr{U}_{n+1} - \{0\}$, and fix $\rho > 1$. Let ρ also denote, as earlier, the dilation ρ: $(z, w) \to (\rho z, \rho^2 w)$. Then the infinite cyclic group $\langle\rho\rangle$ acts freely and properly discontinuously on $\overline{\mathscr{U}}_{n+1} - \{0\}$, and the quotient $\overline{\mathscr{U}}_{n+1} - \{0\}/\langle\rho\rangle$ is a compact complex manifold $\overline{D(\rho)}$ with boundary. Further, the map

$$f : (z, w) \to (z/\sqrt{w}, \exp(\pi i \log w/\log \rho)) \tag{4.1}$$

is well defined on a neighborhood of $\overline{\mathscr{U}}_{n+1} - \{0\}$. f factors through $D(\rho)$, and exhibits $D(\rho)$ as a bounded domain in $\boldsymbol{C}^{n+1}$, namely,

$$D(\rho) = \{(z, w) \in \boldsymbol{C}^{n+1} \mid \sin(-(1/\pi) \log \rho \log |w|) > |z|^2\}. \tag{4.2}$$

First note that the chains on $\partial\mathscr{U}_{n+1}$ given by $z = z_0 \in \boldsymbol{C}^n$ pass down to chains on $\partial D(\rho)$ which are not closed in $\partial D(\rho)$ unless $z_0 = 0$. More interestingly, note that the points (0, 1) and (0, − 1) are not joined by a chain in $\partial\mathscr{U}_{n+1} - \{0\}$ (the unique chain in $\partial\mathscr{U}_{n+1}$ joining them passes through (0, 0)). Hence, the points (0, 1) = $f((0,1))$ and $(0, \exp(-\pi^2/\log \rho)) = f((0, 1))$ on $\partial D(\rho)$ are *not* joined by any chain. Of course, on a compact connected Riemannian manifold, any two points are joined by a geodesic.

We note in passing that by construction, $\partial D(\rho)$ is everywhere locally CR equivalent to the sphere, but $D(\rho)$ is not globally biholomorphic to the ball (since $\partial D(\rho)$ is not homeomorphic to the sphere). Thus the ball is not biholomorphically characterized by the necessary curvature conditions on the boundary. It is simple to see that a Stein manifold D, with ∂D simply connected and having CR curvature identically zero, is biholomorphic to the ball.

(4) Consider a compact Riemann surface C of genus $\geqq 2$ with its usual metric coming from the unit disc. $B(C)$ denotes the unit disc bundle in the tangent bundle of C and $S(C) = \partial B(C)$ is the unit circle bundle. One may make $B(C)$ into a complex manifold with boundary so that $S(C)$ has CR curvature identically 0 and the zero section $C \subset B(C)$ is a totally real submanifold. (This is not the usual complex structure on $B(C)$, but it is obtainable from B_2 as a quotient, as in (3) above.) On $S(C)$ one has the geodesic flow, and it is known that almost all orbits of this flow are dense in $S(C)$. However, these orbits are chains in the given complex structure, and so there exist "Kronecker" chains. We remark that $B(C)$ is a Stein manifold in this structure, but it is not biholomorphic to a domain in C^2. Such dense chains surely exist on the boundaries of bounded domains as well, although these are the only dense examples we know of at present.

A more striking example of the behavior of chains is given by Fefferman in **[14]**, cf. below. Further examples of boundaries with zero curvature and their moduli may be found in [7].

Returning to the general case, the systems of ODE defining chains encountered in the proof of Theorem 3.8 is not accessible for computation, although we remark that the construction of the convergent normal form is the most noteworthy positive application of chains to date. There are at least two other methods for defining the same curves.

The method of Cartan and Chern is to consider, on the structure bundle Y, the differential system:

$$\omega^{(-1)} = \omega^{(1)} = 0, \tag{4.3}$$

notation as in §1. This system is integrable by the structure equation (1.21), and the fact that the curvature tensor Ω has entries in the ideal of forms generated by the entries of $\omega^{(-1)}$ and $\omega^{(-2)}$. The integral manifolds of (4.3) project to curves on M; these curves are the chains. (This system of equations is made more explicit in the appendix to this section.) Define a vector field ξ on Y by:

$$\nu(\xi) \equiv 1, \qquad \omega^j(\xi) \equiv 0, \qquad j = -1, 0, 1, 2. \tag{4.4}$$

The integral curves of ξ project to chains and are transverse to the fibres of Y. Interpreting points in Y as special second-order frames on M, these integral curves give a notion of parallel transport for complex orthonormal frames in $H(M)$ along chains, plus a distinguished family of parametrizations of the chains. (N.B. This transport of frames is the same as that encountered in §3, but the parametrizations *differ* in general.) Unlike the geodesic flow on the orthogonal frame bundle in Riemannian geometry, the vector field ξ generally is not complete, i.e., the distinguished parametrizations along chains generally cannot be continued to $\pm\infty$. (In Example (3) above, the chains coming from $z = z_0 \in C^n$ all have this property.)

The second method for defining chains is by means of the Lorentz geometry introduced by Fefferman. This point of view is also discussed in Diederich's paper in these PROCEEDINGS on the Bergman kernel and its applications **[12]**. We recall the extrinsic approaches discussed there, and indicate an intrinsic approach.

Fefferman's general philosophy is to use the "boundary asymptotics of nontrivial, natural analysis problems" on the bounded s.ψ.c. domain $D \subset C^{n+1}$ to construct CR invariants on the boundary. (This philosophy may, of course, be reversed; cf. the problems at the end of this survey.) In particular, following this philosophy one constructs a conformal class of smooth indefinite metrics (of Lorentz signature) on $\partial D \times S^1$ two ways:

(i) *Bergman kernel.* The metric

$$ds^2_{(i)} = \frac{-\sqrt{-1}}{n+2}(\partial u_D - \bar{\partial} u_D) \cdot d\eta + \sum_{j,k=1}^{n+1} \frac{\partial^2 u_D}{\partial z_j \partial \bar{z}_k} dz_j \cdot d\bar{z}_k \tag{4.5}$$

on $\partial D \times S^1$ is nonsingular, of signature $(2n + 1, 1)$. Here $u_D = K_D^{-1/n+2}$ ($K_D = K_D(z, z)$ is the Bergman kernel function) and η is the angular variable on S^1. This metric is defined globally in terms of K_D, but depends only on the two-jet of u_D along ∂D. Its conformal class is invariant under global biholomorphic transformations, when a biholomorphic F is lifted to $\partial D \times S^1$ by translating η by $-\arg(\det(F'))$. The invariance follows from the global transformation law of the Bergman kernel.

(ii) *Monge-Ampère equation.* The metric

$$ds^2_{(ii)} = \frac{-\sqrt{-1}}{n+2}(\partial \psi_{(2)} - \bar{\partial} \psi_{(2)}) \cdot d\eta + \sum_{j,k=1}^{n+1} \frac{\partial^2 \psi_{(2)}}{\partial z_j \partial \bar{z}_k} dz_j \cdot d\bar{z}_k \tag{4.6}$$

is also nonsingular, of signature $(2n + 1, 1)$, where $\psi_{(2)}$ is Fefferman's second formal approximation to the solution of the Dirichlet problem:

$$(-1)^{n+1} \det\begin{pmatrix} \psi & \psi_{\bar{k}} \\ \psi_j & \psi_{j\bar{k}} \end{pmatrix} \equiv 1 \quad \text{on } D, \qquad \psi | \partial D = 0. \tag{4.7}$$

$\psi_{(2)}$ is locally computable, and the conformal invariance of (4.6) under local biholomorphic transformation follows from the (local) transformation law for the determinant above.

(iii) There is also a local, intrinsic approach, relating this additional structure on ∂D to the invariants already described in this survey.

Consider the bundle $Y \to \partial D$ with group $H = CU(n)/K \ltimes N$, notation as in §1. Set $\hat{E} = Y/(SU(n)/K) \ltimes N$. $\hat{E}$ is a bundle over ∂D, principal with group $CU(n)/SU(n) \simeq C^*$. Factoring $\hat{E}$ by the action of $R^+ \subset C^*$ gives $E =_{\text{def}} \hat{E}/R^+$, a circle bundle over ∂D. $\tilde{E}$ is canonically isomorphic to the circle bundle obtained by restricting the complex line bundle $\Lambda^{n+1,0}$ of forms of type $(n + 1, 0)$ to M, removing the zero-section, and factoring by the action of R^+. The choice of a nonvanishing holomorphic $(n + 1, 0)$-form near ∂D gives an explicit isomorphism of $\tilde{E}$ with $\partial D \times S^1$.

On Y consider the quadratic differential form (notation as in §1)

$$\lambda = \frac{2}{n+2} \nu \cdot \operatorname{tr}(\phi^{\alpha}_{\beta}) + \sum_{\alpha=1}^{n} \omega^\alpha \cdot \bar{\omega}^\alpha. \tag{4.8}$$

The equivariance properties of the connection forms show that λ descends to a form, still denoted λ, on $\hat{E}$. A section σ of $\hat{E}$ over $\tilde{E}$, equivariant with respect to the action of S^1 on $\hat{E}$ and $\tilde{E}$, enables one to pull λ down to $\tilde{E}$. (Such sections are in 1-1 correspondence with sections of E over M, i.e., with choices of 1-form defining $H(M)$.) Then the metric

$$ds^2_{(iii)} = \sigma^*\lambda \tag{4.9}$$

is nonsingular, of signature $(2n + 1, 1)$. As σ runs over all possible equivariant sections, $ds^2_{(iii)}$ runs over a conformal class of S^1-invariant Lorentz metrics on $\partial D \times S^1$. The functoriality of Y and the connection forms shows that $\tilde{E}$ and this conformal class are CR, and hence, biholomorphic invariants.

Each of these three metrics are invariant under the action of S^1. Each vector $X \in T_p(\partial D)$ transverse to $H_p(\partial D)$ has a unique lift to $\tilde{X} \in T_{(p,\zeta)}(\partial D \times S^1)$ where $\zeta \in S^1$ and the length of $\tilde{X} = 0$. The geodesic $\tilde{\gamma}$ through (m, ζ) tangent to $\tilde{X}$ (a "light ray") is a conformal invariant (as smooth curve, ignoring parametrization). The equivariance of the metrics shows that the projection γ of $\tilde{\gamma}$ onto ∂D is independent of $\zeta \in S^1$. One has

THEOREM 4.10. (a) $ds^2_{(i)} = ds^2_{(ii)} = ds^2_{(iii)}$ *(equality of conformal class).*
(b) γ *is the chain tangent to* X.

The first half of (a) follows from recent work of Christoffers and Diederich, and the second half is due to Diederich.

One can also derive the parallel transport of unitary frames along chains from the Lorentz structures above. It seems, in fact, that all the invariants of ∂D arising from Y can be explicitly and directly obtained from conformal invariants of $\tilde{E}$, e.g., parametrization of chains, curvature forms, etc. [**4**].

Theorem 4.10 says that chains are defined by the usual Hamiltonian

$$H_{ds^2} = \sum_{i,j=0}^{2n+1} g^{ij}(x)p_ip_j \tag{4.11}$$

where $(x_0, \cdots, x_{2n+1})$ are coordinates on $\partial D \times S^1$, $(p_0, \cdots, p_{2n+1})$ are the conjugate momenta, g^{ij} is inverse to the matrix g_{ij} expressing ds^2. Taking x_0 as the angular variable in S^1, H_{s^2} is independent of x_0, and p_0 is constant along the solution curves of Hamilton's equations. Thus, H_{s^2} may be rewritten

$$H_{ds^2} = \sum_{i,j\geqq 1} g^{ij}p_ip_j + 2p_0 \sum_{j=1}^{2n+1} g^{0,j}\, p_j + p_0^2 g^{0,0}, \tag{4.12}$$

with p_0 constant. If the quadratic form $B_1 = \sum_{i,j\geqq 1} g^{ij}p_ip_j$ were positive definite, (4.12) together with constancy of H_{s^2} along solution curves would imply that any solution curve $(x(t), p(t))$ with $\lim_{t\to\infty} \bar{x}(t) = \bar{x}_\infty$ $(t_\infty \leqq \infty; \bar{x}(t) = (x_1(t), \cdots, x_{2n+1}(t))$ has

$$0 < c_1 \leqq |p(t)| \leqq c_2 \tag{4.13}$$

as $t \to t_\infty$, and the curve would continue smoothly through $\bar{x}_\infty$. B_1, however, is nonnegative, with one-dimensional null-space, and so this behavior is not guaranteed for chains. We call $p \in \partial D$ a *spiral point* if there is a chain with limit p as above, which does not continue smoothly through p. If ∂D has additional symmetries, other p_i's will be constants along trajectories of H_{s^2}. For example, if ∂D has a CR action of S^1 transverse to the complex subspace $H(\partial D)$, then we can take H_{s^2}

independent of, say, x_0 and x_1, with $B_2 = \sum_{i,j\geq 2} g^{ij} p_i p_j$ positive definite, and there will be no "spiralling".

In **[14]**, Fefferman computes chains with the Hamiltonian derived from $\psi_{(2)}$ for the hypersurface $M \subset C^2$ given by

$$v = |z|^2 + u|z|^4. \tag{4.14}$$

(4.14)[5]

The rotational symmetry of M in z shows as above that there are no spiral points on M near the line $z = 0$, $v = 0$, except possibly on that line itself (which is fixed under the rotational symmetry). Fefferman goes much farther than this to show:

THEOREM 4.15. *There is an infinite family of chains on M as in* (4.14) *which spiral to the origin. The origin is the only spiral point on M.*

The phenomenon described here is global ($t \to \pm\infty$) along the chains, but local about $0 \in M$. It is very difficult to say, in local geometric terms at 0, why this spiralling occurs, since Fefferman actually computes with a simplified Hamiltonian obtained after a very clever ("sneaky" is his word!) canonical transformation. Webster notes that, as a consequence of an old theorem of E. Cartan, the fixed point set of a CR action of S^1 on M ($\dim M = 3$) is a union of chains, each point of which is umbilic. Hence, the line $z = 0$, $c = 0$ in Fefferman's example is a chain of umbilics. Fefferman has asked whether any spiral point must be umbilic, and if so, which umbilics? More specifically, is this determinable by finite order information at the point in question, e.g., if we perturb (4.14) by suitably high order terms in z, u near the origin, is the origin still a spiral point? Are such points stable under more general perturbations? Must such points be isolated, as in (4.14)? The significance of this line of questions is that spiral points on the boundaries are biholomorphic invariants of strictly pseudoconvex domains, and positive answers to the above questions would say that one could make finite local computations at spiral points to obtain strong necessary conditions for global biholomorphic equivalence.

Appendix to §4. We indicate a slightly more explicit form of equations (4.3) for chains. The resulting expression is strictly analogous to the usual equations for geodesics involving the Christoffel symbols. In Riemannian geometry these are the Euler-Lagrange equations for the energy integrand; we do not know an analogous interpretation for chains (other than chasing through the "light rays" construction!).

Pick forms θ, θ^α, $\tilde{\phi}$ locally on M, so that we have

$$d\theta = -i\sum_{\alpha=1}^{n} \theta^\alpha \wedge \overline{\theta^\alpha} + \theta \wedge \tilde{\phi}. \tag{4.16}$$

These define a section : $M \to Y$ such that

$$\begin{array}{lll} \sigma^*\nu = \theta, & \sigma^*\omega^\alpha = \theta^\alpha, & \sigma^*\phi = \tilde{\phi}, \\ \sigma^*\tilde{\phi}^\alpha_\beta = \phi^\alpha_\beta, & \sigma^*\phi^\alpha = \tilde{\phi}^\alpha, & \sigma^*\psi = \tilde{\psi}, \end{array} \tag{4.17}$$

[5] At this time it is not clear if example (4.14) above has spiral chains. See corrections to **[14]**. However it seems that a slight alteration of the argument in **[14]** applies to $v = |z|^2 + u|z|^8$ and shows the existence of spirals.

where the last three equations are definitions: $\tilde{\phi}$, ϕ^α_β, $\tilde{\phi}^\alpha$, $\tilde{\psi}$ are the Christoffel forms for the connection ω.

Suppose we are given a chain $\gamma = \gamma(t)$, $\gamma(0) = 0 \in M$, and assume the parameter t on the interval I chosen so that

$$\theta(\dot{\gamma}(t)) = 1. \tag{4.18}$$

The curve γ must lift locally to a curve $\gamma_1 : I \to Y$ lying in a leaf of (4.3). The curve $\tilde{\gamma} = \sigma \circ \gamma$ is a lift of γ, and we attempt to correct $\tilde{\gamma}$ by right translations in Y by (t-dependent) elements of H in order to satisfy (4.3). Since there will be only $2n$ real functions to determine (because of (4.18)), we consider elements

$$H \ni A(t) = \begin{pmatrix} 1 & 0 & 0 \\ -2ia^* & I & 0 \\ -i|a|^2 & a & 1 \end{pmatrix}, \qquad a(t) \in \mathbf{C}^n, \tag{4.19}$$

and try to make the $\mathfrak{g}^{-1}$ and $\mathfrak{g}^{+1}$ components of the $\mathfrak{g}$-valued 1-form $(R_{A(t)} \cdot \tilde{\gamma}(t))^*\omega$ identically zero on I.[6] Since the connection form ω is H-equivariant, and restricts to the Maurer-Cartan form along the fibers, one computes readily:

$$(R_{A(t)} \cdot \tilde{\gamma}(t))^*\omega = A^{-1}\, dA + \mathrm{Ad}(A^{-1})\gamma^*(\sigma^*\omega). \tag{4.20}$$

Setting the $\mathfrak{g}^{-1}$ component of (4.20) equal to 0 gives

$$\theta^\alpha(\dot{\gamma}(t)) = -2a^\alpha(t). \tag{4.21}$$

Setting the $\mathfrak{g}^{+1}$ component of (4.20) equal to 0 gives, by (4.21),

$$\begin{aligned} -\frac{1}{2}\frac{d\theta^\alpha}{dt}(\dot{\gamma}) &= \frac{i}{2}\left\{\sum_{\lambda=1}^{n} |\theta^\lambda(\dot{\gamma})|^2 + \frac{i}{2}\tilde{\phi}(\dot{\gamma})\right\}\theta^\alpha(\dot{\gamma}) \\ &\quad -\frac{1}{2}\theta^\beta(\dot{\gamma})\,\phi^\alpha_\beta(\dot{\gamma}) - \frac{1}{2}\tilde{\phi}^\alpha(\dot{\gamma}). \end{aligned} \tag{4.22}$$

This is the desired equation.

5. Problems. We collect here several groups of problems, in addition to those scattered throughout the text, in the general area of the geometry of real hypersurfaces (or real submanifolds of higher codimension) in complex manifolds. The *caveat* of the introduction is still in effect!

A. *Interpretation of chains, boundary regularity of maps.* The differential methods for describing chains given in §§3, 4 above require fifth-order information along the hypersurface M (in terms of an arbitrary defining equation). Is there a "$\mathscr{C}^0$-way" to describe chains, analogous to the description of Riemannian geodesics as curves locally realizing the shortest distance between two points? This might be related to an alternate proof of Fefferman's theorem on the boundary regularity of biholomorphic maps between strictly pseudoconvex domains. It seems interesting to pursue further the relation of geometric invariants on the boundary and such regularity properties. One would hope that it is possible to give a proof of general local regularity theorems using strict pseudoconvexity in a more direct way. A

[6] These "nonholonomic" motions along γ are very similar to those used by Moser in **[11**, §3] to derive his chain equations.

related problem is to determine whether, for smooth but not real analytic s.ψ.c. hypersurfaces, chains are still related to the boundary values of holomorphic curves.

B. *Asymptotics.* At the end of §4 we outlined a series of questions coming from Fefferman's paper **[14]**. There are many more very interesting problems in **[14]**, some few of which are answered by the "state of the art" as described here, but most of which (the more interesting ones!) are still quite open. In general, the philosophy which led to the Lorentz structures cries out to be reversed profitably: The boundary asymptotics of natural analysis problems generally have to be given in universal formulae in the local geometric invariants at the boundary. The analogy with the recent work of Gilkey, Patodi and others on the asymptotics of heat kernels in terms of Riemannian and Hermitian invariants would signal that a higher order study of the boundary asymptotics of the Bergman and Szegö kernels, or Monge-Ampère equations, should be viewed in a similar light.

C. *Special examples.* This should cover at least:

(i) Biholomorphic classification of ellipses. It seems as though the domains bounded by ellipses in C^{n+1} should be biholomorphically equivalent by linear maps only. Webster has made very interesting progress on this question in his thesis, especially in the two-dimensional case. In C^2, ellipses modulo "linear equivalence" form a semianalytic set of two real dimensions, and Webster has already shown the existence of one-parameter families of biholomorphically inequivalent ellipses. Thus in C^2 what is needed "generically" is one other global, biholomorphic real invariant of an ellipse (e.g., its volume in the canonical measure of §2) which must be shown independent of Webster's.

(ii) Nonumbilic boundaries in C^2. We just refer to the series of questions raised by Moser and others in §3.

D. *Wave operator,* $\Box_b$ *and chains.* On the circle bundle $\tilde{E}$ of §4, we have a conformal class of (S^1-invariant) Lorentz metrics. Each such metric has a wave operator Δ associated to it, which is closely related to a choice of $\Box_b$ operator on M. The bicharacteristics of Δ are the light rays. Is there a useful interpretation for chains in terms of the analysis of $\Box_b$?

E. *Modular families.* Work of R. Hamilton shows that all small deformations (in a suitable sense) of the complex structure on a bounded s.ψ.c. domain in C^{n+1} are realized by perturbations of the (strictly plurisubharmonic) defining function, i.e., within C^{n+1}. How does one slice the relation of biholomorphic equivalence among the domains defined by the family of all such functions? What structure, if it exists, does such a modular family have: For example, is it a smooth, infinite-dimensional manifold if the original domain has no automorphisms?

F. *Higher codimension problems, degenerate hypersurfaces.* Tanaka has done interesting work on the equivalence problem for higher codimension submanifolds, but with the restrictive assumption that the (vector-valued) Levi form be of constant type. (The equivalence classes of such forms are generally not open in the space of such forms.) It does seem that a mixture of Tanaka's formal prolongation scheme and Moser's technique might give some results in the more general case. (G-structure techniques seem doubtful when the type of the Levi form varies.)

In another direction, D'Angelo and Kohn have been studying the geometry of pseudoconvex, but not s.ψ.c., domains, especially in relation to the regularity properties of the $\bar{\partial}$-Neumann problem and the singularities of the Bergman kernel. Progress here will probably indicate interesting new problems and avenues of approach to degenerate hypersurfaces. That the situation is more complicated than first imagined may be seen in the work of Bloom and Graham (these PROCEEDINGS, part 1, pp. 115–118, 149–152).

NOTES ADDED IN PROOF. (1) The auxiliary structures derived from the 1-form θ_0 of §2 have been used to prove the following:

THEOREM [6]. *Let D be a Stein manifold with compact, smooth* s.ψ.c. *boundary. Then,* Aut°(D) *is compact, unless D is biholomorphic with the ball. If* dim $D \geq 3$, *then* Aut(D) *is compact, unless D is the ball.*

COROLLARY. *The only homogeneous such D is the ball.*

A special case of the corollary has been proved independently by B. Wong [29].

(2) The relationship of the Lorentz structures of §4 to the CR structure is better understood now. In [4] it is shown how to compute the curvature tensor Ω from the conformal curvature tensor of the conformal Lorentz class of §4, and vice versa. The parametrization of chains is retrieved from general considerations of conformal Lorentz geometry. In [28] S. Webster derives somewhat similar results, working directly with Kähler metrics in the ambient space singled out by the higher order approximate solutions of (4.7). This method appears promising for further developments.

REFERENCES

1. H. Alexander, *Proper holomorphic mappings in* C^n (to appear).

2. ———, *Proper holomorphic mappings of bounded domains*, Proc. Sympos. Pure Math., vol. 30, part 2, Amer. Math. Soc. Providence, R. I., 1977, pp. 171–174.

3. L. Boutet de Monvel, *Intégration des équations de Cauchy Riemann induites*, Séminaire Goulaouic-Schwartz, 1974-1975, Exposé IX.

4. D. Burns, Jr., K. Diederich and S. Shnider, *Distinguished curves in the boundaries of strictly pseudoconvex domains* (to appear).

5. D. Burns, Jr. and S. Shnider, *Pseudoconformal geometry*, Lecture Notes, Princeton Univ., Princeton, N. J., 1975.

6. ———, *Geometry of hypersurfaces and some mapping theorems in several complex variables* (to appear).

7. ———, *Spherical hypersurfaces in complex manifolds*, Invent. Math. (to appear).

8. D. Burns, Jr., S. Shnider and R. O. Wells, Jr., *On deformations of strictly pseudoconvex domains* (to appear).

9. E. Cartan, *Sur la géométrie pseudo-conforme des hypersurfaces de deux variables complexes.* I, II, Oeuvres II, 2, 1231–1304; ibid. III, 2, 1217–1238.

10. ———, *Articles on pseudogroups and the equivalence problem*, Oeuvres II, 2.

11. S. S. Chern and J. K. Moser, *Real hypersurfaces in complex manifolds*, Acta Math. **133** (1974), 219–271.

12. K. Diederich, *Some recent developments in the theory of the Bergman kernel function: A survey*, Proc. Sympos. Pure Math, vol. 30, part 1, Providence, R. I., 1977, pp. 127–137.

13. C. Fefferman, *The Bergman kernel and biholomorphic mappings of pseudo-convex domains*, Invent. Math. **26** (1974), 1–65. MR **50** #2562.

14. ———, *Monge-Ampère equations, the Bergman kernel, and geometry of pseudoconvex domains*, Ann. of Math. **103** (1976), 395–416.

15. S. Kobayashi, *Transformation groups in differential geometry*, Springer-Verlag, Berlin, 1972. MR **50** #8360.

16. A. Morimoto and T. Nagano, *On pseudo-conformal transformations of hypersurfaces*, J. Math. Soc. Japan **15** (1963), 289–300. MR **27** #5275.

17. I. Naruki, *The equivalence problem for bounded domains*, Technical Report #118, Res. Inst. Math. Sci, Kyoto Univ., Kyoto, 1973.

18. T. Ochiai, *Geometry associated with semisimple flat homogeneous spaces*, Trans. Amer. Math. Soc. **152** (1970), 159–193. MR **44** #2160.

19. H. Rossi, *Homogeneous strongly pseudoconvex hypersurfaces*, Rice Univ. Studies **59** (1973), 131–145. MR **48** #8851.

20. S. Sternberg, *Lectures on differential geometry*, Prentice-Hall, Englewood Cliffs, N. J., 1964. MR **33** #1797.

21. N. Tanaka, *On the pseudo-conformal geometry of hypersurfaces of the space of n complex variables*, J. Math. Soc. Japan **14** (1962), 397–429. MR **26** #3086.

22. ———, *On generalized graded Lie algebras and geometric structures*. I, J. Math. Soc. Japan **19** (1967), 215–254. MR **36** #4470; erratum, **36**, p. 1568.

23. ———, *On infinitesimal automorphisms of Siegel domains*, J. Math. Soc. Japan **22** (1970), 180–212. MR **42** #7939.

24. S. Webster, *Real hypersurfaces in complex space*, Thesis, Berkeley, 1975.

25. R. O. Wells, Jr., *Deformations of strongly pseudoconvex domains in* $\mathbf{C}^2$, Proc. Sympos. Pure Math., vol. 30, part 2, Amer. Math. Soc., Providence, R. I., 1977, pp. 125–128.

26. T. Bloom and I. Graham, *Geometric characterizations of points of type m on real submanifolds of* $\mathbf{C}^n$, J. Differential Geometry (to appear).

27. ———, *On "type"-conditions for generic real submanifolds of* $\mathbf{C}^n$ (to appear).

28. S. Webster, *Kähler metrics associated to a real hypersurface* (to appear).

29. B. Wong, *Intrinsic measures, metrics, and homogeneous strongly pseudo-convex bounded domains with smooth boundary* (to appear).

PRINCETON UNIVERSITY

THE INSTITUTE FOR ADVANCED STUDY

McGILL UNIVERSITY

SEMINAR SERIES

PROBLEMS IN APPROXIMATION

Proceedings of Symposia in Pure Mathematics
Volume 30, 1977

PROPER HOLOMORPHIC MAPPINGS OF BOUNDED DOMAINS

H. ALEXANDER

1. Survey of the problem. Consider the following general problem: Given two bounded domains D_1 and D_2 in C^n, determine all proper holomorphic mappings $\varphi : D_1 \to D_2$. In this context, "proper" simply means that φ maps sequences in D_1 converging to the boundary of D_1 to sequences in D_2 converging to the boundary of D_2. A direct consequence of the definition is that φ represents D_1 as a branched cover of D_2 of some finite order (v. [**9**]).

For D_1 a general analytic polyhedron and D_2 a domain whose boundary contains a relatively open subset of points of strict convexity (in particular, a bounded domain with $\mathscr{C}^2$ boundary), Henkin [**10**, Theorem 2*] showed that no proper holomorphic map exists. Related results for the more general concept of proper holomorphic covering correspondence were obtained by K. Stein and Rischel [**16**].

The case of holomorphic self-mappings of product domains was treated by Remmert and Stein [**15**]. Rudin [**17**] discussed the special case of the polydisc, where the maps are given by $(z_1, z_2, \cdots, z_n) \mapsto (h_1(z_{i_1}), h_2(z_{i_2}), \cdots, h_n(z_{i_n}))$ where each h_k, $1 \leqq k \leqq n$, is a finite Blaschke product in one complex variable and $(i_1, i_2, \cdots, i_n)$ is a permutation of $(1, 2, \cdots, n)$. Thus, for the polydisc, the only proper holomorphic self-mappings are the "obvious ones."

The same statement holds for the ball, where, for the case of the ball, some reflection should convince the reader that at least the only obvious maps are the automorphisms. This is our main result.

THEOREM. *The only proper holomorphic self-mappings of the ball B_n $(n > 1)$ are the automorphisms of B_n.*

Some special cases were previously obtained by Pelles [**13**].

AMS (MOS) subject classifications (1970). Primary 32H99.

After the ball, the obvious next problem is to show that a proper holomorphic self-mapping of a strictly pseudoconvex domain in C^n $(n > 1)$ is necessarily an automorphism. This conclusion is certainly not generally true for a proper holomorphic mapping between different strictly pseudoconvex domains in C^n. The following example was found independently by S. Pinčuk **[14]** and Y.-T. Siu: Let

$$D_1 = \{(z, w) \in C^2 : |z|^4 + 1/|z|^4 + |w|^2 < 3\}$$

and

$$D_2 = \{(z, w) \in C^2 : |z|^2 + 1/|z|^2 + |w|^2 < 3\}$$

and put $\varphi_0 : D_1 \to D_2$, $\varphi_0(z, w) = (z^2, w)$, to get a proper 2-to-1 map of strictly pseudoconvex domains. Observe that φ_0 is a covering projection (without branching!) in this example. This is what is likely to be true in general. In fact, if one could show that φ extended to be smooth on all of ∂D_1, then a local computation, due independently to J.E. Fornaess **[7]** and S. Pinčuk **[14]**, would apply to show that the derivative φ' is nonsingular on ∂D_1, hence on D_1, and thus φ would be a covering projection; in particular, for D_2 simply connected, φ would be a biholomorphism. Furthermore, even if D_2 is not simply connected, Pinčuk **[14]**, still assuming φ smooth on ∂D_1, has shown that φ is 1-to-1 in the case $D_1 = D_2$; hence φ is a biholomorphism.

Whether or not φ is smooth on all of ∂D_1 will not be settled by the techniques of this paper. Nevertheless, we do show that φ is smooth on a "large" subset of ∂D_1. For the general strictly pseudoconvex domain, this property of φ is not decisive in itself. However, for the ball, where there is a special local characterization of automorphisms, this information on φ is exactly what is needed to verify our theorem. The result on automorphisms of the ball was obtained in **[1**, Proposition 1.1 and Remark, p. 253].

LOCAL CHARACTERIZATION THEOREM. *Let* $p \in \partial B_n$ $(n > 1)$. *Let F be a nonconstant map from a neighborhood N of p in the closed ball* $\overline{B_n}$ *to* C^n, $F : N\ (\subseteq \overline{B_n}) \to C^n$, *such that F is both holomorphic on* $N \cap B_n$ *and* $\mathscr{C}^\infty$ *on N. If* $F(N \cap \partial B_n) \subseteq \partial B_n$, *then F extends to be an automorphism of* B_n.

It appears that this theorem can also be obtained [3] from the general theory developed by Chern and Moser **[4]**. Moreover, by combining this machinery of Chern and Moser with our proof for the ball, D. Burns and S. Shnider [3] have been able to resolve the general problem of showing that a proper self-mapping of a strictly pseudoconvex domain is a biholomorphism. This technique should lead to further new results.

2. Elements of the proof. The full details of the proof of the theorem will appear in **[2]**. Here we shall describe the main lines of the argument. Throughout this section we shall consider a proper holomorphic map $\varphi : D_1 \to D_2$ of strictly pseudoconvex domains D_1 and D_2 in C^n with $\mathscr{C}^\infty$ boundaries. We want to show that φ extends to be $\mathscr{C}^\infty$ on a nonempty open subset of ∂D_1; for if $D_1 = D_2 = B_n$, the local characterization theorem of §1 then implies the theorem.

The first step is to get φ continuous on $\overline{D_1}$. This was proved independently by

Henkin [10] and Pinčuk [14]. In particular, φ induces a continuous map $\partial D_1 \to \partial D_2$.

In order to see that φ is smooth ($\mathscr{C}^\infty$) on part of ∂D_1, we appeal to the recent work of Fefferman [6], who showed that a biholomorphism of strictly pseudoconvex domains extends to be smooth at the boundary. Actually, we utilize a local version of his result.

The reader will observe that Fefferman's theorem refers to a biholomorphism while our φ is merely a proper map. We shall show that there are subdomains $D \subseteq D_1$ such that (a) ∂D meets ∂D_1 in a relatively open nonempty subset, and (b) φ restricted to D is a biholomorphic map to $\varphi(D)$, a domain whose boundary contains a relatively open nonempty subset of ∂D_2. We can then invoke Fefferman's work, which allows us to conclude that φ is smooth on $\partial D \cap \partial D_1$.

In order to construct D of the last paragraph, we need to study the boundary behavior of φ and of certain bounded holomorphic functions. For the latter we have the

GENERALIZED FATOU THEOREM. *Let Ω be a strictly pseudoconvex domain in $\mathbf{C}^n$ and let f be a bounded holomorphic function on Ω. Then f has admissible limits almost everywhere on $\partial\Omega$.*

This theorem was originally due to Korányi [12] for the ball and has been further developed by Hörmander [11], E. Stein [18] and Čirka [5]. We shall not define "admissible" here; suffice it to say that this notion allows approach to a boundary point which is more general than the nontangential approach. This new information (classically, only nontangential limits were known) on admissible limits should have applications. To our knowledge there is, at present, only one, and it stems from the following result of Henkin [10].

LEMMA. *φ maps sequences in D_1 which converge nontangentially to some $p \in \partial D_1$ to sequences in D_2 which converge admissibly to $\varphi(p) \in \partial D_2$.*

A proof of this lemma is given in [2]; it depends on an estimate of Henkin [10] for the Carathéodory metric of a strictly pseudoconvex domain. Graham [8] obtained a sharper form of this estimate and will discuss it elsewhere in these PROCEEDINGS.

Recall that $\varphi : D_1 \to D_2$ is a branched covering map of order λ. By Henkin's theorem, φ extends to give a continuous map $\partial D_1 \to \partial D_2$. The next result shows that this mapping of the boundaries is similar to a branched covering map of order λ.

PROPOSITION. *With the exception of a set of measure zero in ∂D_2, each point $p \in \partial D_2$ has exactly λ φ-preimages in ∂D_1,* i.e., $\#\{\varphi^{-1}(p)\} = \lambda$. *(Here $\#$ denotes the number of elements in a set.)*

The proof of this fact utilizes the generalized Fatou theorem via Henkin's lemma.

Now if $p \in \partial D_2$ is such that $\#\{\varphi^{-1}(p)\} = \lambda$, a point-set argument shows that there exists, for each of the λ points $q \in \{\varphi^{-1}(p)\}$, a subdomain $D \subseteq D_1$ with the properties (a) and (b) mentioned previously and such that $\partial D \cap \partial D_1$ contains a neighborhood of q in ∂D_1. Fefferman's theorem now implies the smoothness of φ

in a neighborhood of q. Thus, φ is smooth on a large subset of ∂D_1, as claimed above.

REFERENCES

1. H. Alexander, *Holomorphic mappings from the ball and polydisc,* Math. Ann. **209** (1974), 249–256. MR **50** #5018.

2. ———, *Proper holomorphic mappings in* C^n, Indiana U. Math. J. (to appear).

3. D. Burns and S. Shnider, Private communication.

4. S. Chern and J. Moser, *Real hypersurfaces in complex manifolds*, Acta Math. **133** (1974), 219–271.

5. E. M. Čirka, *The theorems of Lindelöf and Fatou in* C^n, Mat. Sb. **92 (134)** (1973), 622–644 = Math. USSR Sb. **21** (1973), 619–641. MR **49** #3180.

6. Charles Fefferman, *The Bergman kernel and biholomorphic mappings of pseudo-convex domains*, Invent. Math. **26** (1974), 1–65. MR **50** #2562.

7. John Erik Fornaess, *Embedding strictly pseudoconvex domains in convex domains*, Dissertation, Univ. of Washington, 1974.

8. Ian Graham, *Boundary behavior of the Carathéodory and Kobayashi metrics on strongly pseudo-convex domains in* C^n *with smooth boundary*, Dissertation, Princeton Univ., 1973.

9. R. Gunning and H. Rossi, *Analytic functions of several complex variables*, Prentice-Hall, Englewood Cliffs, N.J., 1965. MR **31** #4927.

10. G. M. Henkin, *An analytic polyhedron is not holomorphically equivalent to a strictly pseudo-convex domain*, Dokl. Akad. Nauk SSSR **210** (1973), 1026–1029 = Soviet Math. Dokl. **14** (1973), 858–862. MR **48** #6467.

11. L. Hörmander, L^p *estimates for (pluri-) subharmonic functions*, Math. Scand. **20** (1967), 65–78. MR **38** #2323.

12. A Korányi, *Harmonic functions on Hermitian hyperbolic space*, Trans. Amer. Math. Soc. **135** (1969), 507–516. MR **43** #3480.

13. D. A. Pelles, *Proper holomorphic self-maps of the unit ball*, Math. Ann. **190** (1971), 298–305; Correction, Math. Ann **202** (1973), 135–136. MR **43** #3501, erratum, **43**, p. 1698.

14. S. J. Pinčuk, *On proper holomorphic mappings of strictly pseudoconvex domains*, Siberian Math. J. **15** (1975), 644–649.

15. R. Remmert and K. Stein, *Eigentliche holomorphe Abbildungen*, Math. Z. **73** (1960), 159–189. MR **23** #A1840.

16. H. Rischel, *Holomorphe Überlagerungskorrespondenzen*, Math. Scand. **15** (1964), 49–63. MR **31** #2418.

17. W. Rudin, *Function theory in polydiscs*, Benjamin, New York, 1969. MR **41** #501.

18. E. Stein, *Boundary behavior of holomorphic functions of several complex variables*, Math. Notes, Princeton Univ. Press, Princeton, N. J., 1972.

UNIVERSITY OF ILLINOIS AT CHICAGO CIRCLE

Proceedings of Symposia in Pure Mathematics
Volume 30, 1977

A PROPERTY OF UNIONS OF ADMISSIBLE DOMAINS

N. KERZMAN*

0. Introduction.

A. *Cones.* Let $D \subset\subset C^n$ be a smooth domain and bD its boundary. A cone of vertex $\xi \in bD$ and aperture $\alpha > 0$ is the set

$$\Gamma_\alpha(\xi) = \{z \in D; \ |z - \xi| < (1 + \alpha)\, \delta(z)\} \tag{0}$$

where $\delta(z)$ is Euclidean distance from z to bD.

Cones play an important role in the theory of boundary behavior of harmonic functions u in D. They are the natural "approach domains". For example a general version of Fatou's theorem [5], [7] states that for a (say) bounded harmonic function u in D

$$\lim_{z \to \xi; z \in \Gamma_\alpha(\xi)} u(z) \quad \text{exists for a.a. } \xi \in bD \tag{0'}$$

where a.a. means almost all ξ in the Lebesgue measure of bD.

Also central to the theory are certain (truncated) unions of cones

$$R = R(E, \alpha, a) = \left[\bigcup_{\xi \in E} \Gamma_\alpha(\xi) \right] \cap D_a$$

where E is arbitrary compact in bD, $\alpha > 0$ is a *fixed* number, $a > 0$ is small and $D_a = \{z \in D; \delta(z) < a\}$. Such compact E usually arise from a measure theoretic argument. However

(L) *No matter how wild E is, the boundary bR is a (piecewise) Lipschitz surface* [6, p. 206].

This property lies at the root of the applications of Stokes' theorem that are repeatedly performed in such regions R.

AMS (MOS) subject classifications (1970). Primary 31B25, 32H99.

*Sloan Fellow.

B. *Admissible domains and E. Stein's results.* Now $D \subset\subset \mathbf{C}^n$ is a smooth domain. If u is holomorphic and bounded in D then (0′) holds even when $\Gamma_\alpha(\xi)$ is replaced by a more generous domain $A_\alpha(\xi)$. See **[5]**. Such "admissible domains" were introduced in **[5]**. See also **[2]**, **[3]** and §1 for the definition of $A_\alpha(\xi)$.

For more recent results see **[1]**.

C. Unions $R = R(E, \alpha, a)$ are defined using $A_\alpha(\xi)$ instead of $\Gamma_\alpha(\xi)$ just as above; bR is still Lipschitz **[5**, p. 65].

Our Theorem 1 shows that bR has an additional property. Roughly stated:

(L′) At $z \in bR$ in the *complex tangential directions*, bR has a Lipschitz constant *dominated* by $(\delta(z))^{1/2}$.

See §1 for the precise statement.

D. In case D is strictly pseudoconvex the analogy with the classical case has been carried further **[5**, Chapter III] by introducing a special Kähler metric in D. The main result is Theorem 12 in **[5]**. However, the proof contains the following gap: In order to use Green's identities the unit normal n_ε on p. 66 of **[5]** should have been the one *in the metric*. Not only its length but also its *direction* will generally disagree with that of the Euclidean normal. The gap is easily bridged by using Theorem 2 of the present work. This is an approximation theorem of extremely familiar type in this subject, but the key property involved is (11) which is a consequence of the special behaviour (L′) of bR.

For another way of approximating R see **[4]** which deals with special domains.

The author lectured on **[5]** at the Massachusetts Institute of Technology in the Spring 1975. He would like to express his thanks to E. Stein for the interest he took in the present questions and to the participants in the course, in particular to Dr. J. Dodziuk, T. Quinto and G. Uhlmann.

1. Unions of admissible domains.

A. *Admissible approach domains.* Let $D \subset\subset \mathbf{C}^n$ be a C^∞ smooth domain and bD its boundary. An admissible approach domain $A_\alpha(\xi)$ with vertex $\xi \in bD$ and aperture $\alpha > 0$ is defined as follows **[5]**:

$$(1) \qquad A_\alpha(\xi) = \{z \in D;\ |\langle z - \xi, \nu_\xi\rangle| < (1+\alpha)\delta_\xi(z),\ |z-\xi|^2 < \alpha\delta_\xi(z)\}$$

where ν_ξ is the unit outer normal to bD at ξ, $\langle\ ,\ \rangle$ is the usual Hermitian product in $\mathbf{C}^n$, and $\delta_\xi(z) = \text{minimum}\{\delta(z), d_\xi(z)\}$. Here $\delta(z)$ is Euclidean distance from z to bD and $d_\xi(z)$ is its distance to T_ξ which is the full $2n-1$ dimensional real plane tangent to bD at ξ. Note that d_ξ *is taken with its sign*, namely $d_\xi(\xi + \nu_\xi t) < 0$ for $t > 0$ and otherwise if $t < 0$. The introduction of δ_ξ and of the second condition in (1), i.e., $|z-\xi|^2 < \alpha\delta_\xi(z)$, only serve to rule out pathological cases when bD has flat or concave parts. For a ball D

$$A'_\alpha(\xi) = \{z \in D;\ |\langle z-\xi, \nu_\xi\rangle| < (1+\alpha)\delta(z)\}$$

essentially coincides with (1).

B. *Their unions.* Let $E \subset bD$ be compact (one should expect a wild E in general) and let $\alpha > 0$ be a *fixed* number. We are interested in the truncated union

$$(2) \qquad R = R(E, \alpha, a) = \Big[\bigcup_{\xi\in E} A_\alpha(\xi)\Big] \cap D_a$$

where $D_a = \{z \in D, \delta(z) < a\}$ and $a > 0$ is small. The "top" of bR is $\{z \in bR,$

$\delta(z) = a\}$. In order to describe the more interesting part $b_0R = \{z \in bR, \delta(z) < a\}$ we recall a special coordinate system frequently used in complex analysis: S_ξ is any unitary coordinate system with origin $\xi \in bD$, such that its $z_2, \cdots, z_n$ directions generate the complex tangent plane to D at ξ (i.e., the special $2n - 2$ dimensional real subspace of T_ξ) and its y_1 direction is the inner normal to D at ξ. Here $z_j = x_j + iy_j$. Finally, let $n(\xi) = -\nu(\xi)$ and $p(\xi, t) = \xi + tn(\xi)$, $t > 0$. U stands for a slight enlargement of E, $U = \{\xi \in bD$; distance from ξ to E is $< \beta\}$. The diameter of E and a are small, and $\beta = \beta(D, \alpha, a)$ is small; a is chosen first and β next in terms of a.

We want to describe the part of bR which "lies above U" in terms of the t which gives $p(\xi, t) \in bR$. The key property is (7), (7′). It has no analogue in the real case.

THEOREM 1 (DESCRIPTION OF bR). *There is a unique function $\phi(\xi)$, $0 \leqq \phi(\xi) < a$, $\xi \in U$ such that*

(3) *For any $\xi \in U$, $p(\xi, \phi(\xi)) \in bR$, $p(\xi, t) \in R$ for $\phi(\xi) < t < a$ and $p(\xi, t) \notin R$ for $t < \phi(\xi)$. (Notice that $\phi(\xi) = \delta[p(\xi, \phi(\xi))]$.)*

This function ϕ has the following properties:

(4) *$\phi(\xi) = 0$ if and only if $\xi \in E$;*

(5) *$\phi(\xi)$ is Lipschitz uniformly in U; and*

(6) *Complex tangential behaviour: Let S_ξ be a special coordinate system at $\xi \in U$, v a unit vector at ξ in a complex tangential direction to bD. For small $s > 0$ set $\xi(s)$ = point of bD whose coordinates in S_ξ are $(y_1(s), sv_2, \cdots, sv_n)$ i.e., its $x_1 = 0$ (but its y_1 cannot be chosen at will if $\xi(s)$ is to be in bD). Then*

$$|\phi(\xi(s)) - \phi(\xi)| \leqq A(\phi(\xi))^{1/2}s \tag{7}$$

if $s < s_0(\xi)$ and $\xi \notin E$. If $\xi \in E$ then $\phi(\xi) = 0$ and we have instead

$$|\phi(\xi(s)) - \phi(\xi)| \leqq As^2, \qquad 0 \leqq s < s_0(\xi). \tag{7′}$$

The constant A is independent of ξ and s.

SKETCH OF THE PROOF OF THEOREM 1. One observes that $p(\xi, \phi(\xi)) \in A_\alpha(\xi_0)$ for some $\xi_0 \in E$. *Restricting attention to* $A_\alpha(\xi_0)$ it is seen that a small change of ξ, followed by a "proportional" increment of t in $p(\xi, t)$, already leads to penetration of $A_\alpha(\xi_0)$ by $p(\xi + \Delta\xi, t + \Delta t)$. Thus $p(\xi + \Delta\xi, t + \Delta t)$ penetrates R. This establishes that ϕ is "Lipschitz semicontinuous". An opposite inequality follows similarly. The required "proportional" increment of t depends on the direction of the change in ξ.

2. Approximations of $R(E, \alpha, a)$ and preferred Kähler metrics. Let $G \subset\subset C^n$ be a smooth C^∞ strictly pseudoconvex domain admitting λ as a smooth strictly plurisubharmonic defining function. Let $ds^2 = \sum_{ij} g_{ij}\, dz_i d\bar{z}_j$ be the preferred Kähler metric introduced in [**5**, p. 56].

The main part of the approximation theorem below is (11). This is the key delicate property one has to use to complete the proof of the central result of [**5**, Chapter III] i.e., Theorem 12 of [**5**]. Our proof of (11) hinges upon the special behaviour of (the interesting part of) bR in complex tangential directions.

THEOREM 2 (APPROXIMATION THEOREM). *There is a family of open sets R_ε defined for $0 < \varepsilon < \varepsilon_0$ having properties* (8) *to* (11) *below:*

(8) *$R_\varepsilon \uparrow R$ when $\varepsilon \downarrow 0$ and $R_\varepsilon \subset\subset R$ for any ε. Here $R_\varepsilon \uparrow R$ means that R is*

the increasing union of the R_ε, $R_{\varepsilon_1} \subset R_{\varepsilon_2}$ *when* $\varepsilon_1 > \varepsilon_2$ *and* $\subset\subset$ *means relatively compact.*

(9) *The boundary* bR_ε *has two parts*: $bR_\varepsilon = b_0R_\varepsilon + b_1R_\varepsilon$ *where* b_0R_ε *is a Lipschitz hypersurface (which does not remain in any* $K \subset\subset D$ *as* $\varepsilon \downarrow 0$) *and* b_1R_ε *(the easy part) is an open set of* $\{z \in D, \delta(z) = a - \varepsilon\}$.

(10) *The divergence theorem holds in each* R_ε *when the functions involved are smooth in a neighborhood of the closure* $\overline{R_\varepsilon}$. *Notice that a Lipschitz function is differentiable a.e. (almost everywhere). Thus* N_ε *(the normal in the metric) and* $d\tau_\varepsilon$ *(the area element in the metric) can be computed a.e. by the same expressions one uses in a smooth surface case. These expressions involve only first derivatives. So, at least the statement makes sense.*

Finally,

$$\int_{b_0(R_\varepsilon)} |\lambda|^n \, d\tau_\varepsilon \leqq C < \infty, \quad C \text{ constant independent of } \varepsilon. \tag{11}$$

SKETCH OF THE PROOF OF THEOREM 2. (See §1 for notation.) One proves the theorem for $R' = R \cap \{z \in D; \delta(z) < a'\}$ where $0 < a' < a$. This suffices.

A. *Construction of* R_ε. $R_\varepsilon = \{z \in D; \delta(z) < a' - \varepsilon$ *and* $z = \xi + (t + \varepsilon)n(\xi)$ for some $\xi \in U, t > \phi(\xi)\}$. Thus R_ε results from pushing R' up in the (changing) normal direction $n(\xi)$ and chopping it off close to the top of R'.

We concentrate on (11). The other proofs are standard.

B. *Remarks on the preferred metric.* In the sequel $\|dz\|$ stands for length in the preferred metric and $|dz|$ for the Euclidean one. For z close to bD, $z \in D$, $\xi(z)$ is the point in bD closest to z. In coordinates S_ξ (see §1) $dz = dz_N + dz_T$, where $dz_N = (dz_1, 0, \cdots, 0)$ and $dz_T = (0, dz_2, \cdots, dz_n)$ is the usual decomposition in complex tangential and normal components. The main properties of the metric are [5, p. 55]

$$\|dz_N\| \approx (1/\delta(z))\,|dz_N|, \tag{12}$$

$$\|dz_T\| \approx (1/\sqrt{\delta(z)})\,|dz_T|. \tag{13}$$

The notation $f \lesssim g$ means $f \leqq cg$, c a constant independent of parameters under consideration; $f \approx g$ means that $f \lesssim g$ and $g \lesssim f$.

PROOF OF (11). Parametrize b_0R_ε by

$$z = \xi + n(\xi)[\phi(\xi) + \varepsilon], \qquad \xi \in U', \tag{14}$$

for some open $U' \subset U$.

We prove (11) by showing

$$d\tau_\varepsilon(z) \lesssim (1/\delta(z))^n d\sigma(\xi), \qquad \xi \in U', \tag{15}$$

where $d\sigma$ is Euclidean measure on bD. (The constants c involved in $\lesssim$ are of course independent of ε.) Now (15) proves (11) because $\lambda(z) \approx \delta(z)$.

To establish (15) consider $d\xi$ and dz connected by (14). For any $d\xi$, $\|dz\| \lesssim (1/\delta(z))\,|dz|$ (by (12) and (13)), and $|dz| \lesssim |d\xi|$ by (5), i.e., by the usual Lipschitz property. Hence

$$\|dz\| \lesssim (1/\delta(z))\,|d\xi| \tag{16}$$

which is not enough to prove (15) since there are $2n - 1$ parameters in U'. But if $d\xi$ *occurs in a complex tangential direction* then

Claim.

$$\|dz\| \lesssim (1/\sqrt{\delta(z)})\,|d\xi|. \tag{17}$$

Applying (17) $2n - 2$ times (once for each "good" real parameter, i.e., the complex tangential ones) and using (16) just once (for the "bad" direction dx_1), we obtain

$$d\tau_\varepsilon(z) \lesssim (1/\sqrt{\delta(z)})^{2n-2}\,(1/\delta(z))\,d\sigma = (1/\delta(z))^n\,d\sigma(\xi)$$

which is (15).

PROOF OF CLAIM. Here enters the special property of bR discussed in §1.

Recall $\phi(\xi) + \varepsilon = \delta(z)$ for $z \in b_0R_\varepsilon$; (14) yields

$$dz = d\xi + (dn)\delta(z) + n(\xi)d\phi. \tag{18}$$

The critical term is $n(\xi)d\phi$. Now $\|nd\phi\| \leqq \|n\|\,|d\phi| \lesssim (1/\delta(z))|d\phi|$. Fortunately $d\xi$ is complex tangential so we have the improved estimate $|d\phi| \lesssim \sqrt{\phi(\xi)}\,|d\xi|$ by (7). (If (7′) applied it would be even simpler.) Hence

$$\|n(\xi)d\phi\| \lesssim (1/\delta(z))\,\sqrt{\delta(z)-\varepsilon} \leqq 1/\sqrt{\delta(z)}.$$

The other terms $\|d\xi\|$ and $\|(dn)\delta(z)\|$ are trivially estimated and we obtain (17). (Note that $n(\xi)$ consistently stands for the unit inner normal to bD at ξ, in the sense of the Euclidean metric.) This completes the sketch of the proof of Theorem 2.

REFERENCES

1. E. M. Čirka, *The theorems of Lindelöf and Fatou in C^n*, Mat. Sb. **92 (134)** (1973), 622–644 = Math. USSR Sb. **21** (1973), 619–641. MR **49** #3180.

2. L. Hörmander, *L^p estimates for (pluri-) subharmonic functions*, Math. Scand. **20** (1967), 65–78. MR **38** #2323.

3. A. Korányi, *Harmonic functions in Hermitian hyperbolic space*, Trans. Amer. Math. Soc. **135** (1969), 507–516. MR **43** #3480.

4. R. Putz, *A generalized area theorem for harmonic functions on hermitian hyperbolic space*, Trans. Amer. Math. Soc. **168** (1972), 243–258. MR **45** #7101.

5. E. M. Stein, *Boundary behaviour of holomorphic functions in several complex variables*, Princeton Univ. Press, Princeton, N. J., 1972.

6. ———, *Singular integrals and differentiability properties of functions*, Princeton Math. Ser., no. 30, Princeton Univ. Press, Princeton, N. J., 1970. MR **44** #7280.

7. K. O. Widman, *On the boundary behaviour of solutions to a class of elliptic partial differential equations*, Ark. Mat. **6** (1966), 485–533. MR **36** #2949.

MASSACHUSETTS INSTITUTE OF TECHNOLOGY

Proceedings of Symposia in Pure Mathematics
Volume 30, 1977

APPROXIMATION THEORY ON CR SUBMANIFOLDS*

JEFFREY NUNEMACHER

This article treats two different questions within the general area of approximation of continuous functions on real submanifolds by holomorphic functions. The first part deals with approximation on noncompact totally real submanifolds, and the second with approximation on compact CR submanifolds. Though the theorem obtained in the latter case is not very satisfactory, we feel that the method of proof deserves mention. For a survey of the broad area to which this study belongs we refer to Wells [9].

1. There is a classical theorem of Carleman [2] which can be thought of as a noncompact generalization of the Weierstrass approximation theorem. It asserts that a continuous function f on the real line $\boldsymbol{R}$ can be approximated by a polynomial to any desired degree of accuracy; more precisely, given a positive continuous function ε on $\boldsymbol{R}$, there exists a polynomial g so that $|f(x) - g(x)| < \varepsilon(x)$ for all x in $\boldsymbol{R}$.

For some time complex analysts have understood that, for many purposes, a good generalization of the real line with its standard embedding in the complex plane is the concept of a totally real submanifold of a complex manifold. This being so, we want to know whether there is an analogue of Carleman's theorem for such a submanifold. We answer this question in the affirmative below.

Let M be a C^1 real submanifold of a complex manifold X. Let $T(X)$ denote the holomorphic tangent bundle of X and J the almost-complex mapping on $T(X)$ induced by the complex structure.

DEFINITION. M is called *totally real* if, for all x in M,

$$J_x T(M)_x \cap T(M)_x = \{0\}.$$

AMS (MOS) subject classifications (1970). Primary 32E30.

*The new results mentioned in this report are contained in the author's Yale dissertation written under the direction of Professor Yum-Tong Siu.

The requirement is that there be no holomorphic directions in the complexified tangent space at any point of M. Thus tangentially M appears as does $\mathbf{R}^k$ embedded in $\mathbf{C}^n$ with the standard embedding as the real part of the first k complex coordinates.

Totally real submanifolds have two significant properties that make them amenable to study by complex analytic techniques.

PROPERTY 1. A totally real submanifold of a complex manifold possesses a neighborhood basis of Stein domains.

PROPERTY 2. If f is a C^k function on a C^k totally real submanifold M ($1 \leqq k \leqq \infty$), then f can be extended to a C^k function F defined on a neighborhood of M with the property that $\bar{\partial}F$ vanishes to order $k - 1$ on M.

Using these two properties together with Hörmander's solution of the $\bar{\partial}$ problem with L^2 bounds, Hörmander and Wermer [5] and Nirenberg and Wells [7] were able to show that a C^∞ function on M could be approximated on any compact set K in the C^∞ topology by a holomorphic function defined on a neighborhood of K. Later work of Harvey and Wells [4] and Range and Siu [8] reduced the order of differentiability required. The final theorem (found in [8]) asserts that a C^k function defined on a C^k totally real submanifold M can be approximated on any compact subset of M in C^k norm by holomorphic functions defined in a neighborhood of M. Essential in these later papers is the solution of the $\bar{\partial}$ problem via integral kernels developed by Henkin and Grauert and Lieb. The main idea is to use Property 2 to extend the given function f to F with $\bar{\partial}F = 0$ on M and then to find a G such that $\bar{\partial}G = \bar{\partial}F$. The function $F - G$ is a holomorphic approximation to f and satisfies very good C^k norm estimates when constructed via kernels.

The work mentioned above finishes the problem of approximation on compact sets. We now give a Carleman-type generalization for the case of supremum norm.

THEOREM 1. *Let M be a connected totally real C^1 submanifold of $\mathbf{C}^n$ and f a continuous function on M. Then given any positive continuous function ε on M, there exists a holomorphic function g defined on a neighborhood of M so that, for all z in M, $|f(z) - g(z)| < \varepsilon(z)$.*

The key to the proof is the observation: It suffices that there exists *some* continuous function a with the property that if f is any continuous function on M then there is some holomorphic function g so that $|f(z) - g(z)| < a(z)$ for all z in M. For assume that such an a exists. We may take $\varepsilon(z)$ less than one for all z in M. Choose b to be a continuous function on M with $b(z) > 2a(z)/\varepsilon(z)$. By assumption there is a holomorphic function h so that $|h(z) - b(z)| < a(z)$. In particular $|h(z) - b(z)| < a(z)/\varepsilon(z)$ so $|h(z)| > a(z)/\varepsilon(z)$. Now approximate the function hf by k holomorphic in a neighborhood of M so that $|h(z)f(z) - k(z)| < a(z)$. But then

$$\left|f(z) - k(z)/h(z)\right| < a(z)/\left|h(z)\right| < \varepsilon(z);$$

thus $g = k/h$ is the holomorphic approximation required.

To prove the theorem we must show the existence of a function a having the above property. This is done by solving a $\bar{\partial}$ problem on a Stein neighborhood of M via the Henkin kernel. Since M is in general noncompact, there are technical complications having to do with the convergence of the integrals involved and the possible wildness of the boundary of a general neighborhood of M. These difficulties we overcome by the use of an averaging technique analogous to that employed in Range and Siu [**8**].

An interesting corollary of Theorem 1 can be drawn having to do with the approximation of mappings.

THEOREM 2. *Let M be a connected totally real C^1 submanifold of a complex manifold X. Let f be a continuous mapping from M into a complex manifold Y, and let the topology on Y be induced by a distance function d. Then if ε is any positive continuous function on M, there exists a holomorphic mapping g from a neighborhood of M into Y so that $d(f(z), g(z)) < \varepsilon(z)$ for all z* in M.

Theorem 2 follows from Theorem 1 in the following manner. We consider the graph of the mapping f as a subset of $X \times Y$. It is a totally real submanifold; hence by Property 1 it has a Stein neighborhood that can be embedded into C^n for some n. The map from M into C^n is now given by a tuple of functions to which we apply Theorem 1. By a well-known theorem of Grauert there is a holomorphic retraction sending some neighborhood of M in C^n back into $X \times Y$. If ε is chosen small enough, the approximating tuple of functions will have its range inside the domain of the retraction. Composing the tuple with the retraction and then with projection onto Y yields the desired approximating map into Y.

For more details in the proofs of Theorems 1 and 2 see the forthcoming article. Whether Theorems 1 and 2 carry over to approximation in C^k norm remains open.

2. A useful generalization of the notion of totally real submanifolds is that of CR submanifold (CR for Cauchy-Riemann).

DEFINITION. A C^1 real submanifold M of a complex manifold is called a *CR submanifold* if the dimension of $J_xT(M)_x \cap T(M)_x$ is constant for x in M.

The requirement is that there be a constant number of independent holomorphic directions in the complexified tangent space to M at any point. Extreme examples of CR submanifolds are provided by totally real submanifolds, having no complex structure, and complex submanifolds, having a maximal amount of complex structure. In between these cases lie parametrized families of complex submanifolds.

The study of CR submanifolds as such was begun by Greenfield in [3]. On a CR submanifold the appropriate object to attempt to approximate is the CR function, i.e., a function annihilated by all antiholomorphic tangent vectors in $T(M)_x$ as x ranges over M. In [**6**] Nirenberg was able to show that under certain geometric conditions an arbitrary CR function can locally be approximated by holomorphic functions. Very little is known about global approximation.

We wish to suggest a technique that can be applied to the global problem, at least on some compact CR submanifolds. That tool is the Kodaira identity from complex differential geometry.

Let V be a holomorphic vector bundle over a Kähler manifold X, and assume that X and V have been given Hermitian metrics. Let U be a smoothly bounded domain in X. Then the Kodaira identity expresses the Dirichlet norm $\|\bar{\partial}\beta\|^2 + \|\bar{\partial}^*\beta\|^2$ for β a V-valued (0, 1) form on U in terms of the curvatures of V and X and the Levi form of the boundary of U (and of course the norm of β). A reference for the identity in this form is Andreotti and Vesentini [**1**, p.155 of the Erratum].

If U is strictly pseudoconvex and if the metrics on X and V satisfy a certain curvature hypothesis, the inequality

$$\|\bar{\partial}\beta\|^2 + \|\bar{\partial}^*\beta\|^2 \geqq \|\beta\|^2 \tag{1}$$

follows. It is just such an inequality which is required in Hörmander's solution of the $\bar{\partial}$ problem.

Now if we start with a C^∞ CR section f of V over M, where M is a CR submanifold of X, f can be extended, analogously as in Property 2, to a C^∞ section F of V over a neighborhood of M with $\bar{\partial}F$ vanishing to high order on M. If hypotheses guaranteeing inequality (1) are imposed, the $\bar{\partial}$ problem can be solved to find a section G of V with $\bar{\partial}G = \bar{\partial}F$. As in Theorem 6.1 of Nirenberg and Wells [7], G satisfies the L^2 estimate near M

$$\|G\|_2 \leqq C\|\bar{\partial}F\|_2$$

for some constant C, and from this follow Sobolev estimates. Hence $F - G$ is a holomorphic section of V closely approximating f.

We state the theorem thus obtained. $\langle\ \rangle$ denotes the pointwise metric for a V-valued (0, 1) form, θ the curvature matrix of V, and R the Ricci form associated to the metric on X.

THEOREM 3. *Let M be a smooth compact CR submanifold of a Kähler manifold X, and let V be a holomorphic vector bundle on X. Assume that M has a neighborhood basis of strictly pseudoconvex domains. Assume also that X and V possess metrics so that the curvature condition*

$$\langle\theta\beta, \beta\rangle + \langle R\beta, \beta\rangle \geqq \langle\beta, \beta\rangle$$

is satisfied on some neighborhood of M for (0, 1) *forms β with coefficients in V. Then if f is a smooth CR section of V over M, f can be approximated in the C^∞ topology by holomorphic sections of V defined on neighborhoods of M in X.*

The many hypotheses of this theorem are oppressive. In particular it would be nice to find weak differential geometric conditions which imply the existence of a strictly pseudoconvex neighborhood basis. Strong ones which include the assumption that M is totally geodesic are known.

BIBLIOGRAPHY

1. A. Andreotti and E. Vesentini, *Carleman estimates for the Laplace-Beltrami equation on complex manifolds*, Inst. Hautes Sci. Publ. Math. No. **25** (1965), 81–130; erratum, ibid., No. **27** (1965), 153–155. MR **30** #5333; **32** #465.

2. T. Carleman, *Sur un théorème de Weierstrass*, Ark. Math. Ast. Fys. **20** (1927), 1–5.

3. S. J. Greenfield, *Cauchy-Riemann equations in several variables*, Ann. Scuola Norm. Sup. Pisa (3) **22** (1968), 275–314. MR **38** #6097.

4. F. R. Harvey and R. O. Wells, Jr., *Holomorphic approximation and hyperfunction theory on a C^1 totally real submanifold of a complex manifold*, Math. Ann. **197** (1972), 287–318. MR **46** #9379.

5. L. Hörmander and J. Wermer, *Uniform approximation on compact sets in $\mathbf{C}^n$*, Math. Scand. **23** (1968), 5–21. MR **40** #7484.

6. R. Nirenberg, *On the H. Lewy extension phenomenon*, Trans. Amer. Math. Soc. **168** (1972), 337–356. MR **46** #392.

7. R. Nirenberg and R. O. Wells, Jr., *Approximation theorems on differentiable submanifolds of a complex manifold*, Trans. Amer. Math. Soc. **142** (1969), 15–35. MR **39** #7140.

8. R. M. Range and Y. -T. Siu, *C^k approximation by holomorphic functions and $\bar{\partial}$-closed forms on C^k submanifolds of a complex manifold*, Math. Ann. **210** (1974), 105–122. MR **50** #2561.
9. R. O. Wells, Jr., *Function theory of differentiable submanifolds*, Contributions to Analysis (A Collection of papers dedicated to Lipman Bers), Academic Press, New York, 1974, pp. 407–441. MR **50** #10322.

UNIVERSITY OF TEXAS AT AUSTIN

Proceedings of Symposia in Pure Mathematics
Volume 30, 1977

UNIFORM APPROXIMATION AND THE CAUCHY-FANTAPPIE INTEGRAL

BARNET M. WEINSTOCK

Let Y be a compact set in C^n. Let $C(Y)$ denote the space of continuous, complex-valued functions on Y, and $P(Y)$ the space of functions on Y which are uniform limits of polynomials in the complex coordinates. A basic approximation problem is to find conditions on Y, in addition to the obvious necessary condition of polynomial convexity, which imply that $P(Y) = C(Y)$. Here we outline a new approach to this problem, based on the Cauchy-Fantappie integral, in the special case when Y is a compact subset of a C^1 manifold of a certain type. Details will appear in **[6]**.

1. Historical background. In 1964 Wermer [9] proved that if X is a closed disc in the complex plane and R is a function such that

$$|R(z) - R(w)| < |z - w|, \qquad z, w \in X,$$

then every continuous function on X is a uniform limit of polynomials in z and $\bar{z} + R$. Wermer's proof, which we sketch in §3 below, was based on the fact that a complex measure with compact support in C is zero if and only if its Cauchy transform is zero almost everywhere on a neighborhood of the support of μ. In the same paper Wermer proved in a similar manner that if f is a C^2 function such that $f_{\bar{z}}(0) \neq 0$ then there exists a compact neighborhood of 0 on which every continuous function is a uniform limit of polynomials in z and f.

This second theorem of Wermer's admits the following geometric formulation. A real C^1 submanifold M of C^m is called totally real if at each point of M the (real) tangent space, considered as a real subspace of C^m, contains no complex line. When $m = 2$ and M is the graph of a function f, then M is totally real if and only if $f_{\bar{z}}$ is nonzero. Furthermore, every totally real M is locally polynomially convex.

AMS (MOS) subject classifications (1970). Primary 32E30, 46J10.

Also, if X is a compact subset of the domain of f and Y is the graph of $f|X$, then the algebra of uniform limits of polynomials in z and f is obviously isomorphic to $P(Y)$, and the conclusion of Wermer's theorem is equivalent to $P(Y) = C(Y)$.

After partial results by Wells [8] and Nirenberg and Wells [5], Hörmander and Wermer [4] proved that if Y is any compact subset of a totally real, r-dimensional submanifold M of $\mathbf{C}^m$, and if M is of class C^p, where $p \geqq r/2 + 1$, then every $f \in C(Y)$ is the uniform limit of holomorphic functions. Thus, by the Oka-Weil theorem, if Y is also polynomially convex, $P(Y) = C(Y)$.

As an application they showed that if X is compact in $\mathbf{C}^n$, N is a neighborhood of X, and $R = (R_1, \cdots, R_n)$ is a $\mathbf{C}^{n+1}$ mapping such that $|R(z) - R(w)| \leqq k\,|z - w|$, $z, w \in N$, for some $k < 1$, then every continuous function on X is the uniform limit of polynomials in the functions $z_1, \cdots, z_n, \bar{z}_1 + R_1, \cdots, \bar{z}_n + R_n$. To do so they were required to use the Lipschitz condition to prove that the graph of R in $\mathbf{C}^{2n}$ is totally real and polynomially convex.

Finally, Harvey and Wells [3] succeeded in reducing the required differentiability of M to the optimal class C^1. It followed that the Hörmander-Wermer perturbation theorem, too, required R to be only of class C^1.

2. Statement of results.

THEOREM 1. *Let X be a compact set in $\mathbf{C}^n$. Let N be a neighborhood of X. Let $R = (R_1, \cdots, R_n)$ be a C^1 mapping defined in N such that, for some $k < 1$,*

$$|R(z) - R(w)| \leqq k|z - w|, \qquad z, w \in N.$$

Then every continuous function on X can be approximated uniformly by polynomials in $z_1, \cdots, z_n, \bar{z}_1 + R_1, \cdots, \bar{z}_n + R_n$.

THEOREM 2. *Let $f = (f_1, \cdots, f_n)$ be an n-tuple of functions of class C^1 on a neighborhood of 0 in $\mathbf{C}^n$. If the matrix $(\partial f_i/\partial \bar{z}_j)$ is nonsingular at 0 there is a neighborhood of 0 on which every continuous function is the uniform limit of polynomials in $z_1, \cdots, z_n, f_1, \cdots, f_n$.*

3. Outline of the proofs. Our proofs of Theorems 1 and 2 contain no reference to the general theory of totally real manifolds or to the Oka-Weil theorem and the notion of polynomial convexity. Rather, they parallel Wermer's original arguments, but with the Cauchy transform replaced by a simple device (the Basic Lemma below) based on the Cauchy-Fantappie integral.

We first recall Wermer's proof of his perturbation theorem. Let A denote the algebra of uniform limits of polynomials in z and $\bar{z} + R$. If $w \in X$ then $G(\cdot, w)$, defined by

$$G(z, w) = (z - w)\,(\bar{z} + R(z) - \bar{w} - R(w)),$$

has the following properties: (i) Re $G(z, w) > 0$ if $z \neq w$; (ii) $G(\cdot, w) \in A$; (iii) $(z - w)^{-1}G(z, w) \in A$ for each $w \in X$. Since, if E is any compact subset of $\{\mathrm{Re}\,\lambda \geqq 0\}$, there exist $P_n(\lambda)$, each the uniform limit on E of polynomials in λ, such that $P_n(\lambda) \to \lambda^{-1}$ for each $\lambda \in E - \{0\}$ and $|P_n(\lambda)| \leqq 2\,|\lambda|^{-1}$, it follows that for each $w \in X$ there is a sequence $F_n \in A$ such that $F_n(z) \to (z - w)^{-1}$ on $X - \{w\}$ and $|F_n(z)| \leqq 2|z - w|^{-1}$.

Now, if μ is any complex measure on X which is orthogonal to A, it follows easily that the Cauchy transform of μ is zero at each $w \in X$ for which

$$\int |z - w|^{-1} d|\mu|(z) < \infty.$$

From this one can conclude that $\mu = 0$, so that $A = C(X)$.

We will use the Cauchy-Fantappie integral to produce a substitute for the Cauchy transform for $n > 1$. Let D be a bounded domain in $\mathbf{C}^n$ with smooth boundary. Given an n-tuple $g_1, \cdots, g_n$ of C^1 functions on $\partial D \times D$, let $G(z, w) = \sum (z_j - w_j) g_j(z, w)$. If G is nonvanishing on $\partial D \times D$ one may consider the associated Cauchy-Fantappie kernel $K(z,w)$ defined by

$$K(z, w) = \frac{(n-1)!}{(2\pi i)^n} \sum_{j=1}^{n} (-1)^{j-1} \frac{g_j(z, w)}{G(z, w)^n} \bigwedge_{k \neq j} \bar{\partial}_z g_j(z, w) \wedge dz$$

where dz stands for $dz_1 \wedge \cdots \wedge dz_n$. For each function f holomorphic in D and continuous on D,

$$f(w) = \int_{\partial D} f(z) K(z, w).$$

When the functions $g_1, \cdots, g_n$ are extended to $\bar{D} \times \bar{D}$ so that $G(z, w) \neq 0$ for $z \neq w$ and so that the coefficients of K are in $L^1(D)$ for fixed $w \in D$, then for each C^1 function f,

$$(*) \qquad f(w) = \int_{\partial D} f(z) K(z, w) - \int_D \bar{\partial} f(z) \wedge K(z, w).$$

Let $K_j(z, w) = g_j(z, w) G(z, w)^{-n}$ and let $\eta_j(z, w) = \bigwedge_{k \neq j} \bar{\partial}_z g_k(z, w)$. Then Fubini's theorem and $(*)$ imply the following

BASIC LEMMA. *Suppose that X is compact in $\mathbf{C}^n$, N is a neighborhood of X, $g_1, \cdots, g_n \in C^1(N \times N)$, and $G(z, w) \neq 0$ for $z \neq w$. Suppose further that η_j is independent of w for each j, $1 \leqq j \leqq n$, and that each set $\{\|K_j(\cdot, w)\|_{L^1(X)} : w \in X\}$ is bounded. Then for each complex measure μ with support in X,*

$$(**) \qquad \int |K_j(z, w)| \, d|\mu|(w) < \infty$$

for almost all $z \in N$ and each j, $1 \leqq j \leqq n$. Moreover, if $\int K_j(z, w)\, d\mu(w) = 0$ for all z satisfying $(**)$, $1 \leqq j \leqq n$, then $\mu = 0$.

The proof of Theorem 1 proceeds by following Wermer's outline, with the Cauchy transform replaced by the Basic Lemma applied to the functions

$$g_j(z, w) = (z_j - w_j)(\bar{z}_j + R_j(z) - \bar{w}_j - R_j(w)).$$

The Lipschitz condition on R guarantees that the hypotheses of the Basic Lemma are satisfied. Theorem 2 is proved in a similar manner, using functions $g_j(z, w)$ which are constructed from the first-order Taylor expansion of f. We refer the reader to **[6]** for the actual details.

4. Concluding remarks. (1) The possibility remains that Theorem 1 is true without the hypothesis that R is of class C^1. [See Appendix below.]

(2) The special case of Theorem 1 when $R \equiv 0$, so that the kernel $K(z, w)$ reduces

to the classical Bochner-Martinelli kernel, provides another proof of the Weierstrass approximation theorem.

(3) The method used to prove Theorems 1 and 2 can be useful in other situations. In a subsequent paper [7] the author plans to apply this technique to prove results like the following:

Let X be compact in C^n, f a C^1 mapping of a neighborhood of X into C^n such that the group of $f \mid X$ is polynomially convex in C^{2n}. If E is the set where $(\partial f_i/\partial \bar{z}_j)$ is singular, then every continuous function on X which vanishes on E can be approximated uniformly by polynomials in $z_1, \cdots, z_n, f_1, \cdots, f_n$.

This result is related to earlier work of Wermer [10], Freeman [2], and Fornaess [1].

(4) There is also the possibility that the methods outlined here could be used to give another proof of the theorem of Harvey and Wells referred to at the end of §2.

Appendix. [Added November 25, 1975.] John Wermer has kindly communicated to the author the following observation which was made jointly by Andrew Browder, Brian Cole and himself:

The proof of Theorem 1 can be varied slighty to yield a proof of the following stronger result.

THEOREM 1*. *Let X be compact in C^n, and let $R = (R_1, \cdots, R_n)$ satisfy*

$$|R(z) - R(w)| \leqq k|z - w|, \qquad z, w \in X,$$

for some $k \leqq 1$. Then every continuous function on X can be approximated uniformly by polynomials in $z_1, \cdots, z_n, \bar{z}_1 + R_1, \cdots, \bar{z}_n + R_n$.

Theorem 1* is a new result, and does not seem to follow from the known approximation theorems on totally real manifolds.

The following is a brief sketch of the proof of Theorem 1*.

Extend R to C^n preserving the Lipschitz bound. If μ is a complex measure on X which annihilates the polynomials in $z_1, \cdots, z_n, \bar{z}_1 + R_1, \cdots, \bar{z}_n + R_n$, the argument outlined above still gives $\int K_j(z, w)\, d\mu(z) = 0$ for almost every w with

$$\int |K_j(z, w)|\, d|\mu|(z) < \infty$$

and hence almost everywhere, for $1 \leqq j \leqq n$.

The only place the C^1 hypothesis on R is used in the proof of Theorem 1 is in the construction of the $(n, n-1)$-form $K(z, w)$ and in the derivation of $(*)$. If R is not C^1, choose sequences $R_j^\nu \in C^1(C^n)$ such that

(1) $R_j^\nu \to R_j$ pointwise,

(2) $\|R_j^\nu\|_{\mathrm{Lip}} \leqq k$.

Define H_j^ν, G^ν and K_j^ν for each ν as above. Then

$$|K_j^\nu(z, w)| \leqq C|z - w|^{1-2n},$$

where C is an absolute constant.

If $\varphi \in C^1(N)$, N a fixed bounded neighborhood of X, then $(*)$ and Fubini's theorem imply that

$$\int \varphi(\zeta)\, d\mu(\zeta) = \int_N \sum (-1)^{j-1} \left[\int_X K_j^\nu(z, \zeta)\, d\mu(\zeta) \right] \sigma^\nu(z)\, dm(z)$$

where $dm(z)$ is Lebesgue measure and σ^ν is a function bounded independent of ν. Passing to the limit using the dominated convergence theorem, one concludes that $\int \varphi(\zeta)\, d\mu(\zeta) = 0$, which completes the proof.

REFERENCES

1. J. E. Fornaess, *Uniform approximation on manifolds*, Math. Scand. **31** (1972), 166–170. MR **49** #9634.

2. M. Freeman, *Uniform approximation on a real-analytic manifold*, Trans. Amer. Math. Soc. **143** (1969), 545–553. MR **40** #1777.

3. F. R. Harvey and R. O. Wells, Jr., *Holomorphic approximation and hyperfunction theory on a C^1 totally real submanifold of a complex manifold*, Math. Ann. **197** (1972), 287–318.

4. L. Hörmander and J. Wermer, *Uniform approximation on compact sets in C^n*, Math. Scand. **23** (1968), 5–21. MR **40** #7484.

5. R. Nirenberg and R. O. Wells, Jr., *Approximation theorems on differentiable submanifolds of a complex manifold*, Trans. Amer. Math. Soc. **142** (1969), 15–35. MR **39** #7140.

6. B. Weinstock, *A new proof of a theorem of Hörmander and Wermer*, Math. Ann. **200** (1976), 59–64.

7. ———, *Uniform approximation on smooth polynomially convex sets* (in preparation).

8. R. O. Wells, Jr., *Holomorphic approximation on real-analytic submanifolds of a complex manifold*, Proc. Amer. Math. Soc. **17** (1966), 1272–1275. MR **34** #832.

9. J. Wermer, *Approximation on a disk*, Math. Ann. **155** (1964), 331–333. MR **29** #2670.

10. ———, *Polynomially convex disks*, Math. Ann. **158** (1965), 6–10. MR **30** #5158.

UNIVERSITY OF KENTUCKY

Proceedings of Symposia in Pure Mathematics
Volume 30, 1977

UNIFORM ALGEBRAS ON PLANE DOMAINS

WILLIAM R. ZAME*

Let Ω be an open subset of the complex plane and let $\mathcal{O}(\Omega)$ denote the Frechét algebra of all holomorphic functions on Ω, with the topology of uniform convergence on compact sets. In this note we announce a complete classification of all the closed subalgebras of $\mathcal{O}(\Omega)$ which contain the polynomials. As one of the applications of this classification, we show that, if Ω has k connected components ($1 \leqq k < \infty$) then every closed subalgebra of $\mathcal{O}(\Omega)$ which contains the polynomials is generated by a set of $k + 1$ functions.

Algebras of analytic functions on plane domains (or on Riemann surfaces) have been studied by a number of authors, including Wermer [8], [9], Bishop [1], [2], [3] and Royden [7]. The topological aspect of a similar classification problem was studied by the author in [10], using the analytic structure techniques of Wermer and Bishop. We will make use of that work here, but the more difficult aspects of the classification scheme require a different collection of techniques.

Our classification of subalgebras of $\mathcal{O}(\Omega)$ will be partly in terms of certain simple subalgebras, and we begin by describing these. We say that a closed subalgebra of $\mathcal{O}(\Omega)$ which contains the polynomials is stable if it is closed under differentiation with respect to the coordinate function z. Stable algebras may be constructed in the following way. Let $\Omega = \bigcup \Omega_\alpha$ be a partition of Ω into (disjoint) open and closed subsets (so that each Ω_α is a union of connected components of Ω). For each α, let Ω'_α be a connected open set which is the union of Ω_α with some of the bounded, connected components of $\boldsymbol{C} \backslash \Omega_\alpha$. Let B be the set of all holomorphic functions f on Ω such that $f \mid \Omega_\alpha$ belongs to $\mathcal{O}(\Omega'_\alpha) \mid \Omega_\alpha$ for each α. It is easy to see that B is a stable subalgebra of $\mathcal{O}(\Omega)$ and that ΔB is the disjoint union of the sets Ω'_α, and is thus a complex-analytic manifold in a natural way. (For a topological algebra F, ΔF denotes the space of continuous, nonzero, complex-

AMS (*MOS*) *subject classifications* (1970). Primary 46E25, 30A98; Secondary 32E25, 32K99.
*Supported in part by NSF grant PO 37961–001.

valued homomorphisms of F, with the Gel'fand topology. For an element f in F, we let $\hat{f}$ denote the Gel'fand transform.) The following result shows that all stable algebras arise in this way.

THEOREM 1. *Let B be a stable subalgebra of $\mathcal{O}(\Omega)$ and let $\delta: \Omega \to \Delta B$ be the evaluation map. Let $\Delta B = \bigcup Y_\alpha$ be the decomposition of ΔB into connected components. Then*

(i) *for each α, $\hat{z} | Y_\alpha$ is a homeomorphism onto a connected open subset of $\boldsymbol{C}$;*

(ii) *for each α, $\Omega'_\alpha = \hat{z}(Y_\alpha)$ is the union of $\Omega_\alpha = \hat{z}(Y_\alpha \cap \delta(\Omega))$ with some of the bounded, connected components of $\boldsymbol{C} \backslash \Omega_\alpha$;*

(iii) $B = \{f \in \mathcal{O}(\Omega): (f | \Omega_\alpha) \in \mathcal{O}(\Omega'_\alpha) | \Omega_\alpha$ *for each* $\alpha\}$.

Now fix a closed subalgebra A of $\mathcal{O}(\Omega)$ which contains the polynomials, and let B be the smallest stable subalgebra which contains A. We show that ΔA is a simple quotient space of ΔB.

THEOREM 2. *Let $\rho: \Delta B \to \Delta A$ be the restriction map. Then ρ is surjective and is a quotient mapping. If x, y are in ΔB and $\rho(x) = \rho(y)$ then $\hat{z}(x) = \hat{z}(y)$. The set $\{(x, y): x, y \in \Delta B, x \neq y, \rho(x) = \rho(y)\}$ is discrete in $\Delta B \times \Delta B$.*

Theorems 1 and 2 are largely contained in a previous paper of the author **[10]** although they were formulated somewhat differently.

Having described in reasonable terms the homomorphism space of A, we proceed to endow it with a "structure sheaf." We first construct analytic sheaves on certain subsets of ΔA.

Recall that a subset Y of ΔB is B-convex if for each compact subset K of Y, the set $\{x \in \Delta B: |\hat{b}(x)| \leqq \sup_{y \in K} | \hat{b}(y) |\}$ is again a compact subset of Y.

THEOREM 3. *Let W be an open, relatively compact, B-convex subset of ΔB. Then $\rho(W)$ admits the structure of a 1-dimensional analytic space $(\rho(W), \mathcal{O}_{\rho(W)})$ in a unique way such that*

(i) *$\rho: W \to \rho(W)$ is holomorphic, and is biholomorphic on each connected component of W;*

(ii) *each irreducible branch of $\rho(W)$ is a manifold on which $\hat{z}$ is biholomorphic;*

(iii) *$\hat{A} | \rho(W)$ is a dense subalgebra of the algebra of all holomorphic functions on $\rho(W)$.*

Theorem 3 is proved by first showing that there is an m-tuple of functions in $\hat{A}$ (for m sufficiently large) which maps $\rho(W)$ homeomorphically onto a local subvariety of $\boldsymbol{C}^m$. We then enlarge this m-tuple, if necessary, so that the analytic structure of the variety, when pulled back to $\rho(W)$, provides $\rho(W)$ with the desired analytic structure.

Now let $\boldsymbol{C}_{\Delta A}$ be the sheaf of germs of continuous functions on ΔA. For each x in ΔA, let ${}_x\mathcal{S}_A$ be the set of elements γ in the stalk ${}_x\mathcal{C}_{\Delta A}$ such that $\gamma | \rho(W)$ belongs to ${}_x\mathcal{O}_{\rho(W)}$ for every open, relatively compact, B-convex subset W of ΔB for which x belongs to $\rho(W)$. It is evident that ${}_x\mathcal{S}_A$ is a subalgebra of ${}_x\mathcal{C}_{\Delta A}$ which contains the identity. Set $\mathcal{S}_A = \bigcup_x \mathcal{S}_A$; $\mathcal{S}_A$ will not generally be a subsheaf of $\mathcal{C}_{\Delta A}$, since it may not be open. Nevertheless, we may still speak of sections of $\mathcal{S}_A$. (In fact, there is a sheaf topology on $\mathcal{S}_A$, which is coarser than the topology as a subspace of $\mathcal{C}_{\Delta A}$, but we shall not make use of this fact.) We denote by $\Gamma(\Delta A, \mathcal{S}_A)$

the space of global sections of $\mathscr{S}_A$. Since $\mathscr{S}_A$ is a subset of $\mathscr{C}_{\Delta A}$, we may of course regard the elements of $\Gamma(\Delta A, \mathscr{S}_A)$ as continuous, complex-valued functions on ΔA.

THEOREM 4. *With the above notation,* $\hat{A} = \Gamma(\Delta A, \mathscr{S}_A)$.

It is evident from the construction that each element of $\hat{A}$ is a global section of $\mathscr{S}_A$; the difficulty lies in establishing the converse. The proof involves a delicate approximation argument.

We have now associated with the algebra A a stable algebra B, a quotient space of ΔB and a sheaf on that space. We complete our classification by showing that for each quotient space and sheaf (of the right sort) we can construct the corresponding algebra.

THEOREM 5. *Let B be a stable subalgebra of $\mathcal{O}(\Omega)$ and let X be a quotient space of ΔB with q: $\Delta B \to X$ the quotient mapping. Assume that $q(x) \neq q(y)$ if $x, y \in \Delta B$ but $\hat{z}(x) \neq \hat{z}(y)$, and that $\{(x, y)\colon x, y \in \Delta B, x \neq y, q(x) = q(y)\}$ is a discrete subset of $\Delta B \times \Delta B$. Let $\mathscr{S}$ be a subset of $\mathscr{C}_X$ such that for each open, relatively compact, B-convex subset W of ΔB we have*

(a) *$\mathscr{S} \mid q(W)$ is a subsheaf of $\mathscr{C}_{\rho(W)} = \mathscr{C}_X \mid q(W)$;*

(b) *$(q(W), \mathscr{S} \mid q(W))$ is a 1-dimensional analytic space and each irreducible branch is a manifold;*

(c) *$q\colon W \to q(W)$ is holomorphic (it will automatically follow that q is biholomorphic on each connected component of W).*

Assume further that if $x \in X$, $\gamma \in {}_x\mathscr{C}_X$ and $\gamma \mid q(W)$ belongs to ${}_x\mathscr{S} \mid q(W)$ for each such W with $x \in q(W)$, then γ belongs to ${}_x\mathscr{S}$. Let A be the set of those functions f in B for which there is a section f_0 in $\Gamma(X, \mathscr{S})$ such that $\hat{f} = f_0 \circ q$. Then

(i) *A is a closed subalgebra of $\mathcal{O}(\Omega)$ which contains the polynomials;*

(ii) *B is the smallest stable subalgebra of $\mathcal{O}(\Omega)$ which contains A;*

(iii) *in the notation of Theorem 4, $X = \Delta A$, $q = \rho$ and $\mathscr{S} = \mathscr{S}_A$.*

The key to the proof of Theorem 5 lies in showing that each section of $\mathscr{S}$ over one of the sets $q(W)$ can be approximated by global sections.

It may seem that the classification we have given is somewhat cumbersome, but in fact it is quite manageable. To illustrate this, we give several applications. The first one follows fairly directly from Theorem 4.

THEOREM 6. *Let A be a closed subalgebra of $\mathcal{O}(\Omega)$ which contains the polynomials. Then every continuous function on ΔA, which can be locally approximated by functions in $\hat{A}$, actually belongs to $\hat{A}$.*

Our next result was suggested by a theorem of Gamelin [**4**]. It is proved by a constructive procedure utilizing Theorem 5.

THEOREM 7. *Let A be a closed subalgebra of $\mathcal{O}(\Omega)$ which contains the polynomials and let B be the smallest stable algebra containing A. Then there is a decreasing sequence $B = B_0 \supset B_1 \supset B_2 \cdots$ of closed subalgebras of $\mathcal{O}(\Omega)$ such that $A = \bigcap B_i$ and such that B_{i+1} is of finite codimension in B_i (for each i).*

Our final application is considerably more difficult. It utilizes Narasimhan's imbedding theorem for Stein spaces [**6**].

THEOREM 8. *Let Ω be an open subset of the complex plane with k connected components* ($1 \leqq k < \infty$). *Then every closed subalgebra of $\mathcal{O}(\Omega)$ which contains the polynomials is generated by a set of $k + 1$ functions.*

Complete details, and some generalizations to Riemann surfaces, will appear elsewhere.

REFERENCES

1. E. Bishop, *Subalgebras of functions on a Riemann surface*, Pacific J. Math. **8** (1958), 29–50. MR 20 #3300.

2. ———, *Analyticity in certain Banach algebras*, Trans. Amer. Math. Soc. **102** (1962), 507–544. MR **25** #5410.

3. ———, *Holomorphic completions, analytic continuation, and the interpolation of semi-norms*, Ann. of Math. (2) **78** (1963), 468–500. MR **27** #4958.

4. T. W. Gamelin, *Embedding Riemann surfaces in maximal ideal spaces*, J. Functional Analysis **2** (1968), 123–146. MR **36** #6941.

5. ———, *Polynomial approximation on thin sets*, Sympos. On Several Complex Variables (Park City, Utah, 1970), Lecture Notes in Math., vol. 184, Springer-Verlag, Berlin and New York, 1971, pp. 50–78. MR **45** #9145.

6. R. Narasimhan, *Imbedding of holomorphically complete complex spaces*, Amer. J. Math. **82** (1960), 917–934. MR **26** #6438.

7. H. Royden, *Algebras of bounded analytic functions on Riemann surfaces*, Acta Math. **114** (1965), 113–142. MR **30** #3972.

8. J. Wermer, *Function rings and Riemann surfaces*, Ann. of Math. (2) **67** (1958), 45–71. MR **20** #109.

9. ———, *Rings of analytic functions*, Ann. of Math. (2) **67** (1958), 497–516. MR **20** #3299.

10. W. Zame, *Algebras of analytic functions in the plane*, Pacific J. Math. **42** (1972), 811–819.

STATE UNIVERSITY OF NEW YORK AT BUFFALO
TULANE UNIVERSITY

PRINCIPAL LECTURE

OTTO FORSTER

POWER SERIES METHODS IN DEFORMATION THEORY

Proceedings of Symposia in Pure Mathematics
Volume 30, 1977

POWER SERIES METHODS IN DEFORMATION THEORY

OTTO FORSTER

Introduction. One of the main problems in deformation theory is the construction of versal deformations (moduli problem). This problem has been solved for the deformation of various objects.

(a) *Riemann surfaces.* Already Riemann stated in his famous memoir *Zur Theorie der abelschen Funktionen* [**23**] that a Riemann surface of genus $p \geqq 2$ depends on $3p - 3$ complex parameters (=moduli). The efforts of Teichmüller [**26**], Rauch [**22**], Ahlfors [**1**], Bers [**3**] and Grothendieck [**15**] led to the construction of the Teichmüller family $\pi\colon \mathfrak{X} \to S$ of all Riemann surfaces of genus p. The parameter space S is a complex manifold of dimension $3p - 3$, the fibres $\pi^{-1}(s)$ are Riemann surfaces of genus p with a distinguished basis of the fundamental group.

(b) *Compact complex manifolds.* Kodaira and Spencer [**16**] were the first to consider families of compact complex manifolds of higher dimensions. Kodaira, Nirenberg and Spencer [**18**] proved for compact complex manifolds X with $H^2(X, \Theta_X) = 0$ the existence of a versal deformation $\mathfrak{X} \to S$, where S is the germ of $\boldsymbol{C}^m$ at the origin, $m = \dim H^1(X, \Theta_X)$. (We denote by Θ_X the sheaf of holomorphic vector fields on X.) Kuranishi [**19**] was able to prove the existence of a versal deformation for a compact complex manifold without the assumption $H^2(X, \Theta_X) = 0$. In this case the parameter space may be singular.

(c) *Compact analytic subspaces of a given complex space.* For this problem Douady [**7**] constructed a universal moduli space: Given a complex space X, there exists a complex space H and an analytic subspace $Z \subset X \times H$, proper and flat over H, with the following universal property: If Y is an analytic subspace of $X \times S$, proper and flat over S, then there exists a uniquely determined holomorphic map $\alpha\colon S \to H$, such that Y is the pull-back of Z with respect to the map

AMS (MOS) subject classifications (1970). Primary 32G13; Secondary 32A05.

id $\times$ α: $X \times S \to X \times H$. By taking S to be a simple point, one sees that H parametrizes all compact analytic subspaces of X.

(d) *Germs of analytic sets with isolated singularities.* The corresponding moduli problem was solved by Tjurina **[27]** for normal singularities with $\mathrm{Ext}^2(\Omega_x, \mathcal{O}_x) = 0$, by Grauert **[13]** for reduced singularities and Donin **[5]** in the general case (see also Pourcin **[21]**).

(e) *Compact complex spaces.* Grauert **[14]** and Douady **[8]** proved as a generalization of Kuranishi's theorem the existence of a versal deformation for an arbitrary compact complex space.

A remarkable feature in most of the proofs for the existence of moduli spaces is that one has to enlarge the category in which one is working. In the case of Riemann surfaces one used the theory of quasi-conformal maps; in the theory of Kodaira-Spencer almost complex structures and potential theory play an important role and Douady developed for his proof the theory of (infinite-dimensional) Banach-analytic spaces.

Another possible method of proof is by power series. By a general theorem of Schlessinger **[24]** it is relatively easy to see that one has formal solutions to all of the moduli problems mentioned above. Thus the "only" thing that remains to be done is to assure convergence of the formal constructions. This has been carried out by Grauert **[13]** for the deformation of isolated singularities. For this purpose he proved a powerful generalization of the Weierstrass division theorem.

We want to present in this article a convergence proof for the Kodaira-Nirenberg-Spencer theorem which has been worked out by K. Knorr and the author **[10]**. This proof does not use potential theory and can be carried over to the deformation problem for arbitrary compact complex spaces. We will also discuss (in §§IV and V) some of the crucial points of this generalization.

I. Calculus of majorants for relative automorphisms.

1. *Norms.* Let A be a Banach space and m a positive integer. We denote by $A\{t\} = A\{t_1, \cdots, t_m\}$ the A-module of all convergent power series

$$f = \sum_{\nu \in \boldsymbol{N}^m} a_\nu t^\nu, \qquad a_\nu \in A.$$

We introduce in $A\{t\}$ a pseudonorm by the following procedure: Consider the Taylor expansion of the function

$$M(t) = \prod_{j=1}^{m} \left(\tfrac{1}{2} - \tfrac{1}{4}\sqrt{1 - t_j}\right) = \sum_{\nu \in \boldsymbol{N}^m} \gamma_\nu t^\nu.$$

It is easy to check that all coefficients γ_ν are positive. For an element $f = \sum a_\nu t^\nu \in A\{t\}$ we set

$$\|f\| := \inf\{c \in \boldsymbol{R}_+ : f \ll cM(t)\}.$$

Here $f \ll cM(t)$ means that f is majorized by the series $cM(t)$, i.e., $\|a_\nu\| \leqq c\gamma_\nu$ for all ν. The set of all elements $f \in A\{t\}$ having finite norm is a Banach space which we denote by $A\langle t\rangle$. This Banach space is a module over $\boldsymbol{C}\langle t\rangle$ and for $\varphi \in \boldsymbol{C}\langle t\rangle$, $f \in A\langle t\rangle$ we have

$$\|\varphi f\| \leqq \|\varphi\| \cdot \|f\|.$$

This inequality follows from the fact that $M(t)^2 \ll M(t)$. If A is a Banach algebra, then $A\langle t\rangle$ is also a Banach algebra.

Now let U be an open set in $\boldsymbol{C}^N$ and B the germ of $\boldsymbol{C}^m$ at the origin. By $\Gamma(U, \mathcal{O}_{\boldsymbol{C}^N\times B})$ we denote the algebra of all germs of holomorphic functions in neighborhoods of $U \times 0$ in $U \times \boldsymbol{C}^m$. Every element $f \in \Gamma(U, \mathcal{O}_{\boldsymbol{C}^N\times B})$ can be represented by a power series

$$f = \sum_{\nu\in \boldsymbol{N}^m} f_\nu t^\nu, \qquad f_\nu \in \Gamma(U, \mathcal{O}_{\boldsymbol{C}^N}),$$

where $t_1, \cdots, t_m$ are the canonical coordinates in $\boldsymbol{C}^m$. We introduce a pseudonorm in $\Gamma(U, \mathcal{O}_{\boldsymbol{C}^N\times B})$ by setting

$$\|f\|_U = \inf\{c \in \boldsymbol{R}_+ : \sum |f_\nu|_U t^\nu \ll cM(t)\}.$$

Here $|f_\nu|_U$ denotes the sup norm on U. The Banach algebra of all elements $f \in \Gamma(U, \mathcal{O}_{\boldsymbol{C}^N\times B})$ of finite norm is denoted by $\Gamma_b(U, \mathcal{O}_{\boldsymbol{C}^N\times B})$.

2. *The sheaf of relative automorphisms.* Let X be a complex manifold, (S, s_0) the germ of a complex space and U an open subset of X. By a relative automorphism of $U \times S$ we understand a holomorphic map

$$g: U \times S \to U \times S$$

which commutes with the projection $U \times S \to S$ and induces the identity map on the distinguished fiber $U \times \{s_0\}$. The set $\mathfrak{G}_S(U)$ of all relative automorphisms of $U \times S$ forms a group (in general nonabelian) under composition of maps. If V is an open subset of U, one has a natural restriction $\mathfrak{G}_S(U) \to \mathfrak{G}_S(V)$. In this way we obtain a sheaf $\mathfrak{G}_S$ of groups on X.

The important role of the sheaf $\mathfrak{G}_S$ in deformation theory is due to the following well-known fact (cf. for example Douady [**6**]):

Let X be a compact complex manifold and denote by Def(X, S) the set of isomorphism classes of deformations of X parametrized by S. Then there is a natural one-to-one correspondence

$$H^1(X, \mathfrak{G}_S) \simeq \mathrm{Def}(X, S).$$

This correspondence can be described in the following way: Let $\xi \in H^1(X, \mathfrak{G}_S)$ be a cohomology class which is represented by a cocycle $(g_{ij}) \in Z^1(\mathfrak{U}, \mathfrak{G}_S)$, where $\mathfrak{U} = (U_i)_{i\in I}$ is an open covering of X. Consider the product $X \times S$, break it up into pieces $U_i \times S$, $i \in I$, and patch them together by the transition functions

$$g_{ij}: (U_i \cap U_j) \times S \to (U_i \cap U_j) \times S$$

to obtain the deformation of X corresponding to ξ.

3. *Norms for relative automorphisms.* Let U be an open subset of $\boldsymbol{C}^N$ and B the germ of $\boldsymbol{C}^m$ at 0. We want to describe explicitly $\mathfrak{G}(U) := \mathfrak{G}_B(U)$. An element $\gamma \in \mathfrak{G}(U)$ is a map

$$U \times B \to U \times B, \qquad (z, t) \mapsto (g(z, t), t),$$

where $g \in \Gamma(U, \mathcal{O}_{\boldsymbol{C}^N\times B})^N$ is an element satisfying $g(z, 0) = z$. In the sequel we identify γ with g. We can write $\mathfrak{G}(U) = \mathrm{id} + T\mathfrak{G}(U)$, where $T\mathfrak{G}(U)$ consists of all elements $\varphi \in \Gamma(U, \mathcal{O}_{\boldsymbol{C}^N\times B})^N$ with $\varphi(z, 0) = 0$. Thus $\mathfrak{G}(U)$ is an affine space over the vector space $T\mathfrak{G}(U)$. The pseudonorm $\| \ \|_U$ on $\Gamma(U, \mathcal{O}_{C^N \times B})$ described in §I.1 induces a pseudonorm on $T\mathfrak{G}(U)$. For $g = \mathrm{id} + \varphi \in \mathfrak{G}(U)$ we write $\|g\|_U = \|\varphi\|_U$. The sets of elements of finite norm will be denoted by $T\Gamma_b(U, \mathfrak{G})$ and $\Gamma_b(U, \mathfrak{G})$, respectively. $\Gamma_b(U, \mathfrak{G})$ is an affine space over the Banach space $T\Gamma_b(U, \mathfrak{G})$. For $\varepsilon > 0$, we denote by $\Gamma_\varepsilon(U, \mathfrak{G})$ the set of all elements of norm $< \varepsilon$ in $\Gamma_b(U, \mathfrak{G})$.

4. *Composition of relative automorphisms.* We keep the notations of the previous paragraph. Let $(z, t) \mapsto (f(z, t), t)$ and $(z, t) \mapsto (g(z, t), t)$ be two relative automorphisms. Then their composition is given by

$$(z, t) \mapsto (h(z, t), t), \quad \text{where } h(z, t) = f(g(z, t), t).$$

By abuse of notation we write $h = f \circ g = f(g)$. The composition of two relative automorphisms of finite norm is not necessarily of finite norm. However we have

(I.4.1) THEOREM. *Let $V \subset\subset U$ be two open subsets of $\mathbf{C}^N$. Then there exists $\varepsilon > 0$ such that for every $f \in \Gamma_b(U, \mathfrak{G})$ and $g \in \Gamma_\varepsilon(V, \mathfrak{G})$ we have $f(g) \in \Gamma_b(V, \mathfrak{G})$. The map*

$$\mu: \Gamma_b(U, \mathfrak{G}) \times \Gamma_\varepsilon(V, \mathfrak{G}) \to \Gamma_b(V, \mathfrak{G}),$$
$$(f, g) \mapsto f(g)$$

is differentiable. Its differential at the point (f, g) is

$$D\mu(f, g): T\Gamma_b(U, \mathfrak{G}) \times T\Gamma_b(V, \mathfrak{G}) \to T\Gamma_b(V, \mathfrak{G}),$$
$$(\varphi, \psi) \mapsto \varphi(g) + (\partial f/\partial z)(g)\psi.$$

By the implicit function theorem for Banach spaces one can deduce the following

(I.4.2) COROLLARY. *Let $V \subset\subset U$ be open subsets of $\mathbf{C}^N$. Then there exists $\varepsilon > 0$ such that the mapping*

$$\iota: \Gamma_\varepsilon(U, \mathfrak{G}) \to \Gamma_b(V, \mathfrak{G}), \qquad g \mapsto g^{-1}$$

is well defined and differentiable. Its differential at the point id *is* $D\iota(\mathrm{id}) = -1$.

REMARK. A differentiable map between open subsets of complex Banach spaces is automatically holomorphic; in particular all derivatives exist and are continuous. From local bounds for the derivative we can get Lipschitz estimates for the map.

In the power series constructions which we will have to make, sometimes the following situation occurs: We have to compose two relative automorphisms f and g which are polynomials of degree $\leqq e$ with respect to the variables t. The composition $f(g)$ is no more a polynomial of degree $\leqq e$. However, the higher terms of $f(g)$ are small of second order. This is made precise by the following lemma.

(I.4.3) EXCESS LEMMA. *Let $V \subset\subset U$ be open subsets of $\mathbf{C}^N$. Then there exists $\varepsilon > 0$ and $K \geqq 1$ such that for every $e \in \mathbf{N}$ the following holds: Let $f \in \Gamma_\varepsilon(U, \mathfrak{G})$, $g \in \Gamma_\varepsilon(V, \mathfrak{G})$ be polynomials of degree $\leqq e$ in t and*

$$\sigma := \max(\|f\|_U, \|g\|_V).$$

Write $f(g)$ as $f(g) = h + \psi$, where h is a polynomial of degree $\leqq e$ and ψ contains only terms of order $\geqq e + 1$. Then $\|\psi\|_V \leqq K\sigma^2$.

An analogous statement is true for the inverse g^{-1} of a polynomial g and for coordinate transformations [**10**, Corollary 2.22].

II. Cohomology with bounds for the sheaf $\mathfrak{G}$.

1. *A Heftungslemma.* We identify $\boldsymbol{C}^N$ with $\boldsymbol{R}^{2N}$. By an open (resp. compact) rectangle in $\boldsymbol{R}^{2N}$ we understand a product of open (resp. compact) intervals.

Let

$$Q' = \prod_{\nu=1}^{2N} I'_\nu, \qquad Q'' = \prod_{\nu=1}^{2N} I''_\nu$$

be two rectangles. (Q', Q'') is said to be a *Cousin pair* if there exists ν such that $I'_\mu = I''_\mu$ for all $\mu \neq \nu$ and $I'_\nu \cap I''_\nu \neq \emptyset$. In this case $Q' \cup Q''$ is again a rectangle.

(II.1.1) HEFTUNGSLEMMA. *Let (Q_1, Q_2) be a Cousin pair of open rectangles in $\boldsymbol{C}^N$ and U be an open neighborhood of $\overline{Q_1 \cap Q_2}$. Then there exists $\varepsilon > 0$ and a differentiable map*

$$\alpha: \Gamma_\varepsilon(U, \mathfrak{G}) \to \Gamma_b(Q_1, \mathfrak{G}) \times \Gamma_b(Q_2, \mathfrak{G})$$

with the following properties:

(i) $\alpha(\mathrm{id}) = (\mathrm{id}, \mathrm{id})$.

(ii) *Let $g \in \Gamma_\varepsilon(U, \mathfrak{G})$ and $\alpha(g) =: (g_1, g_2)$. Then $g = g_1 \circ g_2^{-1}$ on $Q_1 \cap Q_2$.*

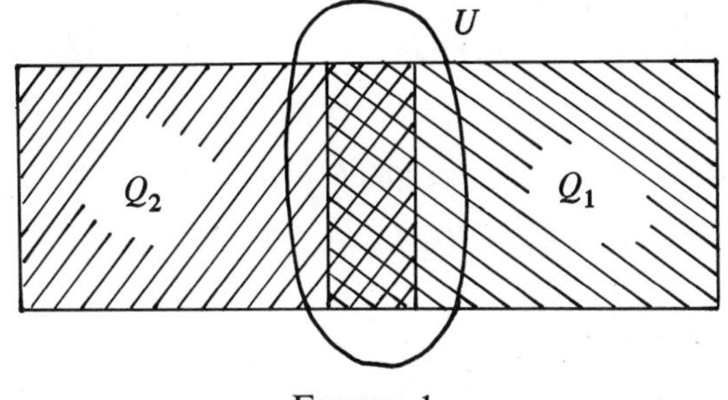

FIGURE 1

PROOF. We will apply the implicit function theorem to the map

$$\Phi: \Gamma_b(U, \mathfrak{G}) \times \Gamma_b(Q_1, \mathfrak{G}) \times \Gamma_{\tilde\varepsilon}(Q_2, \mathfrak{G}) \to T\Gamma_b(Q_1 \cap Q_2, \mathfrak{G}),$$
$$(g, g_1, g_2) \mapsto g(g_2) - g_1$$

which is defined and differentiable for sufficiently small $\tilde\varepsilon > 0$. The partial differential of this map with respect to the second and third argument at the point (id, id, id) is (cf. Theorem I.4.1)

$$L := D_{23}\Phi(\mathrm{id}, \mathrm{id}, \mathrm{id}): T\Gamma_b(Q_1, \mathfrak{G}) \times T\Gamma_b(Q_2, \mathfrak{G}) \to T\Gamma_b(Q_1 \cap Q_2, \mathfrak{G}),$$
$$(\varphi_1, \varphi_2) \mapsto \varphi_2 - \varphi_1.$$

The classical additive Cousin Heftungslemma shows that L admits a continuous linear section

$$\sigma: T\Gamma_b(Q_1 \cap Q_2, \mathfrak{G}) \to T\Gamma_b(Q_1, \mathfrak{G}) \times T\Gamma_b(Q_2, \mathfrak{G}),$$

$L \circ \sigma = 1$. Therefore there exists $\varepsilon > 0$ and a differentiable map

$$\alpha = (\alpha_1, \alpha_2): \Gamma_\varepsilon(U, \mathfrak{G}) \to \Gamma_b(Q_1, \mathfrak{G}) \times \Gamma_\varepsilon(Q_2, \mathfrak{G})$$

with $\alpha(\mathrm{id}) = (\mathrm{id}, \mathrm{id})$ and $\Phi(g, \alpha_1(g), \alpha_2(g)) = 0$ for all g. This is the map we have been looking for.

2. *Theorem* B. We denote by $\mathfrak{m}$ the maximal ideal of $\boldsymbol{C}\{t_1, \cdots, t_m\}$, the local ring of $\boldsymbol{C}^m$ at the origin.

If U is an open set in $\boldsymbol{C}^N$ and $f, g \in \mathfrak{G}(U)$, then $f \equiv g \bmod \mathfrak{m}^{e+1}$ means that the Taylor expansions $f = \sum f_\nu t^\nu$ and $g = \sum g_\nu t^\nu$ coincide up to order e.

Let $\mathfrak{U} = (U_i)_{i \in I}$ be a finite family of open sets in $\boldsymbol{C}^N$. Then we have the cochain groups

$$C^0(\mathfrak{U}, \mathfrak{G}) = \prod_i \Gamma(U_i, \mathfrak{G}), \qquad C^1(\mathfrak{U}, \mathfrak{G}) = \prod_{i,j} \Gamma(U_i \cap U_j, \mathfrak{G}).$$

An element $(g_{ij}) \in C^1(\mathfrak{U}, \mathfrak{G})$ is called a cocycle mod $\mathfrak{m}^{e+1}$, if for every $i, j, k \in I$ we have

$$g_{ij} g_{jk} \equiv g_{ik} \bmod \mathfrak{m}^{e+1} \quad \text{on } U_i \cap U_j \cap U_k.$$

We define

$$C_b^0(\mathfrak{U}, \mathfrak{G}) := \prod_i \Gamma_b(U_i, \mathfrak{G}), \qquad C_\varepsilon^0(\mathfrak{U}, \mathfrak{G}) := \prod_i \Gamma_\varepsilon(U_i, \mathfrak{G}), \quad \text{etc.}$$

Note that $C_b^0(\mathfrak{U}, \mathfrak{G})$ is not a group.

(II.2.1) THEOREM B. *Let $D \subset \boldsymbol{C}^N$ be an open rectangle, $\mathfrak{U} = (U_i)_{i \in I}$ a finite open covering of D and $\mathfrak{V} = (V_i)_{i \in I}$ a family of open subsets of D such that*

$$V_i \subset\subset U_i \quad \text{for all } i \in I.$$

Then there exists $\varepsilon > 0$ and a differentiable map

$$\beta\colon C_\varepsilon^1(\mathfrak{U}, \mathfrak{G}) \to C_b^0(\mathfrak{V}, \mathfrak{G})$$

with the following properties:

(i) $\beta(\mathrm{id}) = \mathrm{id}$.

(ii) *Let $e \in \boldsymbol{N}$ and $(g_{ij}) \in C_\varepsilon^1(\mathfrak{V}, \mathfrak{G})$ be a cocycle* mod $\mathfrak{m}^{e+1}$. *Then $\beta((g_{ij})) =: (h_i)$ satisfies the congruence*

$$g_{ij} \equiv h_i \circ h_j^{-1} \bmod \mathfrak{m}^{e+1} \quad \text{on } V_i \cap V_j.$$

Roughly speaking the theorem states that for an open rectangle D we have $H^1(D, \mathfrak{G}) = 0$, but it gives far more precise information.

The theorem is proved by successive application of the Heftungslemma and the following

(II.2.2) PROJECTION LEMMA. *Let $V \subset \boldsymbol{C}^N$ be an open set and $\mathfrak{U} = (U_i)_{i \in I}$ a finite family of open subsets of $\boldsymbol{C}^N$ such that $V \subset\subset \bigcup U_i$. Then there exists a continuous $\boldsymbol{C}\langle t\rangle$-linear map*

$$\pi\colon C_b^0(\mathfrak{U}, \mathcal{O}_{\boldsymbol{C}^N \times B}) \to \Gamma_b(V, \mathcal{O}_{\boldsymbol{C}^N \times B})$$

with the following property:

Let $e \in \boldsymbol{N}$ and $(f_i) \in C_b^0(\mathfrak{U}, \mathcal{O}_{\boldsymbol{C}^N \times B})$ be an element satisfying

$$f_i \equiv f_j \bmod \mathfrak{m}^{e+1} \quad \text{on } U_i \cap U_j.$$

Then $f := \pi((f_i))$ satisfies

$$f \equiv f_i \bmod \mathfrak{m}^{e+1} \quad \textit{on } V \cap U_i \textit{ for all } i \in I.$$

3. *A smoothing theorem.* Let X be a compact complex N-dimensional manifold. We fix a finite open covering $\mathfrak{U}^* = (U_i^*)_{i \in I}$ of X such that every U_i^* is isomorphic to an open rectangle in $\boldsymbol{C}^N$. For every open covering $\mathfrak{U} \ll \mathfrak{U}^*$ one can then define the sets $C_b^0(\mathfrak{U}, \mathfrak{G})$, $C_b^1(\mathfrak{U}, \mathfrak{G})$, etc.

The following theorem says, roughly speaking, that given two coverings $\mathfrak{V} \ll \mathfrak{U}$, every $\mathfrak{G}$-valued cocycle ξ with respect to the covering $\mathfrak{V}$ may be replaced by a cohomologous cocycle ζ with respect to the covering $\mathfrak{U}$. The cocycles of the covering $\mathfrak{U}$ are considered to be smoother than the cocycles with respect to $\mathfrak{V}$. This accounts for the label "Smoothing Theorem". (This smoothing technique has been invented by Grauert [**12**] in his proof of the direct image theorem.)

(II.3.1) SMOOTHING THEOREM. *Let $\mathfrak{W} \ll \mathfrak{V} \ll \mathfrak{U} \ll \mathfrak{U}^*$ be finite open coverings by rectangles of the compact complex manifold X. Then there exists $\varepsilon > 0$ and a differentiable map*

$$\gamma \colon C_\varepsilon^1(\mathfrak{V}, \mathfrak{G}) \to C_b^1(\mathfrak{U}, \mathfrak{G}) \times C_b^0(\mathfrak{W}, \mathfrak{G})$$

with the following properties:

(i) $\gamma(\mathrm{id}) = (\mathrm{id}, \mathrm{id})$.

(ii) *Let $e \in \boldsymbol{N}$ and $(g_{ij}) \in C_\varepsilon^1(\mathfrak{V}, \mathfrak{G})$ be a cocycle* mod $\mathfrak{m}^{e+1}$. *Set $\gamma((g_{ij})) =: ((G_{ij}), (h_i))$. Then (G_{ij}) is a cocycle* mod $\mathfrak{m}^{e+1}$ *with respect to $\mathfrak{U}$ and one has for all i, j*:

$$G_{ij} \equiv h_i g_{ij} h_j^{-1} \bmod \mathfrak{m}^{e+1} \quad \textit{on } W_i \cap W_j.$$

The proof of the smoothing theorem follows the proof of the classical

(II.3.2) LERAY THEOREM. *Let X be a topological space, $\mathscr{F}$ a sheaf of* (*nonabelian*) *groups on X and*

$$\mathfrak{V} = (V_i)_{i \in I} \ll \mathfrak{U} = (U_i)_{i \in I}$$

open coverings of X. Assume $H^1(U_\alpha \cap \mathfrak{V}, \mathscr{F}) = 0$ for all $\alpha \in I$. Then the restriction map

$$H^1(\mathfrak{U}, \mathscr{F}) \to H^1(\mathfrak{V}, \mathscr{F})$$

is surjective.

PROOF. Let $(f_{ij}) \in Z^1(\mathfrak{V}, \mathscr{F})$ be any cocycle. By hypothesis the cocycle $(f_{ij})|U_\alpha$ is a coboundary for every $\alpha \in I$, i.e.,

$$f_{ij} = \varphi_i^\alpha (\varphi_j^\alpha)^{-1} \quad \text{on } U_\alpha \cap V_i \cap V_j$$

where $\varphi_i^\alpha \in \mathscr{F}(U_\alpha \cap V_i)$. On $U_\alpha \cap U_\beta \cap V_i \cap V_j$ we have

$$\varphi_i^\alpha (\varphi_j^\alpha)^{-1} = \varphi_i^\beta (\varphi_j^\beta)^{-1};$$

hence $(\varphi_i^\alpha)^{-1} \varphi_i^\beta = (\varphi_j^\alpha)^{-1} \varphi_j^\beta$.

Therefore the family $((\varphi_i^\alpha)^{-1} \varphi_i^\beta)_{i \in I}$ defines an element $F_{\alpha\beta} \in \mathscr{F}(U_\alpha \cap U_\beta)$. Clearly $(F_{\alpha\beta})$ is a cocycle with respect to $\mathfrak{U}$. Letting $h_\alpha := \varphi_\alpha^\alpha \in \mathscr{F}(V_\alpha)$, we have on $V_\alpha \cap V_\beta$

$$F_{\alpha\beta} = (\varphi_\alpha^\alpha)^{-1} \varphi_\alpha^\beta = (\varphi_\alpha^\alpha)^{-1} \varphi_\alpha^\beta (\varphi_\beta^\beta)^{-1} \varphi_\beta^\beta = h_\alpha^{-1} f_{\alpha\beta} h_\beta.$$

This shows that class$(F_{\alpha\beta})$ projects onto class$(f_{\alpha\beta})$.

III. Proof of the Kodaira-Nirenberg-Spencer theorem.

Notations. Throughout this section, X will be a compact complex manifold of dimension N. As before, B denotes the germ of $\boldsymbol{C}^m$ at the origin. The local ring of B is $\boldsymbol{C}\{t_1,\cdots,t_m\}$ with maximal ideal $\mathfrak{m}$. The number m will later be chosen to equal $\dim H^1(X, \Theta)$, where Θ is the sheaf of holomorphic vector fields on X.

$\mathfrak{G} = \mathfrak{G}_B$ is the sheaf on X of relative automorphisms with respect to B.

1. *Relative automorphisms and vector fields.* For $e \in \boldsymbol{N}$ and U open $\subset X$ let $\mathfrak{G}_e(U) \subset \mathfrak{G}(U)$ denote the set of all elements $g \in \mathfrak{G}(U)$ satisfying

$$g \equiv \mathrm{id} \mod \mathfrak{m}^e.$$

$\mathfrak{G}_e(U)$ is a normal subgroup of $\mathfrak{G}(U)$. The sheaf $\mathfrak{G}/\mathfrak{G}_{e+1}$ is isomorphic to the sheaf $\mathfrak{G}_{B_e}$, where B_e is the subgerm of B defined by $\mathfrak{m}^{e+1}$.

The sheaf $\mathfrak{G}_1$ coincides with $\mathfrak{G}$.

(III.1.1) LEMMA. *There is a natural isomorphism*

$$\mathfrak{G}_e/\mathfrak{G}_{e+1} \cong \Theta \otimes_{\boldsymbol{C}} (\mathfrak{m}^e/\mathfrak{m}^{e+1}).$$

In particular we have $\mathfrak{G}/\mathfrak{G}_2 = \Theta \otimes (\mathfrak{m}/\mathfrak{m}^2)$. With respect to a local coordinate system $(z_1,\cdots, z_N)$ on X a section g of $\mathfrak{G}_e/\mathfrak{G}_{e+1}$ can be uniquely represented as

$$g(z, t) = \mathrm{id} + \sum_{|\nu|=e} \gamma_\nu(z)t^\nu,$$

where $\gamma_\nu = (\gamma_{\nu 1}, \cdots, \gamma_{\nu N})$ is an N-tuple of holomorphic functions. The elements t^ν, $|\nu| = e$, may be viewed as a basis of $\mathfrak{m}^e/\mathfrak{m}^{e+1}$. The isomorphism

$$\mathfrak{G}_e/\mathfrak{G}_{e+1} \to \Theta \otimes (\mathfrak{m}^e/\mathfrak{m}^{e+1})$$

can now be described by associating to g the section

$$\tau = \sum_{|\nu|=e} \left(\sum_{j=1}^{N} \gamma_{\nu j} \frac{\partial}{\partial z_j} \right) \otimes t^\nu$$

of $\Theta \otimes (\mathfrak{m}^e/\mathfrak{m}^{e+1})$.

2. *The Kodaira-Spencer completeness theorem.* Suppose that $\mathfrak{X} \to B$ is a deformation of X represented by the cocycle $(g_{ij}) \in Z^1(\mathfrak{U}, \mathfrak{G})$. Neglecting all terms of order $\geqq 2$ in t, we get

$$g_{ij} = \mathrm{id} + \sum_{\mu=1}^{m} \gamma_{ij}^{(\mu)} t_\mu \mod \mathfrak{m}^2,$$

where $(\gamma_{ij}^{(1)}), \cdots, (\gamma_{ij}^{(m)})$ are cocycles in $Z^1(\mathfrak{U}, \Theta)$. (We committed some abuse of notation.)

The Kodaira-Spencer *completeness theorem* asserts:

$\mathfrak{X} \to B$ is a complete (resp. versal) deformation of X iff the cohomology classes of $(\gamma_{ij}^{(\mu)})$, $\mu = 1, \cdots, m$, form a system of generators (resp. a basis) of $H^1(X, \Theta)$.

The theorem has been proved by Kodaira-Spencer **[17]** by power series methods (without potential theory). There exist also proofs by Schuster **[25]** and Wavrik **[28]** using the Artin theorem on formal and convergent solutions of analytic equations **[2]**.

From now on, let $m := \dim H^1(X, \Theta)$. (This dimension is finite by the Cartan-Serre finiteness theorem.)

Choose a basis of $H^1(X, \Theta)$ and represent it by cocycles

$$(\gamma_{ij}^{(\mu)}) \in Z^1(\mathfrak{U},\Theta), \qquad \mu = 1, \cdots, m.$$

Let

$$g^1 := \mathrm{id} + \sum_{\mu=1}^{m} \gamma^{(\mu)} t_\mu \in Z^1(\mathfrak{U}, \mathfrak{G}/\mathfrak{G}_2).$$

In order to construct a versal deformation of X it suffices (by the completeness theorem) to construct a cocycle $g \in Z^1(\mathfrak{U}, \mathfrak{G})$ such that $g \equiv g^1 \bmod \mathfrak{m}^2$.

3. *Obstructions to extensions.* We want to construct the cocycle $g \in Z^1(\mathfrak{U}, \mathfrak{G})$ as the limit of a sequence of cocycles $g^e \in Z^1(\mathfrak{U}, \mathfrak{G}/\mathfrak{G}_{e+1})$, $e = 1, 2, 3, \cdots$, such that $g^{e+1} \equiv g^e \bmod \mathfrak{m}^{e+1}$. The cocycle g^e may be viewed as an element of $C^1(\mathfrak{U}, \mathfrak{G})$ which is a polynomial of degree $\leqq e$ with respect to the variables t satisfying the cocycle relations mod $\mathfrak{m}^{e+1}$. The limit $g = \lim_{e\to\infty} g^e$ will exist as a convergent power series if there is a constant $\sigma \in \boldsymbol{R}_+^*$ such that

$$\|g^e\|_{\mathfrak{U}} = \sup_{i,j} \|g_{ij}^e\|_{U_i \cap U_j} \leqq \sigma \quad \text{for all } e \geqq 1.$$

Suppose we have already constructed g^e. Then we have $g_{ij}^e g_{jk}^e = g_{ik}^e \bmod \mathfrak{m}^{e+1}$. If this congruence would hold also mod $\mathfrak{m}^{e+2}$, we could define $g^{e+1} = g^e$. However this is not true in general. Define

$$(*) \qquad \xi_{ijk} = g_{ij}^e g_{jk}^e (g_{ik}^e)^{-1} \bmod \mathfrak{m}^{e+2}.$$

Then $\xi_{ijk} \equiv \mathrm{id} \bmod \mathfrak{m}^{e+1}$; hence ξ_{ijk} is a section of $\mathfrak{G}_{e+1}/\mathfrak{G}_{e+2} \cong \Theta \otimes (\mathfrak{m}^{e+1}/\mathfrak{m}^{e+2})$ over $U_i \cap U_j \cap U_k$. It turns out that the family (ξ_{ijk}) is a cocycle in

$$Z^2(\mathfrak{U}, \mathfrak{G}_{e+1}/\mathfrak{G}_{e+2}).$$

This is the *obstruction cocycle*. We can write.

$$\xi_{ijk} = \mathrm{id} + \sum_{|\nu|=e+1} x_{ijk}^{(\nu)} t^\nu,$$

where $(x_{ijk}^{(\nu)}) \in Z^1(\mathfrak{U}, \Theta)$. If the obstruction cocycle is a coboundary, then there exist cochains $(y_{ij}^{(\nu)}) \in C^1(\mathfrak{U}, \Theta)$ such that

$$(**) \qquad x_{ijk}^{(\nu)} = y_{ij}^{(\nu)} + y_{jk}^{(\nu)} - y_{ik}^{(\nu)}.$$

Now set

$$g_{ij}^{e+1} := g_{ij}^e - \sum_{|\nu|=e+1} y_{ij}^{(\nu)} t^\nu.$$

Then

$$g_{ij}^{e+1} g_{jk}^{e+1} \equiv g_{ik}^{e+1} \bmod \mathfrak{m}^{e+2}.$$

This follows from $(*)$ and $(**)$ together with the following Proposition.

(III.3.1) PROPOSITION. *Let U be an open subset of $\boldsymbol{C}^N$ and $f, g \in \mathfrak{G}(U)$. Suppose $g = \mathrm{id} + \psi$, where $\psi \equiv 0 \bmod \mathfrak{m}^{e+1}$. Then*

$$f \circ g \equiv g \circ f \equiv f + \psi \bmod \mathfrak{m}^{e+2}.$$

This is easily proved by Taylor expansion.

From the above discussion it follows in particular: If we suppose $H^2(X, \Theta) = 0$, then the extension process is always possible and we get at least a formal solution to our problem.

Let us consider the problem of estimates: Let $\tilde{\mathfrak{U}} \ll \mathfrak{U}' \ll \mathfrak{U}$ be shrinkings of $\mathfrak{U}$. If we suppose $\|g^e\|_{\mathfrak{U}} \leqq \sigma$ and σ is sufficiently small, then it follows from the Excess Lemma (I.4.3) that the obstruction cocycle satisfies $\|\xi\|_{\mathfrak{U}'} \leqq K\sigma^2$, where K is a constant depending only on $\mathfrak{U}'$ and $\mathfrak{U}$, but not on g^e. By the Banach open mapping theorem applied to

$$C^1(\mathfrak{U}', \Theta) \xrightarrow{\delta} C^2(\mathfrak{U}', \Theta)$$

there exists a constant K' such that

$$\Big\| \sum_{|\nu|=e+1} {}_i y_j^{(\nu)} t^\nu \Big\|_{\tilde{\mathfrak{U}}} \leqq K' \|\xi\|_{\mathfrak{U}'} \leqq K'K\sigma^2.$$

For $\sigma \leqq \varepsilon_0 := 1/K'K$ we have $K'K\sigma^2 \leqq \sigma$; therefore $\|g^{e+1}\|_{\tilde{\mathfrak{U}}} \leqq \sigma$.

If this estimate would hold with respect to the covering $\mathfrak{U}$ instead of $\tilde{\mathfrak{U}}$, we could iterate the procedure and the construction would be finished. Unfortunately this is not the case. However this difficulty can be overcome by applying the smoothing theorem.

4. *The Kodaira-Nirenberg-Spencer theorem.*

(III.4.1) THEOREM. *Let X be a compact complex manifold with $H^2(X, \Theta) = 0$. Then there exists a versal deformation $\mathfrak{X} \to B$ of X, where B is the germ of $\mathbf{C}^m$ at the origin, $m =$ dim $H^1(X, \Theta)$.*

The proof will proceed along the lines developed in §§III.2 and III.3 and will use the smoothing theorem.

We choose a chain

$$\mathfrak{W} \ll \mathfrak{V} \ll \mathfrak{V}' \ll \tilde{\mathfrak{U}} \ll \mathfrak{U} \ll \mathfrak{U}^*$$

of finite coverings of X by open rectangles. By induction on e, we will construct the following objects:

(a) $g^e \in Z^1(\mathfrak{V}, \mathfrak{G}/\mathfrak{G}_{e+1})$,
(b) $G^e \in Z^1(\mathfrak{U}, \mathfrak{G}/\mathfrak{G}_{e+1})$,
(c) $h^e \in C^0(\mathfrak{W}, \mathfrak{G}/\mathfrak{G}_{e+1})$,
(d) $H^e \in C^0(\mathfrak{V}', \mathfrak{G}/\mathfrak{G}_{e+1})$,
(e) $\tilde{g}^{e+1} \in Z^1(\tilde{\mathfrak{U}}, \mathfrak{G}/\mathfrak{G}_{e+2})$.

These objects have to satisfy the following conditions:

(A_1) $g^{e+1} \equiv g^e \bmod \mathfrak{m}^{e+1}$,
(A_2) $h^{e+1} \equiv h^e \bmod \mathfrak{m}^{e+1}$,
(A_3) $G^{e+1} \equiv G^e \bmod \mathfrak{m}^{e+1}$,
(A_4) $\tilde{g}^{e+1} \equiv G^e \bmod \mathfrak{m}^{e+1}$,
(A_5) $H^{e+1} \equiv H^e \bmod \mathfrak{m}^{e+1}$,
(A_6) $H^e \mid \mathfrak{W} \equiv h^e \bmod \mathfrak{m}^{e+1}$.
(B_1) $G^e_{ij} \equiv H^e_i g^e_{ij} (H^e_j)^{-1} \bmod \mathfrak{m}^{e+1}$ on $V_i \cap V_j$,
(B_2) $g^{e+1}_{ij} \equiv (H^e_i)^{-1} \tilde{g}^{e+1}_{ij} H^e_j \bmod \mathfrak{m}^{e+2}$ on $V_i \cap V_j$.
(C_1) $\|g^e\|_{\mathfrak{V}} \leqq \sigma$,

(C_2) $\max(\|h^e\|_{\mathfrak{W}}, \|G^e\|_{\mathfrak{U}}) \leqq K_1\sigma$,
(C_3) $\|\tilde{g}^{e+1} - G^e\|_{\tilde{\mathfrak{U}}} \leqq K_2\sigma^2$,
(C_4) $\|H^e\|_{\mathfrak{V}'} \leqq K_3\sigma$.

The construction is done according to the following scheme:

	$\mathfrak{W}$	$\mathfrak{V}$	$\mathfrak{V}'$	$\tilde{\mathfrak{U}}$	$\mathfrak{U}$
			H^{e-1}		
				$\tilde{g}^e$	
Induction		g^e			
Step 1	h^e				G^e
Step 2			H^e		
Step 3				$\tilde{g}^{e+1}$	
Step 4		g^{e+1}			
	h^{e+1}				G^{e+1}

We first choose $\tilde{g}^1 \in Z^1(\tilde{\mathfrak{U}}, \mathfrak{G}/\mathfrak{G}_2)$ satisfying the conditions of the completeness theorem. We may assume that $\|\tilde{g}^1\|_{\tilde{\mathfrak{U}}}$ is arbitrarily small. Set $g^1 := \tilde{g}^1|\mathfrak{V}$, $h^0 := \text{id}$, $H^0 := \text{id}$, $G^0 := \text{id}$.

Induction. Step 1. (G^e, h^e) is obtained from g^e by the smoothing theorem. Because the differential of the smoothing map is $C\langle t\rangle$-linear and $g^e \equiv g^{e-1} \bmod \mathfrak{m}^e$ by the induction hypothesis, it follows that

$$G^e \equiv G^{e-1} \bmod \mathfrak{m}^e \quad \text{and} \quad h^e \equiv h^{e-1} \bmod \mathfrak{m}^e.$$

Step 2 (*smoothing of the 2nd kind*). We have by induction hypothesis

$$g^e_{ij} \equiv (H^{e-1}_i)^{-1}\tilde{g}^e_{ij}H^{e-1}_j \bmod \mathfrak{m}^{e+1} \quad \text{on } \mathfrak{V},$$

and

$$g^e_{ij} \equiv (h^e_i)^{-1}G^e_{ij}h^e_j \bmod \mathfrak{m}^{e+1} \quad \text{on } \mathfrak{W};$$

hence

$$G^e_{ij} \equiv (h^e_i(H^{e-1}_i)^{-1})\tilde{g}^e_{ij}(H^{e-1}_j(h^e_j)^{-1}) \bmod \mathfrak{m}^{e+1} \quad \text{on } \mathfrak{W}.$$

By (A_5) and (A_6) we can write

$$h^e_i \equiv H^{e-1}_i + \eta_i \bmod \mathfrak{m}^{e+1} \quad \text{on } W_i,$$

where $\eta_i \equiv 0 \bmod \mathfrak{m}^e$. Then, from Proposition III.3.1, follows

$$H^{e-1}_i(h^e_i)^{-1} \equiv \text{id} - \eta_i \bmod \mathfrak{m}^{e+1} \quad \text{on } W_i$$

and

$$G^e_{ij} - \tilde{g}^e_{ij} \equiv \eta_i - \eta_j \bmod \mathfrak{m}^{e+1} \quad \text{on } W_i.$$

By (A_3) and (A_4)

$$G^e_{ij} - \tilde{g}^e_{ij} =: \xi_{ij} \equiv 0 \bmod \mathfrak{m}^e \quad \text{on } \tilde{U}_i \cap \tilde{U}_j;$$

hence (ξ_{ij}) is a cocycle in $Z^1(\tilde{\mathfrak{U}}, \Theta) \otimes (\mathfrak{m}^e/\mathfrak{m}^{e+1})$. Now we apply the following

(III.4.2) LEMMA. *Let* $(\xi_{ij}) \in Z^1(\tilde{\mathfrak{U}}, \Theta)$ *and* $(\eta_i) \in C^0(\mathfrak{W}, \Theta)$ *satisfy*

$$\xi_{ij} = \eta_i - \eta_j \quad on\ W_i \cap W_j.$$

Then there exists a constant $C > 0$ *(depending only on* $\mathfrak{W} \ll \mathfrak{V}' \ll \tilde{\mathfrak{U}}$*) and* $(\eta'_i) \in C^0(\mathfrak{V}', \Theta)$ *with the following properties*:

(i) $(\eta'_i)|\mathfrak{W} = (\eta_i)$,

(ii) $\xi_{ij} = \eta'_i - \eta'_j$ *on* $V'_i \cap V'_j$,

(iii) $\|\eta'\|_{\mathfrak{V}'} \leqq C \max(\|\xi\|_{\tilde{\mathfrak{U}}}, \|\eta\|_{\mathfrak{W}})$.

In our situation, we get $G^e_{ij} - \tilde{g}^e_{ij} \equiv \eta'_i - \eta'_j \mod \mathfrak{m}^{e+1}$ on $\mathfrak{V}'$. Setting $H^e_i := H^{e-1}_i + \eta'_i$ on V'_i we have

$$g^e_{ij} \equiv (H^e_i)^{-1} G^e_{ij} H^e_j \mod \mathfrak{m}^{e+1} \quad \text{on } \mathfrak{V}.$$

Step 3. $\tilde{g}^{e+1}_{ij}$ is constructed from G^e_{ij} by killing the obstruction cocycle as in §III. 3.

Step 4. Define

$$g^{e+1}_{ij} := (H^e_i)^{-1} \tilde{g}^{e+1}_{ij} H^e_j \mod \mathfrak{m}^{e+2} \quad \text{on } \mathfrak{V}.$$

The estimate is obtained by (C_3), Proposition III.3.1 and the Excess Lemma I.4.3.

This concludes the proof. For more details, cf. **[10]**.

In the case where $H^2(X, \Theta) \neq 0$ one has to use in addition Grauert's division theory **[13]**. Roughly speaking this theory gives canonical representatives $f \in C\{t_1, \cdots, t_m\}$ with estimates for elements $\tilde{f} \in C\{t_1, \cdots, t_m\}/\mathfrak{a}$. See also **[9]**, where this theory has been used to construct versal deformations for vector bundles on compact complex spaces.

IV. Theorem B for deformations of singularities. A small deformation of a manifold is again a manifold. This means that locally a deformation of a manifold is trivial, i.e., a product with the parameter space. This is no longer true for deformations of complex spaces with singularities; in general, singularities admit nontrivial deformations.

In this chapter we describe the analogue of Theorem B (II.2.1) for the deformation of singularities.

1. *Representation of deformations by matrices.* A complex space is locally isomorphic to an analytic subspace X of an open rectangle $D \subset C^N$. We may suppose (after a little shrinking) that there is a free resolution

$$(1) \qquad 0 \longrightarrow \mathcal{O}^{l_N}_{C^N} \xrightarrow{P^0_N} \cdots \longrightarrow \mathcal{O}^{l_1}_{C^N} \xrightarrow{P^0_1} \mathcal{O}_{C^N} \longrightarrow \mathcal{O}_X \longrightarrow 0$$

of the structure sheaf $\mathcal{O}_X$ over D. The morphisms $P^0_n: \mathcal{O}^{l_n}_{C^N} \to \mathcal{O}^{l_{n-1}}_{C^N}$ are given by holomorphic $l_{n-1} \times l_n$ matrices on D.

Now let (S, s_0) be the germ of a complex space and $\pi: Y \to S$ a flat deformation of X over S. Then Y may be embedded as a subspace $Y \subset D \times S$ such that $X \times \{s_0\} = Y \cap (D \times \{s_0\})$ and π is induced by the projection $D \times S \to S$. There exists a resolution of $\mathcal{O}_Y$

$$(2) \qquad 0 \longrightarrow \mathcal{O}^{l_N}_{C^N \times S} \xrightarrow{P_N} \cdots \longrightarrow \mathcal{O}^{l_1}_{C^N \times S} \xrightarrow{P_1} \mathcal{O}_{C^N \times S} \longrightarrow \mathcal{O}_Y \longrightarrow 0$$

which is an extension of (1), i.e., $P_n(0) = P^0_n$ for $n = 1, \cdots, N$. (Here $P_n(0)$ means $P_n | D \times \{s_0\}$.)

Conversely, if we have a complex

$$(3) \qquad 0 \longrightarrow \mathcal{O}^{l_N}_{C^N \times S} \xrightarrow{P_N} \cdots \longrightarrow \mathcal{O}^{l_1}_{C^N \times S} \xrightarrow{P_1} \mathcal{O}_{C^N \times S}$$

over $D \times S$ with $P_n(0) = P_n^0$, $1 \leqq n \leqq N$, then the sequence (3) is exact and $\mathcal{O}_Y = \mathcal{O}_{C^N \times S}/\text{Im } P_1$ is the structure sheaf of a flat deformation of X. Thus every deformation of X is given by a sequence of matrices $(P_N, P_{N-1}, \cdots, P_1)$ with $P_n(0) = P_n^0$ and $P_n P_{n+1} = 0$. (From now on, deformation always means flat deformation.)

However the correspondence of deformations to sequences $(P_N, \cdots, P_1)$ of matrices is not one-to-one. The same deformation $Y \subset D \times S$ of X may admit different resolutions $(P_N, \cdots, P_1)$, $(P'_N, \cdots, P'_1)$. Then there exists a complex isomorphism

$$\begin{array}{ccccccccc} 0 \longrightarrow & \mathcal{O}^{l_N}_{C^N \times S} & \xrightarrow{P_N} & \mathcal{O}^{l_{N-1}}_{C^N \times S} & \longrightarrow \cdots \longrightarrow & \mathcal{O}^{l_1}_{C^N \times S} & \xrightarrow{P_1} & \mathcal{O}_{C^N \times S} \\ & T_N \downarrow & & \downarrow T_{N-1} & & \downarrow T_1 & & \| \\ 0 \longrightarrow & \mathcal{O}^{l_N}_{C^N \times S} & \xrightarrow{P'_N} & \mathcal{O}^{l_{N-1}}_{C^N \times S} & \longrightarrow \cdots \longrightarrow & \mathcal{O}^{l_1}_{C^N \times S} & \xrightarrow{P'_1} & \mathcal{O}_{C^N \times S} \end{array}$$

where T_n is an $l_n \times l_n$ matrix with $T_n(0) = 1$.

Another ambiguity arises because a deformation of X over S may be embedded in two different ways as subspaces $Y \subset D \times S$ and $Y' \subset D \times S$. In this case there exists a relative automorphism $g : D \times S \to D \times S$ that maps Y onto Y'. Let $P = (P_N, \cdots, P_1)$ and $P' = (P'_N, \cdots, P'_1)$ be resolutions of $\mathcal{O}_Y$ and $\mathcal{O}_{Y'}$ respectively. Then the pull-back $P'(g)$ is another resolution of $\mathcal{O}_Y$. Therefore there exists a complex isomorphism $T : P \to P'(g)$, i.e.,

$$T_{n-1} P_n = P'_n(g) T_n \quad \text{for } 1 \leqq n \leqq N$$

$(T_0 = 1)$. We abbreviate this by $T[-1]P = P'(g)T$.

2. *Heftungslemma.* Let B be the germ of C^m at 0 and let $X \subset D$ be as in §IV.1. With respect to the resolution (1) we define a sheaf $\mathfrak{P}$ on D as follows: For U open in D, the set $\mathfrak{P}(U)$ consists of all sequences $(P_N, \cdots, P_1)$, where P_n is a $\mathcal{O}_{C^N \times B}$-module morphism

$$P_n : \mathcal{O}^{l_n}_{C^N \times B} \,|\, U \to \mathcal{O}^{l_{n-1}}_{C^N \times B} \,|\, U$$

with $P_n(0) = P_n^0$, $1 \leqq n \leqq N$.

If $S \subset B$ is a subgerm, we say that $P = (P_N, \cdots, P_1) \in \mathfrak{P}(U)$ is a *complex over* S, if $P_n P_{n+1} | U \times S = 0$, $1 \leqq n < N$. As explained in §IV.1, such a complex $P \in \mathfrak{P}(U)$ over S defines a deformation of $X \cap U$ over S, which we denote by $(P)_S$. Because the coefficients of the matrices P_n are elements of $\Gamma(U, \mathcal{O}_{C^N \times B})$, we can introduce a pseudonorm $\| \ \|_U$ in $\mathfrak{P}(U)$ as in §I.3. The set of elements of finite norm is denoted by $\Gamma_b(U, \mathfrak{P})$. This is an affine space over a Banach space $T\Gamma_b(U, \mathfrak{P})$, which consists of all sequences $(P_N, \cdots, P_1)$ of matrices of the appropriate size satisfying $P_n(0) = 0$ and having finite norm. For $\varepsilon > 0$, $\Gamma_\varepsilon(U, \mathfrak{P})$ denotes the open ball of radius ε in $\Gamma_b(U, \mathfrak{P})$ with center $P^0 = (P_N^0, \cdots, P_1^0)$.

Now consider the following situation: Let (U_1, U_2) be a Cousin pair of open rectangles in D and let $P_i \in \mathfrak{P}(U_i)$ be complexes over $S \subset B$, $i = 1, 2$. Then P_i defines a deformation Y_i of $X \cap U_i$ over S. Suppose further given a relative automorphism $g \in \mathfrak{G}(U_1 \cap U_2)$ which maps $Y_1 | U_1 \cap U_2$ onto $Y_2 | U_1 \cap U_2$. Glueing together Y_1 and Y_2 by means of g, we obtain an abstract deformation of $X \cap$

$(U_1 \cup U_2)$ over S. We want to embed this deformation in $(U_1 \cup U_2) \times S$ and represent it by a complex $\Pi \in \mathfrak{P}(U_1 \cup U_2)$ over S. If we allow a little shrinking, this can be done with estimates simultaneously for all subgerms $S \subset B$. This is the content of the following Heftungslemma.

(IV.2) HEFTUNGSLEMMA. *Let $\mathfrak{U} = (U_1, U_2)$ and $\mathfrak{V} = (V_1, V_2)$ be Cousin pairs of open rectangles in D such that $V_i \subset\subset U_i$. Then there exist $\varepsilon > 0$ and a differentiable map*

$$\alpha: C^0_\varepsilon(\mathfrak{U}, \mathfrak{P}) \times \Gamma_\varepsilon(U_1 \cap U_2, \mathfrak{G}) \to \Gamma_\delta(V_1 \cup V_2, \mathfrak{P}) \times C^0_\delta(\mathfrak{V}, \mathfrak{G})$$

with the following properties:

Let $(P_1, P_2; g) \in C^0_\varepsilon(\mathfrak{U}, \mathfrak{P}) \times \Gamma_\varepsilon(U_1 \cap U_2, \mathfrak{G})$ and let $S \subset B$ be a subgerm such that P_1 and P_2 are complexes over S and g induces an isomorphism $g: (P_1)_S \to (P_2)_S$ on $U_1 \cap U_2$. Set $\alpha(P_1, P_2; g) =: (\Pi; h_1, h_2)$. Then Π is a complex over S and h_i induces an isomorphism $h_i: (P_i)_S \to (\Pi)_S$ on V_i. Moreover the following diagram is commutative on $V_1 \cap V_2$:

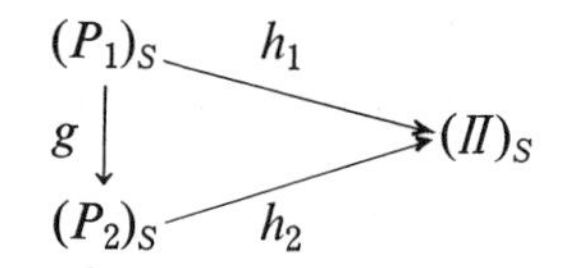

IDEA OF PROOF. By the Heftungslemma for relative automorphisms (II.1.1) we can write $g = h_2^{-1} \circ h_1$. From $(P_1)_S = (P_2(g))_S$ it follows that $(P_1(h_1^{-1}))_S = (P_2(h_2^{-1}))_S$. Therefore there exists a sequence of matrices $T = (T_N, \cdots, T_1)$ on the intersection such that

$$T[-1]P_1(h_1^{-1}) = P_2(h_2^{-1})T \quad \text{over } S.$$

Now by a Heftungslemma for invertible matrices we find $T_i = (T_{iN}, \cdots, T_{i1})$ on V_i with $T = T_2^{-1}T_1$. Then we have on $V_1 \cap V_2$

$$T_1[-1]P_1(h_1^{-1})T_1^{-1} = T_2[-1]P_2(h_2^{-1})T_2^{-1} \quad \text{over } S.$$

Therefore we can define $\Pi \in \mathfrak{P}(V_1 \cup V_2)$ which coincides on V_i with $T_i[-1] \cdot P_i(h_i^{-1})\, T_i^{-1}$ over S. Then $(\Pi; h_1, h_2)$ has the desired properties.

3. *Theorem* B. Let $\mathfrak{U} = (U_i)$ be a finite family of Stein open sets in D. We define

$$C^1(\mathfrak{U}; \mathfrak{P}, \mathfrak{G}) := C^0(\mathfrak{U}, \mathfrak{P}) \times C^1(\mathfrak{U}, \mathfrak{G}).$$

Let $S \subset B$ be a subgerm. An element $(P, g) = (P_i, g_{ij}) \in C^1(\mathfrak{U}; \mathfrak{P}, \mathfrak{G})$ is called a *cocycle over* S, if the following is true:

(i) Every $P_i \in \mathfrak{P}(U_i)$ is a complex over S and defines therefore a deformation $(P_i)_S$ of $X \cap U_i$ over S.

(ii) Every $g_{ij} \in \mathfrak{G}(U_i \cap U_j)$ defines an isomorphism $g_{ij}: (P_j)_S \to (P_i)_S$.

(iii) On every triple intersection we have a commutative diagram

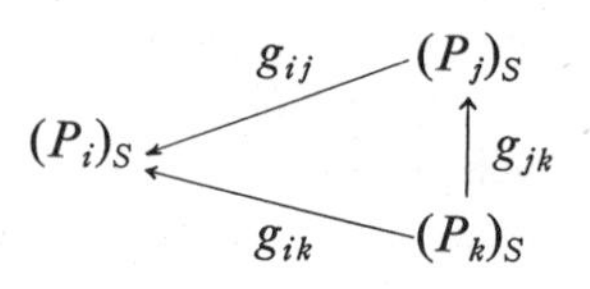

A cocycle (P, g) over S defines a deformation of $X \cap (\bigcup U_i)$ which is obtained by glueing together the deformations $(P_i)_S$ using the isomorphisms g_{ij}. We denote this deformation by $(P, g)_S$.

Again we want to embed this deformation and represent it by a complex $\Pi \in \mathfrak{P}(\bigcup U_i)$. This is done by Theorem B, which is a generalization of (II.2.1).

(IV.3) THEOREM B. *Let $D_1 \Subset D_2 \Subset D$ be open rectangles and*

$$\mathfrak{B} = (V_i)_{i\in I} \ll \mathfrak{U} = (U_i)_{i\in I}$$

finite families of Stein open sets with

$$D_1 = \bigcup_{i\in I} V_i \quad \text{and} \quad D_2 = \bigcup_{i\in I} U_i.$$

Then there exist $\varepsilon > 0$ and a differentiable map

$$\beta\colon C^1_\varepsilon(\mathfrak{U}; \mathfrak{P}, \mathfrak{G}) \longrightarrow \Gamma_b(D_1, \mathfrak{P}) \times C^0_b(\mathfrak{B}, \mathfrak{G})$$

with the following properties:

Let $S \subset B$ be a subgerm and $(P_i, g_{ij}) \in C^1_\varepsilon(\mathfrak{U}; \mathfrak{P}, \mathfrak{G})$ a cocycle over S. Set $\beta(P_i, g_{ij}) =: (\Pi, h_i)$. Then Π is a complex over S and h_i induces an isomorphism h_i: $(P_i)_S \to (\Pi)_S$ on V_i. Moreover the following diagrams are commutative on $V_i \cap V_j$:

$$\begin{array}{ccc} (P_i)_S & \xrightarrow{h_i} & \\ g_{ij}\uparrow & & (\Pi)_S \\ (P_j)_S & \xrightarrow{h_j} & \end{array}$$

Theorem B is proved by successive application of the Heftungslemma.

Using this Theorem B, one can prove a generalization of the Smoothing Theorem II.3.1 for deformations of a compact complex space.

V. Extension of deformations. Whereas in the construction of the versal deformation of a compact complex manifold we had to handle only one obstruction cocycle in $Z^2(\mathfrak{U}, \Theta)$ for the extension by one order, things are more complicated for deformations of complex spaces with singularities. Here we meet already locally obstructions to the extension. This will be described now.

1. *The extension complex.* We consider an analytic subspace X of an open rectangle $D \subset \mathbf{C}^N$ and a fixed free resolution

$$(1) \qquad \cdots \longrightarrow \mathcal{O}^{l_2}_{\mathbf{C}^N} \xrightarrow{P^0_2} \mathcal{O}^{l_1}_{\mathbf{C}^N} \xrightarrow{P^0_1} \mathcal{O}_{\mathbf{C}^N} \longrightarrow \mathcal{O}_X \longrightarrow 0$$

of $\mathcal{O}_X$ (cf. IV.1). We compose P^0_1 with the derivation $d\colon \mathcal{O}_{\mathbf{C}^N} \to \Omega_{\mathbf{C}^N}$ and get

$$(2) \qquad \cdots\ \mathcal{O}^{l_2}_{\mathbf{C}^N} \xrightarrow{P^0_2} \mathcal{O}^{l_1}_{\mathbf{C}^N} \xrightarrow{d\circ P^0_1} \Omega_{\mathbf{C}^N}.$$

The *extension complex* $\mathcal{E}^\cdot$ of X with respect to the resolution (1) is obtained by applying the functor $\mathcal{H}om(-, \mathcal{O}_X)$ to the sequence (2). Because $\mathcal{H}om(\mathcal{O}^l_{\mathbf{C}^N}, \mathcal{O}_X) \cong \mathcal{O}^l_X$, complex $\mathcal{E}^\cdot$ can be written as

$$\mathcal{E}^\cdot\colon \mathcal{O}^N_{\mathbf{C}^N} \xrightarrow{\partial P^0_1/\partial z} \mathcal{O}^{l_1}_X \xrightarrow{P^0_2} \mathcal{O}^{l_2}_X \xrightarrow{P^0_3} \cdots.$$

Here $\partial P^0_1/\partial z$ is the functional matrix of P^0_1 and $\partial P^0_1/\partial z$, P^0_2, P^0_3, etc., operate by multiplication on the right. We define sheaves $\mathscr{E}x^i$ on X by $\mathscr{E}x^i := H^i(\mathscr{E}^\cdot)$. It is easy to see that

$$\mathscr{E}x^0 = \operatorname{Ker}\,(\mathcal{O}_X^N \xrightarrow{\partial P_1^0/\partial z} \mathcal{O}_X^{l_1})$$

is canonically isomorphic to the sheaf Θ_X of holomorphic vector fields on X.

2. *The sheaf* $\mathscr{E}x^1$. In this paragraph we show that $\mathscr{E}x^1$ may be interpreted as the sheaf of isomorphism classes of deformations of X over the double point $S_1 = (0, \boldsymbol{C}[\varepsilon])$, $\varepsilon^2 = 0$.

Let U be a Stein open subset of D and Y a deformation of $X \cap U$ over S_1. We associate to this deformation an element $\xi \in \mathscr{E}x^1(U)$ in the following way: Y can be embedded in $U \times S_1$ and represented by a complex $P = (P_N, \cdots, P_1)$ over S_1. We can write

$$P_1 = P_1^0 + \varepsilon p, \qquad P_2 = P_2^0 + \varepsilon q.$$

We have

$$0 = P_1 P_2 = P_1^0 P_2^0 + \varepsilon(P_1^0 q + p P_2^0);$$

hence $P_1^0 q + p P_2^0 = 0$, which implies $pP_2^0 = 0$ on X because $\mathcal{O}_X = \mathcal{O}_{\boldsymbol{C}^N}/\operatorname{Im} P_1^0$. Therefore

$$p \in \operatorname{Ker}(\mathcal{O}_X^{l_1} \xrightarrow{P_2^0} \mathcal{O}_X^{l_2}) = Z^1(\mathscr{E}^\cdot).$$

We associate to Y the class of p in $\mathscr{E}x^1$. It is easy to see that p is independent of the choice of the resolution P of $\mathcal{O}_Y$.

If $Y' \subset U \times S_1$ is a deformation of $X \cap U$ over S_1 isomorphic to Y, there exists a relative automorphism

$$g = \mathrm{id} + \varepsilon\gamma : U \times S_1 \to U \times S_1$$

which maps Y' onto Y. Then $P' := P(g)$ is a resolution of $\mathcal{O}_{Y'}$. In particular we have

$$\begin{aligned} P_1' &= P_1^0 + \varepsilon p' = (P_1^0 + \varepsilon p) \circ (\mathrm{id} + \varepsilon\gamma) \\ &= P_1^0 + \varepsilon p + \varepsilon\gamma\partial P_1^0/\partial z, \end{aligned}$$

hence $p' = p + \gamma\partial P_1^0/\partial z$. Therefore p' and p differ only by a coboundary $\in B^1(\mathscr{E}^\cdot)$ and define the same element of $\mathscr{E}x^1$. Conversely it is easy to see that every element of $\mathscr{E}x^1(U)$ arises in this way from a deformation of $X \cap U$ over S_1.

3. *Obstructions to extensions.* Let $S \subset S'$ be germs of analytic spaces with local rings $\mathcal{O}_S$ and $\mathcal{O}_{S'}$. The inclusion $S \subset S'$ corresponds to an epimorphism $\mathcal{O}_{S'} \to \mathcal{O}_S$. We call $S \subset S'$ a *small extension* if the kernel $L := \operatorname{Ker}(\mathcal{O}_{S'} \to \mathcal{O}_S)$ is annihilated by the maximal ideal $\mathfrak{m}_{S'}$ of $\mathcal{O}_{S'}$.

Suppose we are given a deformation Y of X over S. The problem is to extend the deformation Y to a deformation Y' of X over S'. We will show that the deformation Y gives rise to a certain obstruction $\xi \in \mathscr{E}x^2 \otimes L$, which vanishes if and only if the deformation can be extended over S'.

For simplicity we consider only the special case

$$S = (0, \boldsymbol{C}[t]/(t^k)), \qquad S' = (0, \boldsymbol{C}[t]/(t^{k+1})).$$

In this case L is a one-dimensional vector space with basis t^k.

The deformation Y can be embedded in $D \times S$ and represented by a resolution $P = (P_N, \cdots, P_1)$ of $\mathcal{O}_Y$. The coefficients of the matrices P_n are polynomials in t of degree $\leqq k - 1$. We have $P_1P_2 = 0 \bmod (t^k)$; hence

$$P_1P_2 = \xi t^k \bmod (t^{k+1}).$$

ASSERTION. (a) $\xi \in Z^2(\mathcal{E}^\cdot)$; *therefore* ξ *defines an element of* $\mathcal{E}x^2$.

(b) $\xi \in B^2(\mathcal{E}^\cdot)$ *if and only if the deformation* Y *can be extended over* S'.

PROOF. (a) We have to show $\xi P_3^0 = 0$ on X. Since $P_2P_3 = 0 \bmod (t^k)$, we have $P_2P_3 = \eta t^k \bmod (t^{k+1})$; hence

$$P_1P_2P_3 = P_1\eta t^k = P_1^0\eta t^k \bmod (t^{k+1}).$$

On the other hand

$$P_1P_2P_3 = \xi t^k P_3 = (\xi P_3^0)t^k \bmod (t^{k+1}).$$

Therefore $\xi P_3^0 = P_1^0\eta \bmod (t^{k+1})$, which vanishes on X.

(b) Suppose $\xi \in B^2(\mathcal{E}^\cdot)$. Then there exists $\pi \in \mathcal{O}_X^{l_1}$ such that $\pi P_2^0 = \xi$ on X. We represent π by an element of $\mathcal{O}_{C^N}^{l_1}$ which we denote by the same letter. Then we have, on D, $\pi P_2^0 = \xi - P_1^0\tau$ with a suitably chosen τ. We set

$$P_1' := P_1 - \pi t^k, \qquad P_2' := P_2 - \tau t^k.$$

Then

$$P_1'P_2' = P_1P_2 - (\pi P_2^0 + P_1^0\tau)t^k = 0 \bmod (t^{k+1}).$$

Therefore defining $\mathcal{O}_{Y'} := \mathcal{O}_{C^N \times S'}/\mathrm{Im}\, P_1'$ we get a flat deformation Y' of X over S' which extends Y.

The converse implication is shown by similar calculations.

4. *Global obstructions.* We keep the notations of V.3.

Let us now consider the problem of uniqueness: Suppose that the deformation Y of X over S can be extended in two ways to deformations Y', Y'' over S', given by complexes P' and P'' with $P' = P'' \bmod (t^k)$. We define ζ by $P_1'' - P_1' = \zeta t^k$.

ASSERTION. $\zeta \in Z^1(\mathcal{E}^\cdot)$; *therefore* ζ *defines an element of* $\mathcal{E}x^1$.

PROOF. We have to show $\zeta P_2^0 = 0$ on X.

Since $P' = P'' \bmod (t^k)$, we can write $P_2'' = P_2' + \tau t^k$.

From $P_1''P_2'' = P_1'P_2' = 0 \bmod (t^{k+1})$ follows

$$0 = (P_1' + \zeta t^k)(P_2' + \tau t^k) = (P_1^0\tau + \zeta P_2^0)t^k \bmod (t^{k+1}).$$

Since $P_1^0\tau$ vanishes on X, we have $\zeta P_2^0 = 0$ on X. Q.E.D.

This gives rise to a global obstruction: Let now X be a compact complex space and $\mathfrak{U} = (U_i)$ a Stein covering of X such that every U_i can be embedded in an open rectangle $D_i \subset C^N$. Suppose that we have a deformation Y of X over S and that the deformations $Y|U_i$ have been extended to deformations Y_i over S'. Then comparing $Y_i|U_i \cap U_j$ and $Y_j|U_i \cap U_j$, we get a cocycle $(\zeta_{ij}) \in Z^1(\mathfrak{U}, \mathcal{E}x^1)$. If its cohomology class in $H^1(X, \mathcal{E}x^1)$ vanishes, we can change Y_i to deformations Y_i' such that for every pair of indices i, j the deformations $Y_i'|U_i \cap U_j$ and

$Y'_j \mid U_i \cap U_j$ are isomorphic. If we want to get a global extension Y' of Y to S' by patching together the deformations Y'_i we meet, as in the case of deformations of manifolds, a last obstruction in $H^2(X, \Theta_X)$.

Apart from several technical complications, the construction of a versal deformation for a compact complex space runs as in the case of manifolds (§III). Details will appear in [**11**].

BIBLIOGRAPHY

1. L. V. Ahlfors, *The complex analytic structure of the space of closed Riemann surfaces*, Analytic Functions, Princeton Univ. Press, Princeton, N. J., 1960. MR **23** #A1798.

2. M. Artin, *On the solutions of analytic equations*, Invent. Math. **5** (1968), 277–291. MR **38** #344.

3. L. Bers, *Uniformization and moduli*, Contributions to Function Theory (Internat. Colloq. Function Theory, Bombay, 1960), Tata Institute of Fundamental Research, Bombay, 1960, pp. 41–49. MR **24** #A1393.

4. M. Commichau, *Deformationen kompakter komplexer Mannigfaltigkeiten*, Math. Ann. **213** (1975), 43–96.

5. I. F. Donin, *Complete families of deformations of germs of complex spaces*, Mat. Sb. (N.S.) **89 (131)** (1972), 390–399, 533 = Math. USSR Sbornik **18** (1972), 397–406. MR **48** #11574.

6. A. Douady, *Obstruction primaire à la déformation*, Séminaire H. Cartan, 13ᵉ Année (1960/61), Exposé 4.

7. ———, *Le problème des modules pour les sous-espaces analytiques compacts d'un espace analytique donné*, Ann. Inst. Fourier (Grenoble) **16** (1966), fasc. 1, 1–95. MR **34** #2940.

8. ———, *Le problème des modules locaux pour les espaces C-analytiques compacts*, Ann. Sci. École Norm. Sup. Sér. IV **4** (1974), 569–602.

9. O. Forster and K. Knorr, *Über die Deformationen von Vektorraumbündeln auf kompakten komplexen Räumen*, Math. Ann. **209** (1974), 291–346.

10. ———, *Ein neuer Beweis des Satzes von Kodaira-Nirenberg-Spencer*, Math. Z. **139** (1974), 257–291. MR **50** #13607.

11. ———, *Konstruktion verseller Deformationen kompakter komplexer Räume* (to appear).

12. H. Grauert, *Ein Theorem der analytischen Garbentheorie und die Modulräume komplexer Strukturen*, Inst. Hautes Études Sci. Publ. Math. No. **5** (1960). MR **22** #12544.

13. ———, *Über die Deformation isolierter Singularitäten analytischer Mengen*, Invent. Math. **15** (1972), 171–198. MR **45** #2206.

14. ———, *Der Satz von Kuranishi für kompakte komplexe Räume*, Invent. Math. **25** (1974), 107–142. MR **49** #10920.

15. A. Grothendieck, *Construction de l'espace de Teichmüller*, Séminaire H. Cartan, 13ᵉ Année (1960/61), Exposé 17.

16. K. Kodaira and D. C. Spencer, *On deformations of complex analytic structures*. I, II, Ann. of Math. (2) **67** (1958), 328–466. MR **22** #3009.

17. ———, *A theorem of completeness for complex analytic fibre spaces*, Acta Math. **100** (1958), 181–294. MR **22** #3010.

18. K. Kodaira, L. Nirenberg and D. C. Spencer, *On the existence of deformations of complex analytic structures*, Ann. of Math. (2) **68** (1958), 450–459. MR **22** #3012.

19. M. Kuranishi, *On the locally complete families of complex analytic structures*, Ann. of Math. (2) **75** (1962), 536–577. MR **25** #4550.

20. S. Lichtenbaum and M. Schlessinger, *The cotangent complex of a morphism*, Trans. Amer. Math. Soc. **128** (1967), 41–70. MR **35** #237.

21. G. Pourcin, *Déformation de singularités isolées*, Astérisque, Soc. Math. France **16** (1974), 161–173.

22. H. E. Rauch, *Variational methods in the problem of the moduli of Riemann surfaces*, Contributions to Function Theory (Internat. Colleq. Function Theory, Bombay, 1960), Tata Institute of Fundamental Research, Bombay, 1960, pp. 17–40. MR **24** #A2020.

23. B. Riemann, *Theorie der abelschen Funktionen*, J. Reine Angew. Math. **54** (1857), Collected Works, Dover Reprint, Dover, New York.

24. M. Schlessinger, *Functors of Artin rings*, Trans. Amer. Math. Soc. **130** (1968), 208–222. MR **36** #184.

25. H.W. Schuster, *Formale Deformationstheorien*, Habilitationsschrift, München, 1971.

26. O. Teichmüller, *Veränderliche Riemannsche Flächen*, Deutsche Math. **7** (1944), 344–359. MR **8**, 327.

27. G. N. Tjurina, *Locally semi-universal flat deformations of isolated singularities of complex spaces*, Izv. Akad. Nauk SSSR Ser. Mat. **33** (1969), 1026–1058 = Math. USSR Izv. **3** (1969), 967–1001. MR **40** #5903.

28. J. J. Wavrik, *A theorem of completeness for families of compact analytic spaces*, Trans. Amer. Math. Soc. **163** (1971), 147–155. MR **45** #3770.

WESTFÄLISCHE WILHELMS-UNIVERSITÄT MÜNSTER

SEMINAR SERIES

VALUE DISTRIBUTION THEORY

Proceedings of Symposia in Pure Mathematics
Volume 30, 1977

SEPARATE ANALYTICITY, SEPARATE NORMALITY, AND RADIAL NORMALITY FOR MAPPINGS

THEODORE J. BARTH

Let X_1, X_2, and Y be complex spaces. A map $f: X_1 \times X_2 \to Y$ is said to be separately analytic if $f(x_1, \cdot): X_2 \to Y$ and $f(\cdot, x_2): X_1 \to Y$ are analytic for every $x_1 \in X_1$ and $x_2 \in X_2$. Osgood **[15]** proved that such a map is analytic provided X_1 and X_2 are complex manifolds and Y is a bounded open subset of $\boldsymbol{C}$; Hartogs **[11]** removed the boundedness assumption, thus allowing $Y = \boldsymbol{C}$; Grauert and Remmert **[9]** improved this by allowing X_1 and X_2 to have singularities. The Hartogs theorem that a separately analytic function is analytic has evolved into the loose statement that "separate analyticity implies analyticity." This statement is false for maps: The map $f: \boldsymbol{C} \times \boldsymbol{C} \to \boldsymbol{C} \cup \{\infty\}$ (the Riemann sphere) defined by

$$\begin{aligned} f(z, w) &= (z + w)^2/(z - w) && \text{if } z \neq w, \\ &= \infty && \text{if } z = w \neq 0, \\ &= 0 && \text{if } z = w = 0, \end{aligned}$$

is separately analytic, but it is not continuous at (0, 0). The first part of this note gives some conditions on Y assuring that every separately analytic map into Y is analytic. Nishino **[14]** and Alexander **[1]** have proved analogs to the Hartogs theorem for separately normal and radially normal families of holomorphic functions. The second part of this note gives generalizations of these results to families of maps; it also contains examples showing some limits of possible generalization.

1. Separate analyticity. Using c-holomorphic maps (instead of holomorphic maps) will yield easier proofs and slightly more general results. Recall that a *c-holomorphic* (or *continuously weakly holomorphic*) *map* is a continuous map $g: X \to Y$ between (reduced, second countable) complex spaces such that the restriction of

AMS (MOS) subject classifications (1970). Primary 32A17, 32C15, 32H99; Secondary 32H20.

g to the set of simple points of X is holomorphic [**16**, (BB), p. 149]. A map $f : X_1 \times X_2 \to Y$ is said to be *separately c-holomorphic* if $f(x_1, \cdot) : X_2 \to Y$ and $f(\cdot, x_2) : X_1 \to Y$ are c-holomorphic for every $x_1 \in X_1$ and $x_2 \in X_2$. Note that f need not be continuous; if f is continuous and separately c-holomorphic, it follows from the classical Osgood theorem that f is c-holomorphic.

By a *complex covering* we mean a locally biholomorphic map $\pi : Z \to Y$ of complex spaces such that (Z, π) is a covering space of Y [**7**, Definition 3, p. 40]. In case Z is a Stein space, we say that *Y has a Stein covering*.

Let d_X denote the Kobayashi pseudodistance on the connected complex space X [**12**, pp. 97–98]. If d_X is an actual distance, X is said to be *hyperbolic*; in this case d_X induces the standard topology on X [**5**]. Every c-holomorphic map $f : X \to Y$ between connected complex spaces is distance decreasing in the sense that $d_X(x_1, x_2) \geqq d_Y(f(x_1), f(x_2))$, for all $x_1, x_2 \in X$.

LEMMA. *Let A be a nowhere dense analytic subset of the complex space X, and let $X_0 = X - A$. Assume that the complex space Y is hyperbolic or has a Stein covering. Let g, g_n $(n = 1, 2, \cdots)$ be c-holomorphic maps from X into Y. If $g_n | X_0 \to g|X_0$ $(n \to \infty)$, then $g_n \to g$ $(n \to \infty)$.*

This lemma is proved by considering successive cases. The hyperbolic case follows from the triangle inequality and the Arzela-Ascoli theorem. The c-holomorphic version of a result of Grauert and Remmert [**9**, Hilfssatz 4, pp. 292–294] reduces the case $Y = \boldsymbol{C}$ to the hyperbolic case. The case in which Y has a Stein covering then follows from the imbedding theorem for Stein spaces [**10**, Theorem 10, p. 224] and the local simple connectivity of X [**13**] using standard lifting arguments [**7**, Proposition 1, p. 50], [**4**, Lemma 1, p. 294].

THEOREM 1. *Let X_1, X_2, and Y be complex spaces. Assume that Y is hyperbolic or has a Stein covering. Then every separately c-holomorphic map from $X_1 \times X_2$ into Y is c-holomorphic.*

By the classical Osgood theorem, it suffices to prove that f is continuous. In case X_1 and X_2 are complex manifolds, the proof of this can be found in [**6**, Lemma 3, p. 183, and Theorem 5(1), p. 186]; a simple argument using the triangle inequality and [**5**] yields the hyperbolic case; the Stein covering case comes from the classical Osgood theorem, the classical Hartogs lemma, the imbedding theorem for Stein manifolds, and standard lifting arguments. Two easy applications of the above lemma reduce the general case to the nonsingular case.

Using the theorem of Grauert and Remmert [**9**, Satz 29, pp. 292–296] mentioned in the introduction, we see that the statement of Theorem 1 remains true if the word "c-holomorphic" is replaced with "holomorphic."

2. Separate and radial normality. A sequence of continuous maps into the complex space Y is said to *diverge compactly* if it converges to the map identically ω, where $Y \cup \{\omega\}$ is the one-point compactification of Y [**17**, p. 197]. A family $\mathscr{F}$ of c-holomorphic maps from the connected complex space X into the complex space Y is said to be *normal* if every sequence in $\mathscr{F}$ has a subsequence that either converges or diverges compactly [**8**, Definition 3, p. 114], [**17**, Definition 1, p. 197].

Let X_1 and X_2 be connected complex spaces. A family $\mathscr{F}$ of c-holomorphic maps from $X_1 \times X_2$ into a complex space Y is said to be separately normal if $\mathscr{F}(x_1, \cdot) = \{f(x_1, \cdot) | f \in \mathscr{F}\}$ and $\mathscr{F}(\cdot, x_2) = \{f(\cdot, x_2) | f \in \mathscr{F}\}$ are normal for every $x_1 \in X_1$ and $x_2 \in X_2$. Nishino [**14**, Théorème II, p. 264] has proved that such a family is normal provided X_1 and X_2 are complex manifolds and $Y = \boldsymbol{C}$.

The most obvious normal families analog to Theorem 1 fails. Take $\mathscr{F} = \{f_j | j = 1, 2, \cdots\}$ where $f_j : \boldsymbol{C} \times \boldsymbol{C} \to \boldsymbol{C}^2$ is defined by $f_j(z, w) = (j^2 z - j, j^2 w - j)$; then $\mathscr{F}$ is separately normal, but it is not normal. The following theorem generalizes Nishino's result. By a *Riemann surface* we mean a connected one-dimensional complex manifold.

THEOREM 2. *Let X_1 and X_2 be connected complex spaces. Assume that Y is either a Riemann surface or a hyperbolic complex space. Then every separately normal family of c-holomorphic maps from $X_1 \times X_2$ into Y is normal.*

The uniformization theorem breaks the proof into three distinct cases:

(1) Y is a hyperbolic complex space;

(2) there is a complex covering $\pi : \boldsymbol{C} \to Y$;

(3) Y is biholomorphically equivalent to the Riemann sphere.

The easy proof of case (1) given in [**6**, Theorem 5(2), p. 186] for nonsingular X_1 and X_2 extends with minor changes to the present situation. The proof of case (2) given in [**6**, Theorem 3, pp. 184–185] for nonsingular X_1 and X_2 uses the nonsingular version of Theorem 1 and a result of Alexander, Taylor, and Ullman [**2**, Theorem 4, pp. 339–340] about sequences of analytic sets; the extension of this to the present situation involves a careful look at the local branching of an analytic set [**16**, Chapters 2 and 3] and another application of the above-mentioned result of Alexander et al. For nonsingular X_1 and X_2, case (3) follows from a theorem of Nishino [**14**, Théorème IV, pp. 279–282] about the normality of a family of meromorphic functions by considering convergence of level sets [**6**, Theorem 2, pp. 181–182]; as was pointed out to me by Joseph A. Becker, an easy argument using a resolution of singularities reduces the present situation to the nonsingular version.

Let B denote the open unit ball centered at the origin in $\boldsymbol{C}^r$. A family $\mathscr{F}$ of holomorphic maps from B into the complex space Y is said to be *radially normal* if the restriction of $\mathscr{F}$ to each complex line through the origin is normal. Alexander [**1**, Theorem 6.2(a), p. 248] has proved that such a family is normal provided $Y = \boldsymbol{C}$.

The theory of radial normality seems to parallel that of separate normality. Again the analog to Theorem 1 fails. Take $r = 2$ and $\mathscr{F} = \{f_j | j = 1, 2, \cdots\}$ where $f_j : B \to \boldsymbol{C}^2$ is defined by $f_j(z, w) = (j^2 z - j, j^2 w^2 - j)$; then $\mathscr{F}$ is radially normal but it is not normal.

THEOREM 3. *Assume that Y is either a hyperbolic complex space or a Riemann surface not biholomorphically equivalent to the Riemann sphere. Then every radially normal family of holomorphic maps from B into Y is normal.*

Invoking the uniformization theorem, we have only cases (1) and (2) above. In case (1), the family is equicontinuous since it consists of distance decreasing maps; normality then follows from the Arzela-Ascoli theorem as in [**3**, Theorem 2, p. 430]. Case (2) follows from Alexander's results using a lifting argument [**4**, Lemma 1, p. 294]. It seems likely to me that the conclusion of Theorem 3 also holds if Y is the Riemann sphere.

REFERENCES

1. H. Alexander, *Volumes of images of varieties in projective space and in Grassmannians*, Trans. Amer. Math. Soc. **189** (1974), 237–249.

2. H. Alexander, B. A. Taylor and J. L. Ullman, *Areas of projections of analytic sets*, Invent. Math. **16** (1972), 335–341. MR **46** #2078.

3. T. J. Barth, *Taut and tight complex manifolds*, Proc. Amer. Math. Soc. **24** (1970), 429–431. MR **40** #5897.

4. ———, *Normality domains for families of holomorphic maps*, Math. Ann. **190** (1971), 293–297. MR **43** #3486.

5. ———, *The Kobayashi distance induces the standard topology*, Proc. Amer. Math. Soc. **35** (1972), 439–441. MR **46** #5668.

6. ———, *Families of holomorphic maps into Riemann surfaces*, Trans. Amer. Math. Soc. **207** (1975), 175–187.

7. C. Chevalley, *Theory of Lie groups*. I, Princeton Math. Ser., vol. 8, Princeton Univ. Press, Princeton, N. J., 1946. MR **7**, 412.

8. H. Grauert and H. Reckziegel, *Hermitesche Metriken und normale Familien holomorpher Abbildungen*, Math. Z. **89** (1965), 108–125. MR **33** #2827.

9. H. Grauert and R. Remmert, *Komplexe Räume*, Math. Ann. **136** (1958), 245–318. MR **21** #2063.

10. R. C. Gunning and H. Rossi, *Analytic functions of several complex variables*, Prentice-Hall, Englewood Cliffs, N. J., 1965. MR **31** #4927.

11. F. Hartogs, *Zur Theorie der analytischen Funktionen mehrerer unabhängiger Veränderlichen, insbesondere über die Darstellung derselben durch Reihen, welche nach Potenzen einer Veränderlichen fortschreiten*, Math. Ann. **62** (1906), 1–88.

12. S. Kobayashi, *Hyperbolic manifolds and holomorphic mappings*, Pure and Appl. Math., 2, Dekker, New York, 1970. MR **43** #3503.

13. B. O. Koopman and A. B. Brown, *On the covering of analytic loci by complexes*, Trans. Amer. Math. Soc. **34** (1932), 231–251.

14. T. Nishino, *Sur une propriété des familles de fonctions analytiques de deux variables complexes*, J. Math. Kyoto Univ. **4** (1965), 255–282. MR **31** #3632.

15. W. F. Osgood, *Note über analytische Funktionen mehrerer Veränderlichen*, Math. Ann. **52** (1899), 462–464.

16. H. Whitney, *Complex analytic varieties*, Addison-Wesley, Reading, Mass., 1972.

17. H. Wu, *Normal families of holomorphic mappings*, Acta Math. **119** (1967), 193–233. MR **37** #468.

UNIVERSITY OF CALIFORNIA, RIVERSIDE

Proceedings of Symposia in Pure Mathematics
Volume 30, 1977

A RESULT ON THE VALUE DISTRIBUTION OF HOLOMORPHIC MAPS $f: C^n \to C^n$

JAMES A. CARLSON

Let f be a holomorphic map from C^n to itself. We assume that f is open, so that a generic fiber $f^{-1}(a)$ is discrete. We measure the growth of f and of $f^{-1}(a)$ by the indicators

$n(a, r) =$ number of points in $f^{-1}(a) \cap \{z: \|z\| < r\}$,

$N(a, r) = \int_0^r n(a, t)\, dt/t$,

$M_f(r) = \max_{\|z\|<r} \|f\|$.

In one complex variable Jensen's theorem gives $N(a, r) \leqq \log M_f(r) + O(1)$. The elementary inequality

$$(\log \theta)\, n(a, r) \leqq \int_r^{\theta r} n(a, t) \frac{dt}{t} = N(\theta r)$$

implies

$$n(a, r) \leqq C_\theta \log M_f(\theta r) + O(1),$$

where $\theta > 1$ and $C_\theta = (\log \theta)^{-1}$.

In several complex variables it is impossible to estimate $N(a, r)$ by $M_f(r)$ for all $a \in C^n$. Indeed, given any $\varepsilon > 0$ and any positive increasing function $\chi(r)$, there is a holomorphic function $t: C^2 \to C^2$ such that $\log M_f(r) < r$ and $n(0, r) \geqq \chi(r) + O(1)$ (see **[4]**).

Let $\mathscr{B}_{\alpha,\delta}$ be the set of points $a \in C^n$ such that no estimate of the form $n(a, r) \leqq r^\alpha[C_\theta \log M_f(\theta r)]^\delta + O(1)$ holds. Here $\theta > 1$, and $C_\theta > 0$. The constant θ can be chosen arbitrarily, but $C_\theta \to \infty$ as $\theta \to 1$. We will call $\mathscr{B}_{\alpha,\delta}$ the singular set for the transcendental Bezout problem.

In **[2]** it was shown that, for any $\varepsilon > 0$, $\mathscr{B}_{1+\varepsilon,n} \subset C^n$ has Lebesgue measure zero. Observe that, for $f(z_1, \cdots, z_n) = (e^{z_1}, \cdots, e^{z_n})$, $n(a, r) \sim r^n$ for most a,

AMS (MOS) subject classifications (1970). Primary 32A10, 32A30; Secondary 30A70.

whereas $\log M_f(r) \sim r$. Thus, from the point of view of the magnitude of the constants α and δ, the above theorem in [2] is not too bad. What we would like to discuss here is the possibility of reducing the size of the singular set.

We first recall the notion of the ρ-capacity of a set in $\boldsymbol{R}^n$. Let ρ be any increasing function such that $\rho(0) = 0$. Given $E \subset \boldsymbol{R}^n$, the ρ-potential of μ is, by definition,

$$V_\mu(x) = \int \frac{1}{\rho(|x - y|)}\, d\mu(y).$$

If the ρ-potential is finite for some choice of μ, then we say that $C_\rho(E) = 0$: E has ρ-capacity zero. If $\rho = r^\alpha$, we call this the α-capacity. The usefulness of the capacity lies in the fact that sets of capacity zero are usually quite small in the measure-theoretic sense. To make this precise, we recall the definition of Hausdorff ρ-measure. Given $E \subset \boldsymbol{R}^n$, consider $\bigwedge_{\rho,\varepsilon}(E) = \inf \sum_{r_i<\varepsilon} \rho(r_i)$, where the infimum is taken over all covers $\bigcup B(x_i, r_i)$ by balls of radius r_i centered at points x_i. Then the Hausdorff ρ-measure is, by definition, $\bigwedge_\rho(E) = \lim_{\varepsilon\to 0} \bigwedge_{\rho,\varepsilon}(E)$. If $\rho = r^\alpha$, we write $\bigwedge_\rho = \bigwedge_\alpha$. If E is a k-dimensional manifold with boundary, then $\bigwedge_k(E)$ is the Euclidean area of E, up to a normalizing constant. Moreover,

$$(*) \qquad \begin{aligned} \textstyle\bigwedge_{k'}(E) &= 0 \quad \text{for } k' > k, \\ \textstyle\bigwedge_{k'}(E) &= \infty \quad \text{for } k' < k. \end{aligned}$$

Given any set E, there is a real number k such that $(*)$ holds. We call k the Hausdorff dimension of E. If $E \subset \boldsymbol{R}^n$, then $n - k$ is by definition the Hausdorff codimension of E. If E has Hausdorff codimension greater than one, then $\boldsymbol{R}^n - E$ is connected. The basic fact relating capacity and measure is the following (see [1]).

If $\int_0^1 d\sigma/\rho < \infty$, then $C_\rho(E) = 0$ implies $\bigwedge_\sigma(E) = 0$. In particular $C_\alpha(E) = 0$ implies $\bigwedge_{\alpha+\varepsilon}(E) = 0$ for all $\varepsilon > 0$.

To state our basic result, let us fix $\rho = r^{-(2n-2)} \log^+ 1/r$. Then we have

THEOREM A. *For any $\varepsilon > 0$, $\mathscr{B}_{1+\varepsilon,\, 2n-1}$ has ρ-capacity zero. Consequently its $(2n - 1 + \delta)$-Hausdorff measure is zero for all $\delta > 0$.*

For most purposes it suffices to say that the singular set $\mathscr{B}_{1+\varepsilon,\, 2n-1}$ has Hausdorff codimension two. Suppose now that the order functions of f (see [31]) satisfy

$$(**) \qquad \overline{\lim} \frac{\{T_{n-1}(r)\}^{1/(n-1)}}{\{T_n(r)\}^{1/n}} < \infty.$$

This is the case if f is balanced in the sense of Wu [6]. Then we have

THEOREM B. *If $(**)$ holds, then for all $\varepsilon > 0$, $\mathscr{B}_{1+\varepsilon,n}$ has ρ-capacity zero.*

The technique of proof is to integrate the First Main Theorem [5] relative to equilibrium measures on the sets $E_m = \{a | N(a, m) > (1 + g(m))T_n(\theta m)\}$, where g is an auxiliary function to be chosen later. $\mathscr{B}$ is defined to be $\bigcap_n \bigcup_{m\geqq n} E_m$, and one concludes that $\mathscr{B}$ has capacity zero if the series $\sum(1/g(l))t_{n-1}(l)/T_n(\theta l)$ converges. On the complement of $\mathscr{B}$ we have

$$N(a, r) = (1 + g(r + 1))T_n(r + 1) + O(1).$$

The results above now follow by making appropriate choices for $\theta \geqq 1$ and $g(r)$:

For Theorem A, $\theta = 1$ and $g(r) = r^{1+\varepsilon} t_{n-1}(r)$; for Theorem B, $\theta > 1$ and $g(r) = r^{1+\varepsilon}$.

The above results are not known to be sharp. The best example I know is the following: Given $\chi(r)$, an increasing positive function, and given $\mathscr{B}$, an arbitrary countable set of points in $\boldsymbol{C}^2$, there is a holomorphic map $f: \boldsymbol{C}^2 \to \boldsymbol{C}^2$ with the properties

(a) f has order zero,

(b) $n(a, r) \geqq \chi(r) + O(1)$,

for all $a \in \mathscr{B}$.

Bibliography

1. Lennart Carleson, *Selected problems on exceptional sets*, Van Nostrand, Princeton, N. J., 1967. MR **37** #1576.

2. James A. Carlson, *A remark on the transcendental Bézout problem*, Value Distribution Theory, Dekker, New York, 1974, pp. 133–143. MR **50** #5029.

3. James A. Carlson and Phillip A. Griffiths, *The order functions for entire holomorphic mappings*, Value Distribution Theory, Dekker, New York, 1974, pp. 225–248.

4. Maurizio Cornalba and Bernard Shiffman, *A counterexample to the "Transcendental Bézout Problem"*, Ann. of Math. (2) **96** (1972), 402–406. MR **47** #499.

5. Phillip A. Griffiths and James R. King, *Nevanlinna theory and holomorphic mappings between algebraic varieties*, Acta Math. **130** (1973), 145–220.

6. Hung-Hsi Wu, *Remarks on the first main theorem in equidistribution theory*. I, II, III, IV, J. Differential Geometry **2** (1968), 197–202, 369–384; ibid. **3** (1969), 83–94, 433–446. MR **43** #2247a, b, c, d.

University of Utah

Proceedings of Symposia in Pure Mathematics
Volume 30, 1977

OPERATOR THEORY AND COMPLEX GEOMETRY*

MICHAEL J. COWEN** AND RONALD G. DOUGLAS

We report on progress we have made recently on several questions in complex geometry, whose original motivation came from the study of bounded operators on complex Hilbert space. A basic problem in operator theory is obtaining useful sets of unitary invariants for operators. One approach to this problem, which can be subsumed under the name "spectral theory", involves associating "local operators" to parts of the spectrum of the operator under investigation. Although this theory has been successful in studying normal operators, finite rank operators, and others, a certain class of operators has proved intractable to this approach. The characteristic property of this class is that the operators in it possess an open set of eigenvalues. A simple example is provided by the backward shift operator U_+^* defined on l^2 such that

$$U_+^*(\alpha_0, \alpha_1, \cdots) = (\alpha_1, \alpha_2, \cdots).$$

Since $U_+^*(1, \lambda, \lambda^2, \cdots) = \lambda(1, \lambda, \lambda^2, \cdots)$ for $|\lambda| < 1$, it follows that the open unit disc lies in the spectrum of U_+^* and consists of eigenvalues. Now a very detailed and deep theory exists for U_+^*, but one which is peculiar to this operator. The problem is how does one study such operators in a systematic fashion.

Let us be a little more precise about the class of operators which we have in mind. For Ω a bounded domain in $\boldsymbol{C}$, let $\mathscr{B}(\Omega)$ denote the collection of bounded

AMS (*MOS*) *subject classifications* (1970). Primary 47A65, 53B35; Secondary 47A70, 47B20, 53A55.

Key words and phrases. Eigenvalue bundles, hermitian vector bundles, curves in Grassmannians.

*Research partially supported by SUNY-Research Foundation Faculty Fellowship and NSF grant MPS7507230 (first author) and NSF grant MP27408062A02 (second author).

We would like to thank our colleagues Mikhail Gromov and Jim Simons for discussing various aspects of the topic of this paper with us.

**Sloan Foundation Fellow.

operators T on a separable complex Hilbert space $\mathscr{H}$ such that

(1) $\Omega \subset$ spectrum of T,

(2) $(T - \omega)\mathscr{H} = \mathscr{H}$ for each ω in Ω,

(3) $\bigvee_{\omega \in \Omega} \ker(T - \omega) = \mathscr{H}$, and

(4) $\dim \ker(T - \omega) = n(T) < \infty$ for all ω in Ω.

We show that such operators can be studied using ideas and techniques from complex geometry.

Let $\mathrm{Gr}(n, \mathscr{H})$ be the complex Grassmann manifold of n-planes through the origin in $\mathscr{H}$ (where we allow $\mathscr{H}$ to be finite dimensional). Thus, $\mathrm{Gr}(n, \mathscr{H})$ is a finite or infinite dimensional complex manifold. Let $S(n, \mathscr{H})$ be the canonical n-plane bundle over $\mathrm{Gr}(n, \mathscr{H})$ with hermitian structure induced by $S(n, \mathscr{H}) \subset \mathrm{Gr}(n, \mathscr{H}) \times \mathscr{H}$; that is, if $\mathscr{V}$ is an n-dimensional subspace of $\mathscr{H}$, and x is the point in $\mathrm{Gr}(n, \mathscr{H})$ determined by $\mathscr{V}$, then the fibre of $S(n, \mathscr{H})$ at x is just $\mathscr{V}$, and the fibre is given the hermitian inner product induced on $\mathscr{V}$ from $\mathscr{H}$.

PROPOSITION 1. *For T in $\mathscr{B}(\Omega)$ the map $t: \Omega \to \mathrm{Gr}(n, \mathscr{H})$ defined by*

$$t(\omega) = \ker(T - \omega)$$

is holomorphic, and thus induces an n-dimensional holomorphic hermitian vector bundle E_T over Ω, by $E_T = t^*S(n, \mathscr{H})$.

That the map t defines a holomorphic bundle has been noted before **[S]**.

It is clear that E_T and any isometric invariants of E_T are unitary invariants for T and to study such invariants we have to know more about bundles which arise from holomorphic maps into $\mathrm{Gr}(n, \mathscr{H})$. Thus we are led to consider the following three questions:

Question I. Let E be a holomorphic hermitian vector bundle over the open domain $\Omega \subset \mathbf{C}$. When does there exist a holomorphic map $f: \Omega \to \mathrm{Gr}(n, \mathscr{H})$ such that $E \cong f^*S(n, \mathscr{H})$, where $\cong$ means holomorphically and isometrically equivalent?

Question II. Given holomorphic maps $f_1, f_2: \Omega \to \mathrm{Gr}(n, \mathscr{H})$, what conditions on $f_i^*S(n, \mathscr{H})$, $i = 1, 2$, guarantee that f_1 and f_2 are *congruent via a unitary transformation* of $\mathscr{H}$; that is, $f_2 = U(f_1)$, where U is unitary on $\mathscr{H}$?

Question III. Given holomorphic hermitian vector bundles E_1, E_2 over Ω, when is $E_1 \cong E_2$ locally?

REMARKS. (1) For a detailed description of holomorphic hermitian vector bundles, see **[B-C]**, **[G$_1$]**.

(2) A holomorphic bundle is locally trivial by definition (locally there is a holomorphic frame) but need not be locally trivial as a hermitian bundle (that is, need not have an *orthonormal* holomorphic frame).

(3) In Question I there is both a local and a global problem. In this paper we discuss only the local problem. Actually, the global problem does not occur in the operator-theoretic problems.

Let E be a holomorphic, n-dimensional vector bundle over Ω with hermitian inner product $(\ ,\ )$. Let $s = \{s_1, \cdots, s_n\}$ be a holomorphic frame, and set

$$N(s) = ((s_i, s_j))_{i,j=1}^n,$$

the matrix of inner products. Let $J(N(s))$ be defined by

$$J(N(s)) = \begin{pmatrix} N(s) & \frac{\partial N(s)}{\partial \bar{z}} & \cdots & \frac{\partial^k N(s)}{\partial \bar{z}^k} & \cdots \\ \frac{\partial N(s)}{\partial z} & \frac{\partial^2 N(s)}{\partial z \partial \bar{z}} & \cdots & & \\ \vdots & \vdots & & & \\ \frac{\partial^k N(s)}{\partial z^k} & & & & \\ \vdots & & & & \end{pmatrix}.$$

Then $J(N(s))$ is an infinite dimensional matrix which gives a hermitian linear form on the total jet bundle of E. Let $J_p(N(s))$ be the upper left $p \times p$ submatrix of $J(N(s))$.

THEOREM I. *In the notation of Question* I, $E \cong f^*S(n, \mathscr{H})$ *locally if and only if* $N(s)$ *is real analytic,* $J(N(s)) \geqq 0$, *the rank of* $J_p(N(s))$ *is constant, except at isolated points, for each* $p = 1, 2, \cdots$, *and the dimension of* $\mathscr{H}$ *equals* $\sup(\text{rank } J_p(N(s)))$.

REMARK. Further, we can characterise when $E \cong E_T$ for some operator T by showing that T, if it exists, acts as the shift operator on the Hilbert space determined by $J(N(s))$.

In the case $n = 1$, that is, for a projective space, Griffiths [**G**$_2$] gave another answer for Question I, equivalent to Theorem I, in terms of the second main theorem of value distribution theory [**W**], [**Ch**], [**C-G**]. This second main theorem has proved very useful in the study of holomorphic maps from Ω into projective space, and we generalize it as follows to $\mathrm{Gr}(n, \mathscr{H})$:

Let $f : \Omega \to \mathrm{Gr}(n, \mathscr{H})$ be holomorphic such that f is spanned by the holomorphic maps $f_1, \cdots, f_n : \Omega \to \mathscr{H}$. Define $\wedge_k^1, \cdots, \wedge_k^n$ for $k = 0, 1, \cdots$ by

$$\wedge_k^i = f_1 \wedge \cdots \wedge f_n \wedge \cdots \wedge f_1^{(k-1)} \wedge \cdots \wedge f_n^{(k-1)} \wedge f_i^{(k)}$$

and define N_k by

$$N_k = ((\wedge_k^i, \wedge_k^j))_{i,j=1}^n.$$

Then $N_k \geqq 0$, and for each k such that $N_k > 0$ except at isolated points, put

$$\Omega_k = i\,\bar{\partial}(\partial N_k \cdot N_k^{-1})/2\pi,$$

the *curvature* of N_k.

PROPOSITION 2. *The following identities hold*:

$$\Omega_0 = -N_1 N_0^{-1} (\det N_0)^{-1} \frac{i}{2\pi}\, dz d\bar{z},$$

and for $k \geqq 1$,

$$\Omega_k = \Big\{ -N_{k+1} N_k^{-1} \prod_{i=0}^{k} (\det N_i)^{\alpha_{k-i}} + (N_k N_{k-1}^{-1} - \mathrm{tr}(N_k N_{k-1}^{-1}) I) \prod_{i=0}^{k-1} (\det N_i)^{\alpha_{k-i-1}} \Big\} \frac{i}{2\pi}\, dz d\bar{z},$$

where $\alpha_0 = -1$ *and* $\alpha_j = n(1-n)^{j-1}$ *for* $j \geqq 1$.

Note: *when* $n = 1$, *this gives*

$$\Omega_k = \frac{|\wedge_{k+1}|^2|\wedge_{k-1}|^2}{|\wedge_k|^4} \frac{i}{2\pi} dzd\bar{z}.$$

REMARK. This gives a deeper necessary condition than Theorem I for $E \cong f^*S(n, \mathscr{H})$ locally; namely, let s be a holomorphic frame and put $N_0 = N(s)$. Define $N_1, \cdots, N_k, \cdots$ inductively so as to satisfy Proposition 2. Then a necessary condition is that $N_k \geqq 0$. This condition is more or less sufficient.

We now turn to Question II. We say that $f_1, f_2 : \Omega \to \mathrm{Gr}(n, \mathscr{H})$ *agree to order* k if and only if for each z in Ω there exists a unitary U on $\mathscr{H}$ such that the kth order jets of f_2 and Uf_1 are equal at z. This can be reformulated in terms of the operator T. For ω in Ω and $k = 1, 2, 3, \cdots$ we define the operators $T_k(\omega)$ and $N_k(\omega)$ by

$$T_k(\omega) = T|_{\ker(T-\omega)^k} = \omega + N_k(\omega).$$

Then $N_k(\omega)$ is a nilpotent operator of order k defined on a finite dimensional Hilbert space.

PROPOSITION 3. *If T, $\tilde{T}$ are two operators in $\mathscr{B}(\Omega)$, then t and $\tilde{t}$ agree to order k as maps of $\Omega \to \mathrm{Gr}(n, \mathscr{H})$ if and only if the operators $N_k(\omega)$ and $\tilde{N}_k(\omega)$ are unitarily equivalent for each ω in Ω.*

Using Proposition 2 it is easy to show the following:

PROPOSITION 4. *The maps f_1, f_2 agree to first order if and only if the eigenvalues of the curvature of $f_1^*S(n, \mathscr{H})$, $f_2^*S(n, \mathscr{H})$ are equal at each point in Ω.*

A theorem of Calabi **[Ca]**, **[G$_2$]** shows that first order agreement implies f_1 and f_2 are congruent when $n = 1$. For $n > 1$, Griffiths conjectured that first order agreement need not imply congruence. This is indeed the case; for example, let $\mathscr{H}$ be $\boldsymbol{C}^4$, $\Omega = \boldsymbol{C}$, $f_1(z)$ be spanned by $(1, 0, \mu_1 e^{\lambda_1 z}, \mu_2 e^{\lambda_2 z})$ and $(0, 1, e^{\lambda_1 z}, e^{\lambda_2 z})$ and $f_2(z)$ be spanned by $(1, 0, -e^{\lambda_2 z}, \mu_2 e^{\lambda_2 z})$ and $(0, 1, e^{\lambda_1 z}, -\mu_1 e^{\lambda_1 z})$. By Propositions 2 and 4, f_1 and f_2 are easily seen to agree to first order, yet are not congruent if 0, λ_1, λ_2, $2\lambda_1$, $2\lambda_2$, $2\lambda_1 + 2\lambda_2$ are distinct, 0, μ_1, μ_2 are distinct, and $\mu_1\bar{\mu}_2 \neq -1$, as can be seen directly.

It is not suprising that agreement to first order is insufficient for congruence, since congruence (for $n > 1$) requires more than equality of the eigenvalues of the curvature:

THEOREM II. *If $f_1, f_2 : \Omega \to \mathrm{Gr}(n, \mathscr{H})$ are holomorphic maps, then f_1 is congruent to f_2 via a unitary transformation of $\mathscr{H}$ if and only if*

$$f_1^*S(n, \mathscr{H}) \cong f_2^*S(n, \mathscr{H}).$$

This has an immediate consequence for operators in $\mathscr{B}(\Omega)$.

COROLLARY 5. *If T_1 and T_2 are operators in $\mathscr{B}(\Omega)$, then T_1 and T_2 are unitarily equivalent if and only if $E_{T_1} \cong E_{T_2}$.*

REMARKS. (1) Since curvature is the only invariant of a holomorphic hermitian line bundle, when $n = 1$ we obtain Calabi's theorem.

(2) The backward shift operator U_+^* can also be represented as the adjoint of multiplication by z on the Hardy space of the unit disc $\boldsymbol{D}$. An analogous operator B^* is obtained as the adjoint of multiplication by z on the Bergman space of $\boldsymbol{D}$.

Both operators are in $\mathscr{B}(\boldsymbol{D})$ and define holomorphic hermitian line bundles over $\boldsymbol{D}$. The curvature of these bundles can be calculated: for U_+^* we have $-\frac{1}{2}\cdot -((i/2\pi)\, d\omega d\bar{\omega}/(1-|\omega|^2)^2)$, while B^* yields $-2\cdot((i/2\pi)d\omega d\bar{\omega}/(1-|\omega|^2)^2)$. There are, of course, easier ways to show U_+^* and B^* are not unitarily equivalent.

(3) Question II is thus a special case of Question III, since if f_1 and f_2 are congruent on any open subset of Ω they are congruent. The bundles in Question II, of course, have real analytic hermitian structures, but real analyticity is not of great help in dealing with Question III, so we only assume C^∞ metrics for Question III.

In order to analyze Question III, we first show:

LEMMA 6. *If E_1, E_2 are holomorphic hermitian vector bundles over $\Omega \subset \boldsymbol{C}$ and $\Phi_1, \Phi_2 : E_1 \to E_2$ holomorphic isometries, then*

(i) *$\Phi_1 \equiv \Phi_2$ if $\Phi_1(z) = \Phi_2(z)$ for some $z \in \Omega$, and*

(ii) *$\Phi_1 = \lambda\Phi_2$ for λ in $\boldsymbol{C}$, $|\lambda| = 1$, if E_1 and E_2 are irreducible.*

By Lemma 6, a holomorphic isometry from $E_1 \to E_2$, if one exists, is determined by its values at one point in Ω, and for irreducible bundles is essentially unique. We should therefore be able to identify the obstructions to constructing a holomorphic isometry. There are several approaches one can take, though the computations always appear to be the same. One possible framework for attacking Question III is the following:

Let E be a holomorphic hermitian vector bundle over $\Omega \subset \boldsymbol{C}$. The curvature is a 2-form valued section of $\mathrm{Hom}(E, E)$, which we will write as $(i/2\pi)\, K dz d\bar{z}$, where K is a section of $\mathrm{Hom}(E, E)$. Relative to a holomorphic frame s of E,

$$K(s) = -\frac{\partial}{\partial \bar{z}}\,(\partial N(s)/\partial z \cdot N(s)^{-1}).$$

Thus a necessary condition for $E_1 \cong E_2$ is that K_1 and K_2 have the same eigenvalues at each z in Ω. For now we make the further assumption that the eigenvalues of K_1, K_2 are distinct at each point, say $\lambda_1(z), \cdots, \lambda_n(z)$. (*Note*: If relative to a holomorphic frame s, $N(s)$ is real analytic (as in the case of $f^*S(n, \mathscr{H})$), then $K(s)$ is real analytic, so generically should have only isolated points where the eigenvalues are not distinct.) Let $E_i^1, \cdots, E_i^n$ be the eigenbundles of E_i, $i = 1, 2$, corresponding to the eigenvalues $\lambda_1(z), \cdots, \lambda_n(z)$ for K_i. Let s_i^j be an orthonormal frame for the line bundle E_i^j. Then $s_i = \{s_i^1, \cdots, s_i^n\}$ is an orthonormal frame for E_i, called a *reduced* frame. Each s_i^j is determined up to multiplication by $\exp\{\sqrt{-1}\,\psi_j\}$, ψ_j real.

LEMMA 7. *The bundles E_1 and E_2 are congruent if and only if there exist reduced frames s_1, s_2 such that $\theta_1(s_1) = \theta_2(s_2)$, where $\theta_i(s_i)$ are the connection* 1*-forms relative to the* (*canonical*) *connection D_i on E_i.*

Now we have very little choice for a reduced frame, so we can read off from Lemma 7 necessary and sufficient conditions. To translate them to invariant form is a little more difficult. We arrive at

THEOREM III. *If the eigenvalues of K_1, K_2 are equal and of constant multiplicity one, then $E_1 \cong E_2$ if and only if the following conditions hold*:

(1) $\mathrm{tr}((K_i)^j(\partial K_i/\partial \bar{z})(K_i)^k(\partial K_i/\partial \bar{z})^*)$ *are equal for* $i = 1, 2$ *for* $0 \leqq j \leqq n-1$,

$0 \leqq k \leqq n - 2$, *where* $n = \dim E_i$, K_i *is the matrix relative to a holomorphic frame and* $*$ *is relative to the hermitian inner product on* E_i;

(2) $\operatorname{tr}((K_i)^{j_1}(\partial K_i/\partial \bar z) \cdots (K_i)^{j_r}\partial K_i/\partial \bar z)$ *are equal for* $i = 1, 2$ *for all* $2 \leqq r \leqq n$, $0 \leqq j_k \leqq n - k$, $j_1 \geqq \cdots \geqq j_r$;

(3) $\operatorname{tr}((K_i)^j(\partial K_i/\partial \bar z)(K_i)^k(\partial K_i/\partial \bar z)^*(K_i)^l\partial K_i/\partial \bar z)$ *are equal for* $i = 1, 2$ *for all* $0 \leqq j \leqq n - 1, 0 \leqq k \leqq n - 2, 0 \leqq l \leqq n - 3$; *and*

(4) $\operatorname{tr}((K_i)^j(\partial K_i/\partial \bar z)(K_i)^k(\partial/\partial \bar z)\{(\partial K_i/\partial \bar z)^*\})$ *are equal for* $i = 1, 2$ *for all* $0 \leqq j \leqq n, 0 \leqq k \leqq n - 1$.

REMARK. Conditions (1)—(4) involve only K, $\partial K/\partial \bar z$, and

$$(\partial/\partial \bar z)(\partial K/\partial \bar z)^* = (\partial/\partial \bar z)\{\partial K/\partial z + [K, (\partial N/\partial z)\, N^{-1}]\}.$$

Thus we have:

COROLLARY 8. *Generic holomorphic maps* $f_1, f_2 : \Omega \to \mathrm{Gr}(n, \mathscr{H})$ *are congruent via a unitary transformation of* $\mathscr{H}$ *if and only if* f_1 *and* f_2 *agree to second order. (We say that* f_1 *and* f_2 *are generic if the eigenvalues of the curvatures have multiplicity one at some point in* Ω.) (*Cf.* [$\mathbf{G}_2$] *for the case* $n = 2$, $\mathscr{H} = \mathbf{C}^4$.)

Using Proposition 2 we can restate Corollary 8 in operator-theoretic terms.

COROLLARY 9. *Operators* T *and* $\tilde T$ *in* $\mathscr{B}(\Omega)$ *are unitarily equivalent if and only if* $N_2(\omega)$ *and* $\tilde N_2(\omega)$ *are unitarily equivalent for each* ω *in* Ω.

CONCLUDING REMARKS. (1) When $n = 2$, Theorem III is true even if the multiplicity of the eigenvalues is constant and equals 2; in fact, the proof is easier. Thus Corollaries 8 and 9 are true without exception when $n = 2$. Indeed, Theorem III suitably modified holds generally, even when the eigenvalues have (constant) multiplicity greater than one. If $E_i^1, \cdots, E_i^p$ (the eigenbundles) have rank > 1, we can reduce these bundles by the eigenbundles of the curvature induced on $E_i^1, \cdots, E_i^p$. Continuing in this fashion, we get a more general reduced frame and the proof still goes through.

(2) Although Corollary 9 may be valid in the generality stated, we can prove it now only in the cases indicated.

(3) We have thus far dealt with one operator and one dimensional Ω. These types of questions clearly generalize to finitely generated commutative algebras of operators where Ω lies in the joint point spectrum and, from the geometric viewpoint, to bundles with base space of dimension > 1.

BIBLIOGRAPHY

[B-C] R. Bott and S. S. Chern, *Hermitian vector bundles and the equidistribution of the zeroes of their holomorphic sections*, Acta Math. **114** (1965), 71–112. MR **32** #3070.

[Ca] E. Calabi, *Isometric imbedding of complex manifolds*, Ann. of Math. (2) **58** (1953), 1–23. MR **15**, 160.

[Ch] S. S. Chern, *Holomorphic curves in the plane*, Differential Geometry (In Honor of K. Yano), Kinokuniya, Tokyo, 1972, pp. 73–94.

[C-G] M. J. Cowen and P. A. Griffiths, *Holomorphic curves and metrics of negative curvature*, J. Analyse Math. **29** (1976), 93–153.

[G₁] P. A. Griffiths, *Hermitian differential geometry, Chern classes, and positive vector bundles*,

Global Analysis (Papers in Honor of K. Kodaira), Univ. Tokyo Press, Tokyo, 1969, pp. 185–251. MR **41** #2717.

[G_2] P. A. Griffiths, *On Cartan's method of Lie groups and moving frames as applied to uniqueness and existence questions in differential geometry*, Duke Math. J. **41** (1974), 775–814.

[S] M. A. Šubin, *Factorization of parameter-dependent matrix functions in normed rings and certain related questions in the theory of Noetherian operators,* Mat. Sb. **73 (113)** (1967), 610–629 = Math. USSR Sb. **2** (1967), 543–560. MR **36** #727.

[W] H. Wu, *The equidistribution theory of holomorphic curves*, Ann. of Math. Studies, no. 64, Princeton Univ. Press, Princeton, N. J., Univ. of Tokyo Press, Tokyo, 1970. MR **42** #7951.

STATE UNIVERSITY OF NEW YORK AT STONY BROOK

Proceedings of Symposia in Pure Mathematics
Volume 30, 1977

A UNICITY THEOREM FOR EQUIDIMENSIONAL HOLOMORPHIC MAPS

S. J. DROUILHET

In 1926, R. Nevanlinna [5], [6] showed that if $f, g: \boldsymbol{C} \to \boldsymbol{P}_1(\boldsymbol{C})$ are nonconstant holomorphic maps and if there exist five points $a_i \in \boldsymbol{P}_1(\boldsymbol{C})$ such that $f^{-1}(a_i) = g^{-1}(a_i)$, then f and g are identical functions. His proof involved an application of his Second Main Theorem. Nevanlinna and H. Cartan [2], [3], [5], [6] subsequently obtained similar theorems showing that if $f^{-1}(a_i) = g^{-1}(a_i)$ for only three or four points a_i, and if wherever f and g assume the value a_i they assume it with the same multiplicity, then f and g are either identical or are related in a manner which can be precisely described. These results were obtained by an application of E. Borel's theorem on the zeros of linear combinations of nonvanishing holomorphic functions. In 1971, E.M. Schmid [7] studied unicity theorems in certain cases for holomorphic maps from open Riemann surfaces to compact Riemann surfaces. Using results developed by Carlson and Griffiths [1], we have generalized the first unicity theorem of Nevanlinna mentioned above to the case of holomorphic mappings from $\boldsymbol{C}^n$ to n-dimensional projective algebraic varieties. Using a generalization of Borel's theorem, H. Fujimoto [4] has obtained extensions of the classical unicity theorems for three and four values to the case of meromorphic mappings from $\boldsymbol{C}^m$ to $\boldsymbol{P}_n(\boldsymbol{C})$. J. Carlson has also studied unicity questions.

Our result is:

1. THEOREM. *Suppose that V is an n-dimensional smooth projective algebraic variety; $f, g: \boldsymbol{C}^n \to V$ are nondegenerate holomorphic maps; L is a positive holomorphic line bundle on V such that $c_1(L) - c_1(K_V^*) > 0$; and $D \in |L|$ is a hypersurface having smooth irreducible components with normal crossings. Further suppose that $f^{-1}(D) = g^{-1}(D)$ as point sets and $f | f^{-1}(D) = g | f^{-1}(D)$. Let $i: V \to \boldsymbol{P}_N(\boldsymbol{C})$ be any holomorphic map of V into some projective space, and let H denote the hyperplane bundle on $\boldsymbol{P}_N(\boldsymbol{C})$. If*

AMS (MOS) subject classifications (1970). Primary 32H25.

$$\left(1 - \left[\frac{c_1(K_V^*)}{c_1(L)}\right] - 2\left[\frac{c_1(i^*H)}{c_1(L)}\right]\right) > 0,$$

then $i \circ f = i \circ g$.

(Here K_V denotes the canonical bundle of V; if $\alpha, \beta \in H^{1,1}(V, R)$, then $[\alpha / \beta] = \inf\{k: k\alpha - \beta > 0\}$.)

The theorem can be used in some specific cases to give explicit conditions which are sufficient for the equality of f and g. The following is such a case.

2. COROLLARY. *Suppose* $f, g: C^n \to P_n(C)$ *are nondegenerate holomorphic maps. Let* D *be a hypersurface in* $P_n(C)$ *of degree at least* $n+4$ *with simple normal crossings. If* $f^{-1}(D) = g^{-1}(D)$ *as point sets, and if* $f | f^{-1}(D) = g | f^{-1}(D)$, *then* $f \equiv g$.

There are examples which show that the condition $f | f^{-1}(D) = g | f^{-1}(D)$ cannot be removed in the theorem or in the corollary. In the absence of this condition, it is unknown whether an explicit description can be given of all the ways in which f and g can be related.

BIBLIOGRAPHY

1. J.A. Carlson and P.A. Griffiths, *A defect relation for equidimensional mappings between algebraic varieties*, Ann. of Math. (2) **95** (1972), 557–584. MR **47** #497.

2. H. Cartan, *Sur quelques théorèmes de M.R. Nevanlinna*, C.R. Acad. Sci. Paris **185** (1927), 1253–1254.

3. ———, *Un nouveau théorème d'unicité relatif aux fonctions méromorphes*, C. R. Acad. Sci. Paris **188** (1929), 301–303.

4. H. Fujimoto, *The uniqueness problem of meromorphic maps into the complex projective space*, Nagoya Math. J. **58** (1975), 1–23.

5. R. Nevanlinna, *Einige eindeutigkeitsätze in der theorie der meromorphen funktionen*, Acta Math. **48** (1926), 367–391.

6. ———, *Le théorème de Picard-Borel et la théorie des fonctions méromorphes*, Gauthier-Villars, Paris, 1929.

7. E.M. Schmid, *Some theorems on value distributions of meromorphic functions*, Math. Z. **120** (1971), 61–92. MR **44** #1808.

UNIVERSITY OF UTAH

Proceedings of Symposia in Pure Mathematics
Volume 30, 1977

MEROMORPHIC MAPPINGS INTO COMPACT COMPLEX SPACES OF GENERAL TYPE*

PETER KIERNAN

1. Introduction. In [2], Griffiths studied holomorphic mappings into canonical algebraic manifolds. He showed that many of the results from the function theory of Riemann surfaces of genus $\geqq 2$ generalize to this case if the mappings are equidimensional and are not totally degenerate. Kobayashi and Ochiai [4], Kodaira [6] and Kwack [7] have obtained further generalizations for projective algebraic manifolds of general type. The proofs of all the theorems have depended upon two basic facts concerning a manifold M of general type:

(1) There exist sections $\phi_0, \cdots, \phi_N$ of a sufficiently high power K^m of the canonical bundle which determine a projective imbedding of M. This means that nondegenerate holomorphic mappings can be defined in terms of the pull-backs of $\phi_0, \cdots, \phi_N$.

(2) The pseudovolume form $\mu_M = (\sum_{j=0}^N \phi_j \bar{\phi}_j)^{1/m}$ has positive Ricci curvature. Thus the Ahlfors-Chern-Kobayashi lemma can be applied to get a volume decreasing theorem for mappings into M.

In this paper we use two simple observations to generalize the results of [2] to the case of meromorphic mappings into compact algebraic spaces of general type. The first observation, as pointed out by Yau in [8], is that the volume decreasing theorem is valid for meromorphic mappings into manifolds of general type. The second observation is that (1) and (2) imply a basic "lifting principle" which precisely explains the behavior of the mappings under consideration. This principle says that if f is an equidimensional meromorphic mapping from the unit polydisk P into M which is not totally degenerate, then there exists a holomorphic lift $\hat{f}$: $P \to C^{N+1}$ such that

AMS (*MOS*) *subject classifications* (1970). Primary 32C15, 32H25.

*This work was done partially under the program "Sonderforschungsbereich Theoretische Mathematik" at the University of Bonn and partially with the support of National Research Council grant A-8497.

(i) $\pi \circ \hat{f} = \phi \circ f$ where $\pi: C^{N+1} \to P_N(C)$ is the meromorphic projection and $\phi = [\phi_0, \cdots, \phi_N]$.

(ii) $|\hat{f}(z)|^2 \leqq (1/2^n n!) \prod_{j=1}^n (1/(1-|z_j|^2)^2)$.

Since (ii) gives a uniform estimate on the growth of $\hat{f}$, the lift behaves like a mapping into a bounded domain. This fact can be used to explain the function theory of M in the same way as the uniformization theorem explains the function theory of Riemann surfaces of genus $\geqq 2$. However, there are some subtleties which occur in the higher dimensional case due to the fact that π is not holomorphic at 0.

2. The lifting principle. Let M be an n-dimensional projective algebraic manifold of general type and let K denote its canonical bundle. Then for some $m > 0$, there exist sections $\phi_0, \cdots, \phi_N$ of K^m such that the mapping $\phi = [\phi_0, \cdots, \phi_N]$ of M into $P_N(C)$ is a holomorphic imbedding. We define a pseudovolume element μ_M on M by setting $\mu_M = (\lambda \sum_{j=0}^n \phi_j \bar{\phi}_j)^{1/m}$. Here λ is a constant chosen so that if $\phi_j = S_j (dw_1 \cdots dw_n)^m$ in a coordinate neighborhood W, then

$$\mu_M = \left(\sum_{j=0}^n |s_j|^2\right)^{1/m} i^n \, dw_1 \wedge d\bar{w}_1 \wedge \cdots \wedge d\bar{w}_n \wedge d\bar{w}_n \quad \text{in } W.$$

Now the Ricci form is given by

$$\operatorname{Ric} \mu_M = i\partial\bar{\partial} \log \left(\sum_{j=0}^N |s_j|^2\right)^{1/m} = \frac{2}{m} \phi^* \omega$$

where ω is the Kähler form associated to the Fubini-Study metric on $P_N(C)$. Since ϕ is an imbedding, Ric μ_M is a positive definite (1, 1)-form on M. Thus $(\operatorname{Ric} \mu_M)^n \geqq c\mu_M$ for some constant $c > 0$. Since multiplying all the sections $\phi_0, \cdots, \phi_N$ by the same constant changes μ_M but not Ric μ_M, we can assume that the ϕ_j were originally chosen so that $(\operatorname{Ric} \mu_M)^n \geqq \mu_M$. Note that the set of base points for ϕ, that is, the set $\{p \in M | \phi_0(p) = \cdots = \phi_N(p) = 0\}$ is, in general, a nonempty divisor on M. For further details concerning ϕ and the volume form μ_M see the addendum to [**4**].

Throughout this paper ϕ': $M' \to M$ will be a bimeromorphic mapping of a Moišezon space M' of general type onto a projective algebraic manifold M, where M and ϕ are as above. A point $p \in M'$ is called a regular point for the imbedding $\psi = \phi \circ \phi'$ if ϕ' is biholomorphic in a neighborhood of p and $\phi'(p)$ is not a base point for ϕ. Otherwise p is called a base point for ψ.

We shall use the following notation. Let $P(r) = \{z \in C^n \,|\, \max |z_j| < r\}$ and $P^*(r) = \{z \in P(r) \,|\, z_1 z_2 \cdots z_n \neq 0\}$. If X is an n-dimensional complex manifold, let $\operatorname{Mer}_n(X, M') = \{f: X \to M' | f$ is meromorphic and $f(X)$ contains an open set$\}$. If $\omega \in H^0(M, K^m)$ and $f \in \operatorname{Mer}_n(X, M)$, then $f^*\omega$ is holomorphic outside a set of codimension 2 and therefore, by Hartogs' theorem, $f^*\omega \in H^0(X, K_X^m)$.

THEOREM 1 (LIFTING PRINCIPLE). *Let $f \in \operatorname{Mer}_n(P(r), M')$. Define holomorphic functions f_j on $P(r)$ by $(\phi' \circ f)^* \phi_j = f_j (dz_1 \wedge \cdots \wedge dz_n)^m$ and let $\hat{f} = (f_0, \cdots, f_N)$. Then $\pi \circ \hat{f} = \psi \circ f$ as meromorphic mappings where $\pi : C^{N+1} \to P_N(C)$ is the projection. Furthermore*

$$|\hat{f}(z)|^2 = \sum_{j=0}^N |f_j(z)|^2 \leqq a^m \left(\prod_{j=0}^n \frac{r}{r^2 - |z_j|^2}\right)^{2m} \tag{1}$$

where $a = (2^n n!)^{-1}$.

PROOF. Let μ_r be the Poincaré volume form on $P(r)$. Since $(\text{Ric}\ \mu_M)^n \geqq \mu_M$, the Alhfors-Chern-Kobayashi lemma (see [3, p. 28]) implies that $f^*\mu_M \leqq \mu_r$. Now (1) follows immediately by comparing coefficients of these two volume forms. Q.E.D.

THEOREM 2. *Let X be an n-dimensional complex manifold and let $E = K_X^m \oplus \cdots \oplus K_X^m$ ($n + 1$ times). For each $f \in \text{Mer}_n(X, M')$ define $\hat{f} \in H^0(X, E)$ by $\hat{f} = ((\phi' \circ f)^* \phi_0, \cdots, (\phi' \circ f)^*\phi_N)$. Let $\rho : E \to \boldsymbol{P}_N(\boldsymbol{C})$ be defined by $\rho(v_0, \cdots, v_N) = [v_0, \cdots, v_N]$. Then $\rho \circ \hat{f} = \phi \circ f$ and there exists a hermitian metric $\langle\ ,\ \rangle_E$ on E such that*

$$\langle \hat{f}(p), \hat{f}(p)\rangle_E < 1 \tag{2}$$

for every $f \in \text{Mer}_n(X, M')$.

PROOF. This follows immediately from Theorem 1 by taking a suitable covering of X by polydisks. Q.E.D.

The last result gives a natural imbedding of $\text{Mer}_n(X, M')$ into $H^0(X, E)$. Since $H^0(X, E)$ has the topology of uniform convergence on compact sets, we get a topology on $\text{Mer}_n(X, M')$. If $\{f^k\} \subset \text{Mer}_n(X, M')$, then (2) implies that there exists a subsequence which converges to $f \in H^0(X, E)$. If f is not the zero section, then it is easy to see that $f \in \text{Mer}_n(X, M')$. This says that $\text{Mer}_n(X, M')$ is a normal family in the sense that if $\{f^k\}$ does not contain a convergent subsequence then $(\phi' \circ f^k)^*\mu_M \to 0$ uniformly on compact subsets of X. If X is compact, then

$$\int_X (\phi' \circ f^k)^* \mu_M = (\deg f^k) \int_M \mu_M$$

where $\deg f^k$ is a positive integer. Therefore $(\phi' \circ f^k)^*\mu_M$ cannot converge to 0 and $\text{Mer}_n(X, M')$ is compact. Kobayashi and Ochiai [5] have obtained this result independently and have used it to show that $\text{Mer}_n(X, M')$ is finite.

3. Applications. We point out that weaker versions of all the corollaries are contained in [2], [4], [6] or [7].

COROLLARY 1 (BIG PICARD THEOREM). *Let X be a complex space of pure dimension n and let A be a subvariety of codimension $\geqq 1$. Then every $f \in \text{Mer}_n(X - A, M')$ extends to an element f in* $\text{Mer}_n(X, M')$.

PROOF. By Hironaka's resolution of singularities we can assume that $X = P$ and $X - A = P^*$. Using the Poincaré metric on P^*, a slight modification of Theorem 1 gives a lift $\hat{f} : P^* \to \boldsymbol{C}^{N+1}$ such that

$$|\hat{f}(z)|^2 \leqq a^m \left(\prod_{j=1}^n \frac{1}{|z_j| \log|z_j|^2}\right)^{2m}.$$

Thus $h(z) = (z_1 z_2 \cdots z_n)^m \hat{f}(z)$ is holomorphic on P^* and locally bounded on P and therefore holomorphic on P. This implies that f is meromorphic since $f = \phi^{-1} \circ \pi \circ h$. (Note that since $\hat{f}$ is the canonically determined lift of f, $\hat{f}$ itself is holomorphic on P.) Q.E.D.

COROLLARY 2 (LITTLE PICARD THEOREM). *Let X and A be as above. If X is compact and the Kodaira dimension $\kappa(X) < n$, then $\text{Mer}_n(X - A, M')$ is empty. In particular,* $\text{Mer}_n(\boldsymbol{C}^n, M')$ *is empty.*

PROOF. By Corollary 1, it suffices to show that $\mathrm{Mer}_n(X, M')$ is empty. But this follows immediately from the definition of $\kappa(X)$. Q.E.D.

COROLLARY 3 (SCHOTTKY-LANDAU THEOREM). *Let*

$$\mathscr{F}_c(r) = \{f \in \mathrm{Mer}_n(P(r), M') \mid |\hat{f}(0)| \geqq c\}.$$

If $c > 0$ *and* $r > (a^{2m}/c)^{1/mn}$, *then* $\mathscr{F}_c(r)$ *is empty.*

This follows immediately from Theorem 1. To see that this is the standard Schottky-Landau theorem, let $p \in M'$ be a regular point for ψ and let $W = \{w \in C^n \mid |w| < 1\}$ be a coordinate neighborhood of p. Let $\mathscr{M}(r, p, W)$ be the set of meromorphic mappings of $P(r)$ into M' satisfying

(a) f is holomorphic at 0 and $f(0) = p$,

(b) $|Jf(0)| = |\det(\partial w_j/\partial z_k|_0)| \geqq 1$.

Since p is a regular point for ψ, $(w_1, \cdots, w_n)$ define coordinates near $\phi'(p)$. Let $\phi_j = S_j(w)\,(dw_1 \wedge \cdots \wedge dw_n)^m$ in these coordinates. If $f \in \mathscr{M}(r, p, W)$, then

$$|\hat{f}(0)|^2 = |Jf(0)|^{2m} \sum_{j=0}^{N} |S_j(\phi'(p))|^2 \geqq \sum_{j=0}^{N} |s_j(\phi'(p))|^2 > 0$$

where the last inequality follows from the fact that p is not a base point for ψ. Thus $r \leqq (a^{2m}/c)^{1/mn}$ where $c^2 = \sum_{j=0}^{N} |s_j(\phi'(p))|^2$. This is the usual form of the Schottky-Landau theorem. It should be pointed out that if $\phi'(p)$ is a base point, then $\sum_{j=0}^{N} |s_j(\phi'(p))|^2 = 0$ and $|\hat{f}(0)| = 0$ for every f with $f(0) = p$. In this case we do not get an upper bound for r. Finally, we point out that Corollary 3 has the advantage of giving a uniform bound for r in terms of a global normalization of the mapping.

Assume that p is not a base point for the imbedding ψ and let $\mathscr{M}(p, W) = \mathscr{M}(1, p, W)$. Since $\mathrm{Mer}_n(P, M')$ is a "normal family", it is easy to see that there exists $r > 0$ such that each $f \in \mathscr{M}(p, W)$ maps $P(r)$ holomorphically into W. Since W is a polydisk in $\boldsymbol{C}^n$, this reduces many questions about mappings in $\mathscr{M}(p, W)$ to the corresponding question for holomorphic mappings of polydisks into bounded domains. For example, this technique can be used to generalize Bloch's theorem.

COROLLARY 4 (BLOCH'S THEOREM). *Using the notation above, there exists a neighborhood* U *of* p *such that each* $f \in \mathscr{M}(p, W)$ *maps a neighborhood of* 0 *biholomorphically onto* U.

4. Examples. The spaces we have been studying can be listed in increasing order of generality as follows: canonical algebraic manifolds (i.e., K very ample), manifolds with negative first Chern class (i.e., K ample), projective algebraic manifolds of general type and Moišezon spaces of general type. As far as the function theory (for nondegenerate mappings) of these spaces is concerned, the only important distinction seems to be the existence or nonexistence of base points. This is true essentially because the projection $\pi\colon \boldsymbol{C}^{N+1} \to \boldsymbol{P}_N(\boldsymbol{C})$ is not holomorphic at 0 and therefore statements about the lifting $\hat{f}$ do not easily translate to statements about f at points where $\hat{f}$ vanishes.

Let $M_1 \subset \boldsymbol{P}_3(\boldsymbol{C})$ be the Fermat surface of degree 5, that is, $M_1 = \{[z_0, z_1, z_2, z_3] \mid z_0^5 + z_1^5 + z_2^5 + z_3^5 = 0\}$. M_1 is a canonical algebraic variety so it has the best behavior possible of the spaces we are considering. Let $S \subset M_1$ be the curve $S = \{(z_0, z_1, z_2, z_3) | z_1 = \omega z_0 \text{ and } z_3 = \omega z_2\}$ where $\omega^5 = -1$. Then S is a nonsingularly imbedded rational curve. Using the adjunction formula and the fact that the canonical divisor is given by a hyperplane section, it is easy to compute $S \cdot S = -3$. Thus the normal bundle of S is L^3 where L is the tautological line bundle over $\boldsymbol{P}_1(\boldsymbol{C})$. Let $g : L \to L^3$ be the natural mapping and let $\rho : \boldsymbol{C}^2 \to L$ be the meromorphic mapping which blows up the origin. By a theorem of Grauert there exists a neighborhood $U \subset L^3$ which is biholomorphic to a neighborhood of $S \subset M_1$. Let $V = g^{-1}(U)$ and without loss of generality assume $P \subset \rho^{-1}(V)$. Consider $h = g \circ \rho$ as a mapping into a neighborhood of S. Then $h : P^* \to M_1$ is a nondegenerate holomorphic mapping into M which does not extend holomorphically to P. This shows that even if K is very ample, the extension in Corollary 1 will only in general be meromorphic. Define $F^k : P \to P - \{0\}$ by $F^k(z, w) = (z, zw/2 + 1/2k)$, and let $f^k = h \circ F^k$. Then $\{f^k\}$ converges in $\mathrm{Mer}_n(P, M_1)$ to some f which is holomorphic in P. However, $\{f^k\}$ does not converge uniformly on compact subsets of P to f.

Let M_2 be obtained from a canonical algebraic surface by blowing up a point. There exists a point $p \in M_2$ which has a coordinate neighborhood $U = \{(z_1, z_2) | |z_1| < 1 \text{ and } |z_1 z_2| < 1\}$ with $p = (0, 0)$. Define $f^k : P \to U \subset M_2$ by $f^k(z_1, z_2) = (z_1/k, kz_2)$. Clearly, there does not exist a neighborhood V of p such that every f^k maps some neighborhood of 0 biholomorphically onto V. Since each f^k is in $\mathscr{M}(p, U)$, M does not satisfy Bloch's theorem at p.

Let M_3 be obtained from a canonical algebraic manifold of dimension 4 by blowing up a point. There exists a point $q \in M_3$ which has a coordinate neighborhood $U = \{(z_1, z_2, z_3, z_4) | |z_1| < 1, |z_1 z_2| < 1, |z_1 z_3| < 1 \text{ and } |z_1 z_4| < 1\}$ with $q = (0, 0, 0, 0)$. Define $g^k : P(k) \to U \subset M_3$ by $g^k(z_1, z_2, z_3, z_4) = (z_1/k^3, kz_2, kz_3, kz_4)$. Then g^k is in $\mathscr{M}(k, p, U)$ for each k and therefore M_3 does not satisfy the Schottky-Landau property at q. Another example of a manifold of general type which does not satisfy the Schottky-Landau property at some points is the surface M_4 obtained by blowing up $p \in M_2$. Finally, we note that it is easy to construct a sequence $\{f^k\}$ in $\mathrm{Hol}(P, M_4)$ such that $f^k \to f$ in $\mathrm{Mer}_2(P, M_4)$ and f is not holomorphic.

In this paper we have used elementary facts to show that nondegenerate mappings into a Moišezon space M behave very nicely. The examples indicate the only "bad" behavior occurs near subvarieties of M which are not of general type. In order to study arbitrary mappings into M it will be necessary to have a much deeper knowledge of the structure of M. I feel that these mappings also behave nicely. However, in order to prove this, it is probably necessary to solve the following extremely difficult problem.

Problem. Let M be a Moišezon space and let d_M denote the Kobayashi pseudodistance on M. Show that $d_M(x, y) > 0$ unless there exists a subvariety $A \subset M$ which is not of general type and which contains x and y. In particular, if every subvariety of M is of general type, show that M is hyperbolic.

References

1. D. Eisenman (D. A. Pelles), *Intrinsic measures on complex manifolds and holomorphic mappings*, Mem. Amer. Math. Soc. No. 96 (1970). MR **41** #3807.

2. P. A. Griffiths, *Holomorphic mapping into canonical algebraic varieties*, Ann. of Math (2) **93** (1971), 439–458. MR **43** #7668.

3. S. Kobayashi, *Hyperbolic manifolds and holomorphic mappings*, Pure and Appl. Math., 2, Dekker, New York, 1970. MR **43** #3503.

4. S. Kobayashi and T. Ochiai, *Mappings into compact complex manifolds with negative first Chern class*, J. Math. Soc. Japan **23** (1971), 137–148. MR **44** #5514.

5. ———, *Meromorphic mappings onto compact complex spaces of general type* (to appear).

6. K. Kodaira, *Holomorphic mappings of polydiscs into compact complex manifolds*, J. Differential Geometry **6** (1971/72), 33–46. MR **46** #386.

7. M. H. Kwack, *Holomorphic mappings into complex spaces* (to appear).

8. S.-T. Yau, *Intrinsic measures of compact complex manifolds* (to appear).

UNIVERSILY OF BRITISH COLUMBIA

Proceedings of Symposia in Pure Mathematics
Volume 30, 1977

REAL AND SEMIREAL ZEROS OF ENTIRE FUNCTIONS IN C^n

PIERRE LELONG

1. The indicatrices which occur in the value distribution theory for entire functions are plurisubharmonic (psh) functions. Therefore a part of the theory extends to a conic set γ with vertex at the origin in C^n, if we assume that γ is not C^n-polar nor C^n-negligible. In the space C^n of the coordinates $z = (z_k)$, it is a natural problem to take for γ the real subspace R^n of C^n, or the cone SR^n of the semireal points in C^n.

DEFINITION. *A point $z = (z_k)$ in C^n is called semireal if there exists $u \in C$, and $x = (x_k) \in R^n$ such that $z_k = ux_k$, $1 \leqq k \leqq n$.*

SR^n is the image of the real projective space $\mathrm{Pr}(R^n)$ in $\mathrm{Pr}(C^n)$ by the complexification $(x_k) \to (ux_k)$.

In the following $P(G)$ denotes the convex cone of the psh functions $f \not\equiv -\infty$ defined in a domain G. We put $f^*(z) = \limsup_{y \to z} f(y)$ (upper regularization of f). A set $A \subset G$ is called G-polar if it is contained in $[z \in G;\ V(z) = -\infty]$ for some $V \in P(G)$; A is called G-negligible if $A \subset [z \in G;\ W(z) < W^*(z)]$, $W = \sup_q V_q$ and $V_q \in P(G)$ is a locally upper bounded sequence.

Problem 1. *Does there exist a connection between the value distribution on SR^n and on C^n*

(i) *for an entire function F on C^n?*

(ii) *for a plurisubharmonic function f on C^n?*

Because $R^n \subset SR^n \subset C^n$, Problem 1 gives a way to get an upper bound on the real zeros of F.

Problem 2. *If a function $\phi(r)$ dominates $M(r) = \sup_{\|z\| \leqq r} \log|F(z)|$ and if $F(0) = 1$, does there exist a majorant $g(r)$ of the $n - 1$ real dimensional area of the set $[x \in R^n, F(x) = 0, \|x\| \leqq r]$?*

2. For the first problem we recall (see [**3a**, p. 537]):

AMS (MOS) subject classifications (1970). Primary 32A15, 32C25.

PROPOSITION 1. *The restriction of a plurisubharmonic function to* $\boldsymbol{R}^n$ *belongs to* $L^1_{\mathrm{loc}}(\boldsymbol{R}^n)$.

PROPOSITION 2. *In a domain* $G \subset \boldsymbol{C}^n$, *the sets* $\boldsymbol{R}^n \cap G$, *and* $S\boldsymbol{R}^n \cap G$ *are not* G*-polar nor* G*-negligible* (*see* [**3a**, *p.* 537] *and* [**3b**, *p.* 276]).

3. Method of the inverse function. For Problem 1, we use a method of inverse function as given in [**3c**, Chapter 6]. If $f(z, \xi) \in P(\Gamma)$, where Γ is the cylinder $\Gamma = [z \in d; \xi \in \boldsymbol{C}]$, and d is a domain of $\boldsymbol{C}^n$, we put $M(z, r) = \sup_\theta f(z, re^{i\theta})$, and define

$$\delta(z, m) = [\sup r,\ r > 0,\ M(z, r) < m]. \tag{2}$$

Then $-\log \delta(z, m)$ is a psh function in each domain $d_1 \subset\subset d$ for $m > \sup_{z \in d_1} f(z, 0)$.

4. Example of indicatrices. A general result is: The behaviour of the classical indicatrices is the same on $S\boldsymbol{R}^n$ and in $\boldsymbol{C}^n$. We consider some examples.

THEOREM 1. *For* $f \in P(\boldsymbol{C}^n)$ *let* $c(z) = \lim_{r=\infty}(M(rz)/\log r)$. *Then* $-[c^{-1}(z)] = \lim_{m \to +\infty} m^{-1}[-\log \delta(z, m)]$ *is the limit of an increasing sequence of negative psh functions in* $\boldsymbol{C}^n$, *and only the following situations occur*:

(I) $c^*(z) = +\infty$ *and the set* $[z \in \boldsymbol{C}^n; c(z) < +\infty]$ *is a polar cone in* $\boldsymbol{C}^n$.

(II) $c^*(z)$ *is a constant* c_0, $0 < c_0 < +\infty$, *and the set* $[z \in \boldsymbol{C}^n;\ c(z) < c_0]$ *is a polar cone in* $\boldsymbol{C}^n$.

(III) $c(z) \equiv 0$ *and* f *is the constant* $f(0)$.

We give some consequences of Theorem 1.

COROLLARY 1. *If* $F(z)$ *is an entire function in* $\boldsymbol{C}^n$ *with* $|F|$ *bounded from above on* $S\boldsymbol{R}^n$, *then* F *is a constant. The same conclusion holds if* $|\varphi_z(u)| = |F(uz)|$, $u \in \boldsymbol{C}$, *is bounded from above by* $m(z)$, $m(z) < +\infty$, *for* $z \in \gamma$, *and* γ *is a conic set which is not* $\boldsymbol{C}^n$*-polar.*

COROLLARY 2. *If* $F(z)$ *is entire and* $F(z) \neq 0$ *for* $z \in S\boldsymbol{R}^n$, *then* $F = e^G$, G *entire. More generally if we suppose that* $\varphi_z(u) = F(uz)$, $u \in \boldsymbol{C}$, *has a finite number* $n(z)$ *of zeros for* $z \in \gamma$, *if* γ *is* $S\boldsymbol{R}^n$ *or is not a* $\boldsymbol{C}^n$*-polar conic set, then the conclusion holds*: $F(z) = P(z)e^{G(z)}$ *where* P *is a polynomial.*

COROLLARY 3. *If* t *is a positive and closed* $(1, 1)$*-current in* $\boldsymbol{C}^n$ *and* $\operatorname{supp} t \cap S\boldsymbol{R}^n = \varnothing$, *then* t *vanishes.*

For the proof of Corollary 2, apply Theorem 1 to the indicatrix

$$N(z, r) = \frac{1}{2\pi} \int_0^{2\pi} \log |F(rze^{i\theta})|\, d\theta. \tag{3}$$

As a consequence of Proposition 1, the constant c_0 in Theorem 1 is determined if the value distribution on $S\boldsymbol{R}^n$ is known for F.

5. Similar results hold for the "order" $\rho(z)$ defined by

$$M(z, r) = \sup_\theta f(rze^{i\theta}), \qquad f \in P(\boldsymbol{C}^n),$$

and

$$\rho(z) = \limsup_{r=\infty} \frac{\log M^+(z, r)}{\log r},$$
$$-\rho^{-1}(z) = \limsup_{m=\infty} (\log m)^{-1} [-\log \delta(z, m)];$$

$\rho^*(z)$ is a constant $\rho, 0 \leqq \rho \leqq +\infty$; the set $\rho(z) < \rho$ is a polar cone; ρ is determined by the value distribution of F on SR^n. If $0 \leqq \rho < \infty$, we consider

$$\begin{aligned} g(z) &= \limsup_{r=\infty} r^{-\rho} M(z, r); \\ \log g(z) &= \limsup_{m} U_m(z); \\ U_m(z) &= \log m - \rho \log \delta(z, m) \in P(C^n); \\ U_m(vz) &= U_m(z) + \rho \log v \quad \text{for all } v \in C. \end{aligned}$$

Then only three situations are possible and the decision can be made using the value distribution on SR^n:

(I) $g^*(z) = +\infty$ and $g(z) < \infty$ is a polar cone in C^n (maximal typus).

(II) $g^*(z) \in P(C^n)$ and $g(z) < g^*(z)$ is a negligible cone.

(III) $g(z) \equiv 0$ (minimal typus, which holds if $g(z) = 0$ on a not negligible conic set in C^n).

(For the proof of (I): if $\sup_{\|z\|\leqq 1} U_m(z) = a_m$ and $a_m \to +\infty$, consider $a_m^{-1} U_m(z) = V_m(z)$, and define $p(z) = \limsup_{m\to+\infty} V_m(z)$; then $p^*(z) \equiv 1$ in C^n, and $p(z) = 0$ if $g(z) < \infty$.)

The preceding results can be applied to the indicatrix (3) and give order and typus for the distribution of the zeros of F in C^n, if they are given on SR^n.

6. The preceding method remains valid, using the notion of the "relative order" as defined in [**3c**]. Let

$$\mu(r) = \sup_{\|z\|\leqq 1} f(rz) \quad \text{and} \quad \varphi(m) = [\sup r, \mu(r) < m].$$

For $z \in C^n$, we define $h(r, z)$ by the implicit relation $\sup_\theta f(re^{i\theta}z) = M(z, r) = \mu[rh(r, z)]$.

Then, using the inverse functions $\varphi(m)$, $\delta(r, m)$, we have

$$\log h(r, z) = \log \varphi(m) - \log \delta(z, m) = S_m(z) \in P(C^n)$$

and we define $c(z)$ by

$$\log c(z) = \limsup_{r=\infty} \log h(r, z) = \limsup_{m} S_m(z).$$

$c^*(z)$ and $\log c^*(z)$ are plurisubharmonic functions. The set $c(z) < c^*(z)$ is a negligible one in C^n. The results of §5 given for $g(z)$ hold for $c(z)$ and such results in C^n are determined as in §5, by the value distribution on SR^n.

7. An upper bound for the $(n-1)$-dimensional area of the real zeros of an entire function F. We assume $|F(0)| = 1$; let M_{n-1} be the component of maximal dimension $n-1$ of $M = [x \in R^n; F(x) = 0]$; M_{n-1} is a real analytic set in R^n and has locally finite area of dimension $n-1$ (cf. [**2**]). To calculate an upper bound of its area $s(r_1, \cdots, r_n)$ in the real domain $Q(r_i) = [x \in R^n; |x_i| \leqq r_i, 1 \leqq i \leqq n]$, we start from an upper bound of the number $n'_1(r_1, x_2, \cdots, x_n)$ of the zeros of F for given real x_i, $2 \leqq i \leqq n$, and complex z_1, $|z_1| \leqq r_1$. For $\sigma > 1$ and almost all $(x_2, \cdots, x_n) \in R^{n-1}$,

$$\begin{aligned} &n'_1(r_1, x_2, \cdots, x_n) \\ (4) \qquad &\leqq (\log \sigma)^{-1}\left[\frac{1}{2\pi}\int_0^{2\pi} \log|F(\sigma r_1 e^{i\theta_1}, x_2, \cdots, x_n)|\, d\theta_1 - \log|F(0, x_2, \cdots, x_n)|\right]. \end{aligned}$$

In the following we write $M(r_1, \cdots, r_n) = \sup \log |F(z_1, \cdots, z_n)|$ for $z_i \in C$, $|z_i| \leqq r_i$. For given $x_2, \cdots, x_n$, the number $n_1(r_1, x_2, \cdots, x_n)$ of the real zeros of F with $|x_1| \leqq r_1$ is smaller than $n_1'(r_1, x_2, \cdots, x_n)$ and the method gives an upper bound for the projection $s_1(r_1, \cdots, r_n)$ of $M_{n-1} \cap Q(r_i)$ on $R^{n-1}(x_2, \cdots, x_n)$: from (4), we deduce

$$(5) \quad s_1(r_1, \cdots, r_n) \leqq M(\sigma r_1, r_2, \cdots, r_n)\, r_2 \cdots r_n - \int \log|F(0, x_2 \cdots x_n)|dx_2 \cdots dx_n.$$

In (5), the integral is extended over the set $|x_2| \leqq r_2, \cdots, |x_n| \leqq r_n$. To obtain an upper bound, we use the following lemma of W. Stoll (cf. **[5]**) which can be given for a psh function $V(z)$ and applied to $V(z) = \log|F(z)|$.

LEMMA (W. STOLL). *Given* $f \in P(C^n)$, $M(r) = \sup_{\|z\| \leqq r} f(z)$, *and* $\sigma > 1$, *there exists* $C(\sigma) > 0$ *such that*

$$(6) \qquad \int_0^1 f(tz)\, dt \geqq f(0) - C(\sigma)M(\sigma r).$$

By repeated application of (5), we obtain:

THEOREM 2. *For given* $k > 1$, *there exists* $C(k) > 0$ *such that, if* $F(z)$ *is an entire function and* $F(0) = 1$, *then the* $(n-1)$*-dimensional area of the set of the real zeros of* F *in* $Q(r_1, \cdots, r_n) = [x \in R^n, |x_i| \leqq r_i, 1 \leqq i \leqq n]$ *has the upper bound*

$$(7) \qquad s(r_1, \cdots, r_n) \leqq C(k)M(kr_1, \cdots, kr_n)\, r_1 \cdots r_n \sum_1^n r_i^{-1}.$$

The proof uses $ds \leqq \sum ds_k$ and (5).

8. A consequence of the preceding theorem is

COROLLARY 4. *Let* $F(z)$ *be an entire function in* C^n *and* $F(0) = 1$ *and* $M(r) = \sup|F(z)|$, $\|z\| \leqq r$, $z \in C^n$. *Then the* $(n-1)$*-dimensional area* $s(r)$ *of the real zeros of* F *in a ball* $\|x\| \leqq r$ *in* R^n, *has the property*:

$$(8) \qquad s(r) \leqq C'(k, n)\, r^{n-1} M(kr\sqrt{n}).$$

CONJECTURE. *In* (8), $M(kr\sqrt{n})$ *can be replaced by* $M(kr)$.

Let us define an indicatrix $\nu_s(r)$ for M_{n-1}:

$$(9) \qquad \nu_s(r) = r^{1-n} s(r)$$

and compare with the indicatrix $\nu(r) = (n-1)!\pi^{1-n}\sigma(r)$ of $M = [z \in C^n; F(z) = 0]$; $\sigma(r)$ is the $(2n-2)$-dimensional area of $M \cap [\|z\| \leqq r]$. Now instead of considering an entire function F, we give Cousin data M of zeros in C^n. The Cousin problem has a "minimal" solution which is an entire function F whose indicatrix $M(r)$ is calculated using $\nu(r)$. Therefore, there exists an estimate of an upper bound for $\nu_s(r)$ using $\nu(r)$ instead of $M(r)$. Roughly speaking, if a bound is given for the increasing of the indicatrix $\nu(r)$ of M on C^n, then there exists a bound for the increasing of the indicatrix $\nu_s(r)$ of M_{n-1}, which is the variety of maximal dimension of the real zeros.

EXAMPLE. If $q < \lambda \leqq q + 1$, $\nu(t)$ is of finite order λ, and $\int \nu(t)dt/t^{q+2} < \infty$, then $M(r)$ in (8) can be replaced by the upper bound (see **[3d]**)

$$M(r) \leqq A(n, q)\left[r^q \int_0^r \frac{\nu(t)}{t^{q+1}}\, dt + r^{q+1} \int_r^\infty \frac{\nu(t)\, dt}{t^{q+2}}\right] \tag{10}$$

and a majorant of $\nu_s(r)$ is obtained using only $\nu(r)$.

A consequence of (10) is

COROLLARY 5. *If M is a Cousin data of zeros of finite order λ in C^n, then the indicatrix $\nu_s(r)$ of the real part $M \cap R^n$ of dimension $n-1$ in R^n is of finite order $\lambda' \leqq \lambda$.*

The results of §8 are due to F. Fages and the author.

9. Now, we consider the zeros of an entire function $F(z_1, \cdots, z_n) = F(z)$ located on a real line $x_k = \nu\alpha_k$, $\alpha = (\alpha_k) \in R^n$, $\|\alpha\| = 1$, $\nu \in R$, and denote by $n(\alpha, r)$ the number (finite or infinite) of such zeros in the ball $B(r) = [z \in C^n, \|z\| \leqq r]$. We assume $F(0) = 1$, and we write $n'(\alpha, r)$ for the number of the zeros of F in $B(r)$ which lie on the *complex* line $z_k = u\alpha_k$, $u \in C$; $n(\alpha, r) \leqq n'(\alpha, r)$. Let us consider the indicatrices:

$$N(\alpha, r) = \int_0^r n(\alpha, t)\, t^{-1}\, dt; \qquad N'(\alpha, r) = \int_0^r n'(\alpha, t) t^{-1}\, dt$$

and the mean value of $\log|F|$ on the sphere $S^{2n-1}(r)$:

$$N(r) = \omega_{2n-1}^{-1} \int \log|F(r\beta)\, d\omega_{2n-1}(\beta), \qquad \|\beta\| = 1,\ \beta \in C^n. \tag{11}$$

By a result of Ronkin ([**4**], cf. also [**1**]), for each $k > 1$, there exists $C_1(k) > 0$ such that we have $N(\beta, r) \leqq C_1(k)N(kr)$. On the other hand, $N(r)$ is given by the indicatrix $\nu(r)$ (cf. [**3d**])

$$N(r) = \int_0^r \nu(t)\, t^{-1}\, dt; \qquad \nu(r) = \omega_{2n-1}^{-1} \int n(\beta, r)\, d\omega_{2n-1}(\beta). \tag{12}$$

Then, using $N(\alpha, r) \leqq N'(\alpha, r)$, (11) and (12), we obtain for the number $n'(\alpha, r)$ of the real zeros in $B(r)$ on the real line $x_k = \alpha_k\nu$, α_k, $\nu \in R$:

$$\int_0^r n'(\alpha, t)\, t^{-1}\, dt = N(\alpha, r) \leqq C(k) \int_0^{kr} \nu(t) t^{-1}\, dt. \tag{13}$$

Define a projective indicatrix $\nu_p(r)$ for the real zeros of F as a mean value of $n(\alpha, r)$ for $\alpha \in R^n$, $\|\alpha\| = 1$:

$$\nu_p(r) = \omega_{n-1}^{-1}\, n(\alpha, r)\, d\omega_{n-1}(\alpha), \qquad \|\alpha\| = 1,\ \alpha = (\alpha_k) \in R^n. \tag{14}$$

Then we obtain an upper bound for $\nu_p(r)$:

$$\int_0^r \nu_p(t)\, t^{-1}\, dt \leqq C_1(k)\, N(kr) = C_1(k) \int_0^{kr} \nu(t) t^{-1}\, dt. \tag{15}$$

A consequence of (15) is the following: *If the projective indicatrix ν of the zeros of F in C^n is of finite order λ, then the projective indicatrix ν_p of the real zeros is of finite order $\lambda' \leqq \lambda$.*

The results of the paragraphs 1—6 give only information on the indicatrices; the bounds in §§7 and 8 hold for finite r.

BIBLIOGRAPHY

1. F. Fages, *Croissance maximale de certaines fonctions plurisousharmoniques positives*, C. R. Acad. Sci. Paris Sér. A-B **273** (1971), A1040–A1043. MR **45** #3738.

2. M. E. Herrera, *Integration on a semianalytic set*, Bull. Soc. Math. France **94** (1966), 141–180. MR **35** #4837.

3a. P. Lelong, *Fonctions plurisousharmoniques et fonctions analytiques de variables réelles*, Ann. Inst. Fourier (Grenoble) **11** (1961), 515–562. MR **26** #358.

3b. ———, *Fonctions entières de type exponential dans C^n*, Ann. Inst. Fourier (Grenoble) **16** (1966), fasc. 2, 269–318. MR **35** #1827.

3c. ———, *Fonctions entières et fonctionnelles analytiques*, Cours publié aux Presses de l'Université de Montréal, Chap. 6, 1967.

3d. ———, *Fonctions entières (n variables) et fonctions plurisousharmoniques d'ordre fini dans C^n*, J. Analyse Math **12** (1964), 365–407. MR **29** #3668.

4. L. I. Ronkin, *Fonctions de plusieurs variables*, Chap. 4, §2, Moscow, 1971.

5. W. Stoll, *A Bezout estimate for complete intersections*, Ann. of Math. (2) **96** (1972), 361–401. MR **47** #2091.

UNIVERSITÉ DE PARIS VI

Proceedings of Symposia in Pure Mathematics
Volume 30, 1977

A DEFECT RELATION ON STEIN MANIFOLDS

JOHN MURRAY

1. Introduction. We will be concerned throughout with the following situation: M is a connected Stein manifold of dimension m with a positive strictly pseudoconvex exhaustion function τ; V is a Hermitian vector space of dimension $n + 1$, and

$$f: M \to \boldsymbol{P}(V)$$

is a generic meromophic map (so $f(M)$ is not contained in any hyperplane in $\boldsymbol{P}(V)$).

Under these conditions, the first main theorem of value distribution theory takes the form [3]:

$$T_f(r) = N_f(r, a) + m_f(r, a) - D_f(r, a)$$

(for any regular value r of τ and any $a \in \boldsymbol{P}(V^*)$) where

$$T_f(r) = \int_{G_r} \psi_r f^*(\omega) \wedge (dd^c\tau)^{m-1} \qquad \text{(characteristic function)},$$

$$N_f(r, a) = \int_{\gamma(\nu(f,a)) \cap G_r} \nu(f, a)(dd^c\tau)^{m-1} \qquad \text{(valency function)},$$

$$m_f(r, a) = \frac{1}{2\pi} \int_{\Gamma_r} \log \frac{1}{\|f; a\|} d^c\tau \wedge (dd^c\tau)^{m-1} \qquad \text{(compensation function)},$$

$$D_f(r, a) = \int_{G_r} \log \frac{1}{\|f; a\|} (dd^c\tau)^m \qquad \text{(deficit)};$$

and where we use the following notation:

$G_r = \tau^{-1}([0, r])$,
$\Gamma_r = \tau^{-1}(\{r\})$,
$\psi_r = \max(r - \tau, 0)$,

AMS (MOS) subject classifications (1970). Primary 32A30, 32A70.

a denotes a hyperplane in $\boldsymbol{P}(V)$,
$\|z; a\|$ is the usual distance from a point z to the hyperplane a,
ω is the Fubini-Study metric on $\boldsymbol{P}(V)$,
$\nu(f, a)$ is the intersection divisor of the map f with the hyperplane a, and
$\gamma(\nu(f, a))$ denotes the support of this divisor.

Griffiths [2] (in a special case) and Stoll [3] (in the general case) have given estimates of the growth of $D_f(r, a)$, and hence $N_f(r, a)$, in terms of functions independent of a. In this note we give a corresponding defect relation, together with a slight generalization of one of Stoll's estimates.

2. Some definitions. The estimates given by Stoll involve the idea of the associated maps of the given map f. These are defined as follows: Let B be a fixed holomorphic form of bidegree $(m - 1, 0)$ and let

$$\alpha = (\alpha_1, \cdots, \alpha_m): U_\alpha \to \boldsymbol{C}^m$$

be a local coordinate system on M. Let $\tilde{f}: U_\alpha \to \boldsymbol{C}^{n+1}$ be a local irreducible representation of f (that is, $\boldsymbol{P} \circ \tilde{f} = f | U_\alpha$ and the components of $\tilde{f}$ have no nontrivial common factor), and define $\tilde{f}'$ by

$$\tilde{f}' d\alpha_1 \wedge \cdots \wedge d\alpha_m = df \wedge B.$$

$\tilde{f}'', \tilde{f}''', \tilde{f}^{(4)}$, etc., are defined by iteration. Then the pth associated map, $f_p: M \to \boldsymbol{P}(\bigwedge^{p+1} V)$, is defined locally by

$$f_p = \boldsymbol{P} \circ (\tilde{f} \wedge \tilde{f}' \wedge \cdots \wedge \tilde{f}^{(p)}).$$

This makes sense only if $\tilde{f} \wedge \cdots \wedge \tilde{f}^{(p)} \not\equiv 0$; it can be shown that there are always forms B on M such that $\tilde{f} \wedge \tilde{f}' \wedge \cdots \wedge \tilde{f}^{(n)} \not\equiv 0$ [3]; in what follows we always assume that one such fixed form B has been chosen. It can then be shown that f_p is in fact a well-defined meromorphic map on M, depending only on f and B. We also define the nth greatest common divisor, d_n, of f to be the zero divisor of $\tilde{f} \wedge \tilde{f}' \wedge \cdots \wedge \tilde{f}^{(n)}$ locally. (For a more precise and detailed description of these ideas, see [3].)

One next defines ρ and h_p by

$$\rho (dd^c\tau)^m = (1/2\pi)\, d\tau \wedge d^c\tau \wedge (dd^c\tau)^{m-1}$$

and

$$h_p (dd^c\tau)^m = (1/2\pi)^{m-1}(-1)^{(m-1)(m-2)/2}\, B \wedge \bar{B} \wedge f_p^*(\omega_p)$$

(where ω_p is the Fubini-Study metric on $\boldsymbol{P}(\bigwedge^{p+1}(V))$), and $z(r)$ by

$$z(r) = \operatorname{Inf}\{c \in [1, \infty) | (1/2\pi)^{m-1}(-1)^{(m-1)(m-2)/2}\, B \wedge \bar{B} \leqq (c dd^c\tau)^{m-1} \text{ on } G_r \cup \Gamma_r\}.$$

Finally, we define

$$S_f^{+p}(r) = \frac{1}{2} \int_{G_r} \log^+(1/h_p)(dd^c\tau)^m \geqq 0,$$

$$J^+(r) = \frac{1}{4\pi} \int_{\Gamma_r} \log^+ \rho\, d^c\tau \wedge (dd^c\tau)^{m-1} \geqq 0,$$

$$\Phi(r) = \int_{G_r} (dd^c\tau)^m,$$

$$N_{d_n}(r) = \int_{\gamma(d_n)\cap G_r} \psi_r d_n (dd^c\tau)^{m-1},$$

$$\mathrm{Ric}^+(r) = \int_{G_r} \psi_r (\mathrm{Ric}(dd^c\tau)^m \wedge (dd^c\tau)^{m-1})^+,$$

where Ric $(dd^c\tau)^m$ denotes the usual Ricci form of the volume form $(dd^c\tau)^m$.

3. Statement of results. As a generalization of one of Stoll's estimates, we have the following:

PROPOSITION 1. *Let the set $A \subseteq \boldsymbol{P}(V^*)$ be in general position, and assume that all our previous conditions are satisfied. Then, for any $\varepsilon > 0$,*

$$\sum_{a\in A} D_f(r,a) \lesssim (1+\varepsilon)n \sum_{j=0}^{n-1} \{S_f^{+j}(r) + \Phi(r)\log(1 + T_{f_j}(r)/\Phi(r)\} + (1+\varepsilon)(n^2/2)\Phi(r)\{\log^+ \Phi(r) + (m-1)\log^+ z(r)\}$$

($\lesssim$ *means that the inequality holds on the complement of a set of finite measure*).

The proof of this is a modification of the proof given by Griffiths and Stoll, using the lemma of Ahlfors **[1]** and Weyl **[4]** on hyperplanes in general position.

Our main result is an analogous estimate for the compensation function $m_f(r, a)$ (that is, a second main theorem):

THEOREM 2. *Under the same hypotheses as Proposition* 1, *there exists a constant C_1 such that*

$$\begin{aligned}\sum_{a\in A} m_f(r,a) + N_{d_n}(r) \lesssim{}& (n+1)T_f(r) + n(n+2)\sum_{p=0}^{n-1} S_f^{+p}(r) \\ &+ 2n(n+1)\{\mathrm{Ric}^+(r) + J^+(r) + \\ &\qquad \Phi(r)[(m-1)\log^+ z(r) + \log^+\Phi(r) + C_1]\} \\ &+ n(n+1)\Phi(r)\log(1 + T_f(r)/\Phi(r)).\end{aligned}$$

The proof follows the same lines as the classical Ahlfors-Weyl proof, using several of the results in **[3]**.

The defect of f at a is defined in the usual way:

$$\delta_f(a) = \liminf_{r\to\infty}(m_f(r,a)/T_f(r)) \geqq 0.$$

Then $\delta_f(a) \geqq 0$, and if f does not intersect the hyperplane a, $\delta_f(a) \geqq 1$. However, we cannot say in this situation that $\delta_f(a) \leqq 1$.

We can obtain a defect relation from Theorem 2 by putting several restrictions on the growth of f.

THEOREM 3. *In addition to our previous assumptions, suppose that the map f also satisfies the following conditions*:

(i) $\mathrm{Ric}^+(r) = o(T_f(r))$,
(ii) $J^+(r) = o(T_f(r))$,
(iii) $\Phi(r)\log^+\Phi(r) = o(T_f(r))$,
(iv) $\log^+ z(r) = o(T_f(r)/\Phi(r))$,
(v) $\sum_{p=0}^{n-1} S_f^{+p}(r) = o(T_f(r))$.

Then $0 \leqq \delta_f(a) \leqq 1$, $\forall\, a \in \boldsymbol{P}(V^*)$ *and* $\sum_{a\in A} \delta_f(a) \leqq n + 1$.

The proof of the second statement is obvious; the first can be proved by using the Plücker difference formula [3] to eliminate the characteristic functions of the associated maps from the right-hand side of the estimate in Proposition 1.

Finally, if the map f is *steady* in the sense of Griffiths and Stoll [2], [3] then $\sum_{p=0}^{n-1} S_f^{+p}(r) = O(\Phi(r))$ and so Theorem 3 applies to steady maps which satisfy conditions (i) through (iv).

References

1. L. V. Ahlfors, *The theory of meromorphic curves*, Acta Soc. Sci. Fenn. Nova Ser. **A3** (1941), no. 4, pp. 1–31. MR **2**, 357.

2. P. Griffiths, *On the Bezout problem for entire analytic sets*, Ann. of Math. (2) **100** (1974), 533–552.

3. W. Stoll, *Deficit and Bezout estimates*, Part B of Value-Distribution Theory, edited by R. O. Kujala and A. L. Vitter III. Dekker, New York, 1973.

4. H. Weyl, *Meromorphic functions and analytic curves*, Ann. of Math. Studies, no. 12, Princeton Univ. Press, Princeton, N. J., 1943. MR **5**, 94.

Texas Tech University

Proceedings of Symposia in Pure Mathematics
Volume 30, 1977

ON HOLOMORPHIC CURVES IN ALGEBRAIC VARIETIES WITH AMPLE IRREGULARITY

TAKUSHIRO OCHIAI*

1. Introduction and remarks on a paper of Bloch. This is a report on some results which will be published elsewhere as a joint work with P. A. Griffiths.

Let X be an irreducible quasi-projective algebraic variety over the field of complex numbers. We write X_{an} for its associated complex analytic space. By a *holomorphic curve* in X, we mean a holomorphic mapping $f: C \to X_{\mathrm{an}}$ from the Gaussian plane C into X_{an}. We call f *algebraically degenerate* if there exists a closed algebraic subvariety Y of X such that $Y \neq X$ and $f(C) \subset Y$.

In his paper [2] written in 1926, A. Bloch makes the following remarkable assertion.

(1-1) *Let X be an irreducible projective algebraic variety whose irregularity*[1] *is strictly greater than the dimension of X. Then a holomorphic curve is always algebraically degenerate.*

He only gives a sketchy proof to (1-1) when X is a surface, and the general case is left to the reader. However, even his rough proof for surfaces, in itself, contains serious gaps. Notably (i) he himself admits in his paper that his Lemma III is not proved in enough generality to support his argument (cf. [2, p. 22]). In fact we have proved this Lemma III using results in [3]. The next objection (ii) is that one of the essential steps in his proof (i.e., that (3-1) below is true) is overlooked and not proved at all. As we see this last step, it is so highly nontrivial that we do not know yet the validity of this step in general, except the case when X is a surface. There-

AMS (MOS) subject classifications (1970). Primary 32C10, 30A70; Secondary 14K20.

Key words and phrases. Algebraically degenerate, irregularity, Albanese variety, abelian variety, holomorphic curve.

*Partially supported by Sonderforschungsbereich "Theoretische Mathematic", University of Bonn, and by University of Notre Dame under NSF MPS71-03140 A04.

[1]The irregularity of X is the dimension of the space of regular rational 1-forms on X.

fore, if we could have nothing other than his incomplete arguments, then (1-1) must remain a conjecture at most.

Thus we have been trying to understand Bloch's paper **[2]**. Now we would like to report some of the results which we have obtained in this direction. We emphasize here that (1-1) in general is still a conjecture.

2. Reduction of the problem. Let X be an irreducible quasi-projective variety. For convenience, we call X a *degenerate variety* if any holomorphic curve in X is algebraically degenerate. Now let X be as in (1-1). Consider a resolution of singularities of X, $\rho: \hat{X} \to X$, where $\hat{X}$ is a nonsingular projective algebraic variety and ρ is a surjective regular rational mapping. Let $A(\hat{X})$ be the Albanese variety of $\hat{X}$ and $\alpha: \hat{X} \to A(\hat{X})$ be the Albanese mapping from $\hat{X}$ into $A(\hat{X})$. It is clear that the irregularity of the image variety $\alpha(\hat{X})$ is equal to that of X, and that if $\alpha(\hat{X})$ is a degenerate variety, so is X. Thus, in order to prove (1-1), we may assume that X is a proper closed subvariety of an abelian variety A and that X is *in good position* in A, defined as follows.

DEFINITION. An algebraic subvariety X of an abelian variety A is said to be *in good position* if, for a nonzero regular rational 1-form ω on A, the restriction $\omega | X_{\text{reg}}$ is nonzero, where X_{reg} is the set of regular points of X.

Therefore our problem to prove becomes

(2-1) *Let X be a proper closed subvariety of an abelian variety A which is in good position. Then X is a degenerate variety.*

3. Statements of results. We have established the following theorem which should be a revised form of Bloch's main theorem in **[2]**.

THEOREM 1. *Let X be an m-dimensional irreducible algebraic subvariety (closed or not) of an abelian variety A. Suppose X satisfies the following property:*

(3-1) *There exist regular rational* 1-*forms $\omega_1, \cdots, \omega_{m+1}$ on A such that the restriction of the system of m-forms*

$$\{\omega_1 \wedge \cdots \wedge \hat{\omega}_a \wedge \cdots \wedge \omega_{m+1}\}_{1 \leqq a \leqq m+1}$$

onto X_{reg} is linearly independent.[2]

Then X is a degenerate variety.[3]

Roughly speaking, Bloch claims that property (3-1) is automatically satisfied in Theorem 1 when X is closed, and hence that assertions (2-1) and (1-1) are true in general. This is the gap mentioned in (ii) (cf. §1). It seems to be difficult to verify (3-1) in general and this is the reason that assertion (1-1) is still an open problem.

Before going further, let us see that a further reduction of problem (2-1) is possible. So let X be as in (2-1). Let B be the closed algebraic subgroup of A defined by $\{a \in A; a(X) \subset X\}$. Let B_0 be the neutral irreducible component of B. Let $\sigma: A \to A/B_0$ be the natural projection. Then of course $\sigma(X)$ is in good position in the abelian variety A/B_0, and if $\sigma(X)$ is a degenerate variety, so is X. Therefore, in order to prove (2-1), we may assume that X in (2-1) is *simple* in A, defined as follows.

---

[2] $\hat{\omega}_a$ means that ω_a is deleted.

[3] It can be shown that a holomorphic curve is either in the singular locus of X or a canonical divisor of the form $\{x \in X; \sum_{a=1}^{m+1} C_a(\omega_1 \wedge \cdots \wedge \hat{\omega}_a \wedge \cdots \wedge \omega_{m+1} | X)(x) = 0\}$, where $c_1, \cdots, c_{m+1}$ are constants.

DEFINITION. Let X be a closed algebraic subvariety of an abelian variety A. X is said to be *simple* in A if there is no irreducible closed subgroup of A other than zero which leaves X invariant.

Thus our problem to prove becomes

(3-2) *Let X be an irreducible closed subvariety of an abelian variety A, which is in good position and simple in A. Then X is a degenerate variety.*

Now we have established the following

THEOREM 2. *Let X be an irreducible closed subvariety of an abelian variety A which is in good position and simple in A. Then property* (3-1) *is satisfied if X is one of the following cases:*

(a) *X is a curve:*
(b) *X is a surface;*
(c) *X is a hypersurface;*
(d) *X is nonsingular and the Euler number of X is nonzero.*

Combining Theorems 1 and 2, we can see that assertion (3-2) (and so assertion (1-1)) is true for many interesting cases.

COROLLARY 1. *Let X be a projective algebraic surface whose irregularity is strictly greater than two. Then any holomorphic curve in X is algebraically degenerate.*

Thus we have shown that Bloch's conjecture (1-1) is true in the case when X is a surface.

COROLLARY 2. *Let X be an irreducible closed hypersurface of a complex torus, which is not a subcomplex torus. Let $f: \mathbf{C} \to X$ be a holomorphic mapping. Then there exists a proper closed analytic subvariety Y of X such that $f(\mathbf{C}) \subset Y$.*

Let X be a compact Riemann surface of genus $g \geqq 2$. Let r be $1 \leqq r < g$. The permutation group S_r of degree r naturally operates on $X \times \cdots \times X$ (r times). We write $X_{(r)}$ for the quotient space of $X \times \cdots \times X$ (r times) with respect to S_r. It is well known that $X_{(r)}$ is a projective algebraic manifold.

COROLLARY 3. *Any holomorphic curve in $X_{(r)}$ $(1 \leqq r < g)$ is algebraically degenerate.*

Let M be a compact Kähler manifold. We call M *weakly ample* if the Albanese mapping of M is an immersion into the Albanese torus of M (cf. **[6]**). Some of the results in **[4]**, **[5]** and **[6]** together with Theorems 1 and 2 imply the following corollaries, among others.

COROLLARY 4. *Let M be a compact Kähler manifold which is weakly ample. Suppose the Euler number of M is nonzero. Then for any holomorphic mapping $f: \mathbf{C} \to M$, there exists a proper closed analytic subset Y of M such that $f(\mathbf{C}) \subset Y$.*

COROLLARY 5. *Let M be a compact Kähler surface which is weakly ample. Then for any holomorphic mapping f: $\mathbf{C} \to M$ there exists a closed analytic curve Y in M such that $f(\mathbf{C}) \subset Y$.*

As was hinted by Bloch in **[2]**, a slight modification of the proof for Theorem 1 gives

THEOREM 3. *Let A be an abelian variety. Let X and Y be distinct closed hypersurfaces of A which are linearly equivalent. If a homomorphic curve in A does not meet X and Y, then it is algebraically degenerate.*

This is a special case of the conjecture raised by S. Lang (cf. Bull. Amer. Math. Soc. **77** (1971), 635–677) which claims that the above theorem should be true even for $X = Y$. This conjecture is true if a holomorphic curve in A is given by a group homomorphism [**1**].

REFERENCES

1. J. Ax, *Some topics in differential algebraic geometry.* II: *On the zeros of theta functions*, Amer. J. Math. **94** (1972), 1025–1213.

2. A. Bloch, *Sur les systèmes de fonctions uniformes satisfaisant à l'équation d'une variété algèbrique dont l'irrégularité dépasse la dimension*, J. Math. **5** (1926), 19–66.

3. P. A. Griffiths, *Function theory of finite order on algebraic varieties*, J. Differential Geometry **6** (1972), 285–306; ibid. **7** (1972), 45–66. MR **46** #3829; **48** #4345.

4. A. Howard and Y. Matsushima, *Weakly ample vector bundles and submanifolds of complex tori*, Séminaire de Mathématique Supérieures, Les Presses de l'Université de Montréal.

5. Y. Matsushima, *Holomorphic immersions of a compact Kähler manifold into complex tori*, J. Differential Geometry **9** (1974), 309–328. MR **50** #2550.

6. Y. Matsushima and W. Stoll, *Ample vector bundles on compact complex spaces*, Rice Univ. Studies **59** (1973), no. 2, 71–107.

7. T. Ochiai, *Some remarks on the defect relations of holomorphic curves*, Osaka J. Math. **11** (1974), 483–510.

THE INSTITUTE FOR ADVANCED STUDY

Proceedings of Symposia in Pure Mathematics
Volume 30, 1977

VALUE DISTRIBUTION ON PARABOLIC SPACES*

WILHELM STOLL

In [3], the theory of Carlson and Griffiths [1] and Griffiths and King [2] is extended to parabolic spaces. Here an outline shall be given. Let M and N be irreducible complex spaces of dimensions m and n respectively. M is noncompact and N is compact. Let $f: M \to N$ be a holomorphic map. Let $\pi: L \to N$ be a holomorphic line bundle. Then L is said to be nonnegative (respectively positive), if there exists a hermitian metric κ along the fibers of L such that the associated Chern form $c(L, \kappa)$ is nonnegative (respectively positive) on N. Assume κ is given with $c(L, \kappa) \geqq 0$. The vector space $V = \Gamma(L, N)$ of global holomorphic sections of L over N has finite dimension $k + 1$. Assume $k \geqq 0$. If $\mathfrak{v} \in V$, define $\boldsymbol{P}(\mathfrak{v}) = \boldsymbol{C}\mathfrak{v}$. Then $\boldsymbol{P}(V) = \{\boldsymbol{P}(\mathfrak{v}) \mid 0 \neq \mathfrak{v} \in V\}$ is the complex projective space associated to V. Take a hermitian metric l on V. Let φ be the exterior form of the Fubini-Study-Kähler metric on $\boldsymbol{P}(V)$ associated to l. The volume element φ^k gives $\boldsymbol{P}(V)$ the total volume one.

If $\mathfrak{v} \in V$, then $Z(\mathfrak{v}) = \{x \in M \mid \mathfrak{v}(x) = 0_x\}$ is the zero set of $\mathfrak{v}$. If $a \in \boldsymbol{P}(V)$, then $\boldsymbol{P}(\mathfrak{v}) = a$ with $0 \neq \mathfrak{v} \in V$ and $E_L[a] = Z(\mathfrak{v})$ is well defined. For each $x \in M$, also

$$0 \leqq \|a, x\|_\kappa = |\mathfrak{v}(x)|_\kappa / |\mathfrak{v}|_l$$

is well defined with $E_L[a] = \{x \in M \mid \|a, x\|_\kappa = 0\}$. If $x \in N - E_L[a]$, then

$$c(L, \kappa) = - dd^c \log \|a, x\|_\kappa^2$$

where $4\pi d^c = i(\bar{\partial} - \partial)$. Now κ is said to be distinguished if $0 \leqq \|a, x\|_\kappa \leqq 1$ for all $(a, x) \in \boldsymbol{P}(V) \times N$. A constant $\lambda > 0$ exists such that $\lambda\kappa$ is distinguished. Since $c(L, \lambda\kappa) = c(L, \kappa)$, without loss of generality κ can be assumed to be distinguished. The base point set $E_L[\infty] = \bigcap_{a \in \boldsymbol{P}(V)} E_L[a]$ is thin analytic. Define $N_\infty = N - E_L[\infty]$.

AMS (MOS) subject classifications (1970). Primary 32H25, 32F15.

*This research was partially supported by the National Science Foundation NSF grant MPS 75–07086.

If $x \in N_\infty$, the linear map $\eta_x \colon V \to L_x$ defined by $\eta_x(\mathfrak{v}) = \mathfrak{v}(x)$ is surjective. Then l on V defines a quotient metric l_x on L_x, which defines a distinguished hermitian metric l along the fibers of $L \mid N_\infty$. Then $c(L, l) = c(L \mid N_\infty, l) \geqq 0$ on N_∞.

Let $\tau \geqq 0$ be a nonnegative function of class C^∞ on M. For $r \geqq 0$ define

$$\begin{aligned} M[r] &= \{x \in M \mid \tau(x) \leqq r^2\}, \\ M(r) &= \{x \in M \mid \tau(x) < r^2\}, \\ M\langle r\rangle &= \{x \in M \mid \tau(x) < r^2\}, \\ \omega &= dd^c \log \tau && \text{on } M - M[0], \\ \sigma &= d^c \log \tau \wedge \omega^{m-1} && \text{on } M - M[0], \\ \upsilon &= dd^c \tau && \text{on } M. \end{aligned}$$

Then τ is called a *parabolic exhaustion* and (M, τ) a *parabolic space*, if $M[r]$ is compact for all $r \geqq 0$, if $M[0]$ has measure zero, if $\omega \geqq 0$ (hence $\upsilon \geqq 0$), if $\upsilon^m \not\equiv 0$ and if $\omega^m \equiv 0$. Then $M[0] \neq \varnothing$ and

$$\int_{M[r]} \upsilon^m = \int_{M(r)} \upsilon^m = \varsigma r^{2m}$$

where $\varsigma = \int_{M\langle r\rangle} \sigma > 0$ is constant. If $A \subseteq M$, define $A[r] = A \cap M[r]$.

(C^m, τ) with $\tau(\mathfrak{z}) = |\mathfrak{z}|^2$ is parabolic. If (M, τ) is parabolic, and if $\tilde{M}$ is an irreducible, noncompact complex space with $\dim \tilde{M} = \dim M$, if $\beta \colon \tilde{M} \to M$ is a surjective, proper holomorphic map, then $(\tilde{M}, \tau \circ \beta)$ is parabolic. Hence each affine algebraic variety is parabolic. If (M_j, τ_j) are parabolic for $j = 1, 2$, and if $\pi_j \colon M_1 \times M_2 \to M_j$ is the projection, then $(M_1 \times M_2, \tau_1 \circ \pi_1 + \tau_2 \circ \pi_2)$ is parabolic. A noncompact Riemann surface is parabolic if and only if M is in the class $\mathcal{O}_g$, i.e., each subharmonic function bounded above is constant. If B is a compact, irreducible, complex space of dimension $m - 1$, if $\pi \colon M \to B$ is a holomorphic line bundle, if τ is a hermitian metric along the fibers of M with $c(M, \tau) \leqq 0$, and if an open subset $U \neq \varnothing$ of B exists such that $c(M, \tau) \mid U < 0$, then (M, τ) is parabolic. $(C - Z) \times C^{m-1}$ is parabolic, but not affine algebraic.

Let ν be a nonnegative divisor on M with $A = \operatorname{supp} \nu$. The *counting function* of ν is defined by

$$n_\nu(r) = r^{2-2m} \int_{A[r]} \nu \upsilon^{m-1} \geqq 0$$

where ν also denotes the multiplicity function of ν. Then n_ν increases. The *valence function* of ν is defined for $0 < s < r$ by

$$N_\nu(r, s) = \int_s^r n_\nu(t) \frac{dt}{t} \geqq 0.$$

Consider the holomorphic map $f \colon M \to N$. Assume $f(M) \nsubseteq E_L[a]$ for all $a \in \boldsymbol{P}(V)$. For each $a \in \boldsymbol{P}(V)$ a nonnegative divisor $\theta_f^a[L]$ is defined with support $f^{-1}(E_L[a])$. Its counting function and valence function are denoted by $n_f^a(r, L)$ and $N_f^a(r, s, L)$ respectively. The *compensation function* is defined for almost all $r > 0$ by

$$m_f^a(r, L, \kappa) = \int_{M\langle r\rangle} \log \frac{1}{\|a, f\|_\kappa} \sigma \geqq 0.$$

For all $r > 0$, the spherical image is defined by

$$A_f(r, L, \kappa) = r^{2-2m} \int_{M[r]} f^*(c(L, \kappa)) \wedge \upsilon^{m-1} \geqq 0$$

which increases with r. For $0 < s < r$, the *characteristic* is defined by

$$T_f(r, s, L, \kappa) = \int_s^r A_f(t, L, \kappa) \frac{dt}{t} \geqq 0.$$

If $c(L, \kappa) > 0$, then $T_f(r, s, L, \kappa) \to \infty$ for $r \to \infty$. Although $c(L, l)$ is only defined outside $E_L[\infty]$, the integrals $A_f(r, L, l)$, $T_f(r, s, L, l)$ and $m_f^a(r, L, l)$ exist. For $0 < s < r$, the *First Main Theorem* holds:

$$T_f(r, s, L, \kappa) = N_f^a(r, s, L) + m_f^a(r, L, \kappa) - m_f^a(s, L, \kappa),$$
$$T_f(r, s, L, l) = N_f^a(r, s, L) + m_f^a(r, L, l) - m_f^a(s, L, l),$$

and shows that the compensation function extends to a continuous function for all $r > 0$. The *Mean Value Theorem* holds for l only:

$$T_f(r, s, L, l) = \int_{P(V)} N_f^a(r, s, L) \varphi^k.$$

The defects

$$0 \leqq \delta_f^0(a, L) = \liminf_{r \to \infty} \frac{m_f^a(r, L, l)}{T_f(r, s, L, l)} \leqq 1,$$
$$0 \leqq \delta_f(a, L) = \liminf_{r \to \infty} \frac{m_f^a(r, L, \kappa)}{T_f(r, s, L, \kappa)} \leqq 1,$$

do not depend on s and κ. If $E_L[\infty] = \varnothing$, then $\delta_f^0(a, L) = \delta_f(a, L)$. If $f^{-1}(E_L[a]) = \varnothing$, then $\delta_f^0(a, L) = 1 = \delta_f(a, L)$. The Mean Value Theorem implies

$$\int_{P(V)} \delta_f^0(a, L) \varphi^k = 0.$$

Hence $f^{-1}(E_L[a]) \neq \varnothing$ for almost all $a \in \boldsymbol{P}(V)$. These results can be extended to meromorphic maps and to higher codimensions (see **[3]**).

Now, assume that M and N are complex manifolds. To each positive form Ω of degree $2m$ on M, a Ricci form Ric Ω is associated. For $0 < s < r$, the Ricci function

$$\operatorname{Ric}(r, s, \Omega) = \int_s^r t^{1-2m} \int_{M[t]} \operatorname{Ric} \Omega \wedge \upsilon^{m-1} \, dt$$

is defined. Define $v \geqq 0$ by $\upsilon^m = v\Omega$. For almost all $0 < s < r$, the *Ricci function* of τ is defined by

$$\operatorname{Ric}_\tau(r, s) = \operatorname{Ric}(r, s, \Omega) + \frac{1}{2} \int_{M\langle r\rangle} \log v\sigma - \frac{1}{2} \int_{M\langle s\rangle} \log v\sigma$$

and does not depend on the choice of Ω. If $\upsilon^m > 0$, then $\operatorname{Ric}_\tau(r, s) = \operatorname{Ric}(r, s, \upsilon^m)$. If $\beta\colon M \to \boldsymbol{C}^m$ is surjective, proper and holomorphic with $\tau = |\beta|^2$, the Jacobian of β defines a nonnegative divisor θ on M with $\operatorname{Ric}_\tau(r, s) = N_\theta(r, s)$. If $\beta(\operatorname{supp} \theta)$ is affine algebraic, then $N_\theta(r, s) = O(\log r)$ for $r \to \infty$. In the general situation, the *Ricci defect of f* is defined by

$$R_f = \limsup_{r\to\infty} \frac{\mathrm{Ric}_\tau(r, s)}{T_f(r, s, L, \kappa)}.$$

The projective sections $a_1, \cdots, a_q$ in $\boldsymbol{P}(V)$ are said to be in *general position* if and only if for each $x \in E_L[a_1] \cap \cdots \cap E_L[a_q]$ the following condition holds: "There are integers $1 \leqq \mu(1) < \cdots < \mu(t) \leqq q$ such that $x \in E_L[a_j]$ if and only if $j = \mu(\lambda)$ for exactly one λ. An open neighborhood U of x and a holomorphic section ν: $U \to L$ with $Z(\nu) = \emptyset$ exist. Take $0 \neq \mathfrak{v}_j \in V$ with $a_i = \boldsymbol{P}(\mathfrak{v}_j)$ for $j = 1, \cdots, q$. Then $\mathfrak{v}_j = w_j\nu$ on U where w_j: $U \to \boldsymbol{C}$ is holomorphic. Then $(dw_{\mu(1)} \wedge \cdots \wedge dw_{\mu(t)})(x) \neq 0$."

Let K_N and K_M be the canonical bundles of N and M respectively. Let K_N^* be the dual bundle of K_N. Then f: $M \to N$ pulls back K_N and K_N^* to holomorphic line bundles $K_{N,f}$ and $K_{N,f}^*$ on M which are duals. A global holomorphic section $F \not\equiv 0$ of the *Jacobian bundle* $K(f) = K_M \otimes K_{N,f}^*$ is called a *Jacobian section*. Let ν_F be the divisor of F. The *ramification defect* of F is defined by

$$\Theta_F = \limsup_{r\to\infty} \frac{N_{\nu_F}(r, s)}{T_f(r, s, L, \kappa)} \geqq 0.$$

If $U \neq \emptyset$ is open in N, let $\Omega_N^n(U) = \Gamma(U, K_N)$ be the vector space of holomorphic forms of degree $(n, 0)$ on U. If $\tilde{U} = f^{-1}(U) \neq \emptyset$, then F acts as a linear map F: $\Omega_N^n(U) \to \Omega_M^m(\tilde{U})$. If $\psi \in \Omega_N^n(U)$, the section $\psi : U \to K_N$ pulls back to a section ψ_f: $\tilde{U} \to K_{N,f}$. Since $K_{N,f}$ is dual to $K_{N,f}^*$, the inner product $F[\psi] = (F, \psi_f)$ defines a holomorphic form of bidegree $(m, 0)$ on $\tilde{U}$. If $p \geqq 0$ is an integer, define

$$i_p = (-1)^{p(p-1)/2}\, p!\, (i/2\pi)^p.$$

Let $A_N^{2n}(U)$ be the vector space of forms of degree $2n$ on U. Then F acts as a linear map F: $A_N^{2n}(U) \to A_M^{2m}(\tilde{U})$ such that

$$F[i_n\varphi \wedge \bar{\chi}] = i_m F[\varphi] \wedge \overline{F[\chi]}$$

for all φ and χ in $\Omega_N^n(U)$. If $0 \leqq \psi \in A_N^{2n}(U)$, then $F[\psi] \geqq 0$. If $0 < \psi \in A_N^{2n}(U)$, then $\mathrm{Ric}\, F[\psi] = f^*(\mathrm{Ric}\, \psi)$ on $f^{-1}(U) - Z(F)$.

Under reasonable assumptions a Jacobian section F exists. The Second Main Theorem of **[1]** and **[2]** can be extended to parabolic manifolds without dimension restriction.

Define $M^+(r) = \{x \in M(r) \mid \upsilon(x) > 0\}$ if $r > 0$. Let $A_N^{p,q}(U)$ be the vector space of forms of bidegree p, q on the open subset U of N. The Jacobian section F is said to be dominated by τ if and only if for each $r > 0$ a minimal constant $Y(r) \geqq 1$ exists such that

$$n(F[\psi^n]/\upsilon^m)^{1/n} \leqq Y(r) f^*(\psi) \wedge \upsilon^{m-1}$$

on $f^{-1}(U) \cap M^+(r)$ whenever U is open in M and $0 \leqq \psi \in A_N^{1,1}(U)$. Then Y is called the *dominator* of F. If F is dominated by τ, then $m \geqq n = \mathrm{Rank}\, f$. Under reasonable assumptions a Jacobian section F dominated by τ exists to the given f. In some cases, even $Y \equiv m$ can be shown. Let F be a Jacobian section dominated by τ with dominator Y. Then the dominator defect is defined by

$$Y_f = \limsup_{r\to\infty} \frac{\log Y(r)}{T_f(r, s, L, \kappa)} \geqq 0.$$

The *Main Defect Relation* holds under the following assumptions. Let N be a compact, connected complex manifold of dimension n. Let L be a positive holomorphic line bundle on N. Let K_N be the canonical bundle on N. A smallest integer $p \geqq 0$ exists such that $L^p \otimes K_N$ is nonnegative. (Here $L^p = L \otimes \cdots \otimes L$.) Let $V = \Gamma(N, L)$ be the vector space of global holomorphic section of L. Let l be a hermitian metric on V. Take $a_1, \cdots, a_q$ in $\boldsymbol{P}(V)$ in general position with $q > p$. Let M be a noncompact, connected, complex manifold of dimension m carrying a parabolic exhaustion τ. Let f: $M \to N$ be a holomorphic map with $f(M) \not\subseteq E_L[a]$ for all $a \in \boldsymbol{P}(V)$. Assume a Jacobian section F dominated by τ is given. Let Y be the dominator. The *Main Defect Relation* holds:

$$\Theta_F + \sum_{j=1}^{q} \delta_f(a_j, L) \leqq p + R_f + \varsigma n Y_f.$$

If $R_f = \infty$ or $Y_f = \infty$, this is meaningless. If $R_f = Y_f = 0$, then $\sum_{j=1}^{q} \delta_f(a_j, L) \leqq p$. Since $q > p$ and $0 \leqq \delta_f(a_j, L) \leqq p$, this implies $f(M) \cap E_L[a_j] \neq \emptyset$ for at least one index j. The Picard-Borel theorem is obtained.

Several applications can be given. For instance if (M, τ) is a parabolic manifold define

$$\operatorname{Ord} \tau = \limsup_{r \to \infty} \frac{\log^+ \operatorname{Ric}_\tau(r, s)}{\log r}$$

as the *order of* τ. It is independent of s. Let N be a compact, connected projective algebraic manifold of dimension n. Let L be a holomorphic line bundle on N with a hermitian metric κ along L such that $c(L, \kappa) > 0$. Define

$$\operatorname{Ord} f = \limsup_{r \to \infty} \frac{\log^+ \operatorname{Ric}_\tau(r, s)}{\log r}$$

as the order of f. Then $\operatorname{Ord} f$ does not depend on s, L, κ. Assume $m = n =$ Rank f. Suppose N is of general type, meaning

$$\limsup_{k \to \infty} k^{-n} \dim \Gamma(N, K_N^k) > 0.$$

Then $\operatorname{Ord} f \leqq \operatorname{Ord} \tau$. This extends a result of [2] and of Kodaira to parabolic spaces.

References

1. J. Carlson and Ph. Griffiths, *A defect relation for equidimensional holomorphic mappings between algebraic varieties*, Ann. of Math. (2) **95** (1972), 557–584. MR **47** #497.

2. Ph. Griffiths and J. King, *Nevanlinna theory and holomorphic mappings between algebraic varieties*, Acta. Math. **130** (1973), 145–220.

3. W. Stoll, *Value distribution theory on parabolic spaces* (to appear).

University of Notre Dame

Proceedings of Symposia in Pure Mathematics
Volume 30, 1977

ON THE EQUIDISTRIBUTION THEORY OF HOLOMORPHIC MAPS

CHIA-CHI TUNG

In [7] a general First Main Theorem (F.M.T.) for an admissible family of analytic sets is proved. A formulation of this result is given in §2 of this note. For spaces with suitable exhaustions, the F.M.T. yields several Casorati-Weierstrass type results (see §3) generalizing theorems of Chern [1], [2], Cowen [3], Griffiths and King [4], Stoll [5] and Wu [8].

1. Admissible family of analytic sets. In the following, all complex spaces are reduced, pure dimensional and have a countable base of open sets. Let Y be a complex space. A collection $\mathfrak{A} = \{S_t\}_{t \in N}$ is said to be an admissible family in Y iff:

(A_1) The parameter set N is a locally irreducible complex space.

(A_2) There exist a complex space M and holomorphic maps $\pi: M \to N$, $h: M \to Y$, such that π is surjective, open, and h is proper, locally trivial at every point of M.

(A_3) For each $t \in N$, the restriction $h: \pi^{-1}(t) \to Y$ is injective, and $S_t = h(\pi^{-1}(t))$.

(A_4) S_t contains no branch of Y for all $t \in N$.

It follows that each S_t in $\mathfrak{A}$ is an analytic subset of Y of pure positive codimension s (independent of t). Consider the following example. Let $0 \leqq p \leqq q \leqq k - 1$ be integers and V a complex vector space of dimension $k + 1$. Let $G_p(V)$ be the Grassmann manifold of all projective p-planes in $\boldsymbol{P}(V)$. If $y \in G_p(V)$, let $E(y)$ denote the p-plane spanned by y in $\boldsymbol{P}(V)$. Then the associated flag manifold

$$F_{pq} = \{(y, z) \in G_p(V) \times G_q(V) \mid E(y) \subseteq E(z)\}$$

and the natural projections $\pi: F_{pq} \to G_q(V)$, $h: F_{pq} \to G_p(V)$ define an admissible family in $G_p(V)$.

AMS (MOS) subject classifications (1970). Primary 30A70, 32H99; Secondary 32C30.

Another example is provided by the set of all Schubert varieties in $G_p(V)$ associated to a given symbol $(a_0, a_1, \cdots, a_n)$ (see Cowen [3]).

Let X be a complex space of dimension m, $\mathfrak{A} = \{S_t\}_{t\in N}$ an admissible family in Y of codimension $s \leqq m$, and $f\colon X \to Y$ a holomorphic map. Let $(\tilde{f}, \tilde{h})$ be the fiber product of (f, h).

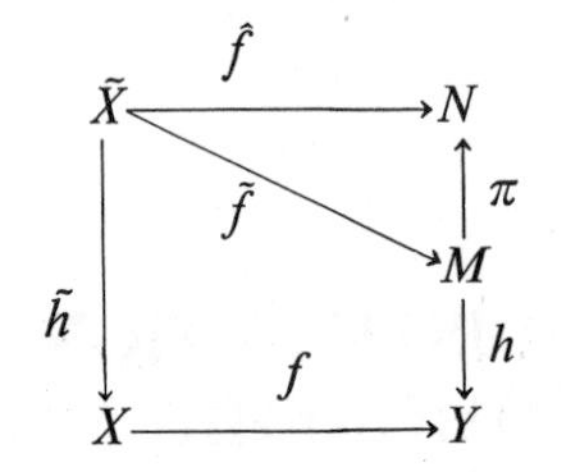

Then f is said to be almost adapted to $\mathfrak{A}$ iff $\hat{f} = \pi \circ \tilde{f}$ has strict maximal rank, i.e., the restriction of $\hat{f}$ to every branch of $\tilde{X}$ has rank $k = \dim N$. Let $(a, x) \in N \times X$: f is said to be adapted to a at x (rel. to $\mathfrak{A}$) iff there exist open neighborhoods U of a and V of x such that either $f^{-1}(S_t) \cap V$ has pure codimension s for all $t \in U$ or the set is empty for all $t \in U$. If $G \subseteq X$ let $N_G(f, \mathfrak{A})$ be the set of all $t \in N$ to which f is adapted at every $x \in G$. If f is almost adapted to $\mathfrak{A}$, then $N - N_G(f, \mathfrak{A})$ has measure zero in N.

THEOREM 1 (CROFTON FORMULA). *Let X, $\mathfrak{A}$ and $f\colon X \to Y$ be as above. Assume f is almost adapted to $\mathfrak{A}$. Let ζ, resp. ω, be a continuous differential form on X of degree $2(m-s)$, resp. on N of degree $2k$. Define $\Omega = h_*\pi^*(\omega)$. (Here h_* denotes integration along fibers of h.) Let G be an open set in X such that $K = \bar{G} \cap \operatorname{supp} \zeta$ is compact. For each $t \in N_K(f, \mathfrak{A})$, define $n(f(G) \cap S_t) = \int_{f^{-1}(S_t)\cap G} \nu_f^t \zeta$. Then*

$$\int_G f^*(\Omega) \wedge \zeta = \int_N n(f(G) \cap S_t)\, \omega .$$

REMARK. The intersection multiplicity ν_f^t of f with S_t is inserted because of its appearance in the F.M.T. (see [7]).

2. The first main theorem. A relatively compact open set G in a complex space X of dimension m is called a Stokes domain iff (1) $\Gamma = \partial G$ has locally finite (Hausdorff) $(2m-1)$-measure; (2) Γ contains a C^∞ boundary manifold Γ_0 of $G \cap X_{\mathrm{reg}}$ such that $\Gamma - \Gamma_0$ has zero $(2m-1)$-measure. A triple (g, G, ψ) is called a bump in X iff (1) $\phi \subseteq g \subset\subset G$, where G and g (if nonempty) are Stokes domains in X; (2) $\psi\colon X \to \boldsymbol{R}_+ = [0, \infty)$ is a continuous function such that (i) $\operatorname{supp} \psi \subseteq \bar{G}$, (ii) $\psi \mid \bar{G} - g$ is of class C^2, (iii) $\psi \mid \bar{g} = \operatorname{Max} \psi \mid \bar{G} > 0$ (if $g \neq \phi$).

THEOREM 2 (F.M.T.). *Let $f\colon X \to Y$ be a holomorphic map from a complex space X of dimension m into a complex space Y, and $\mathfrak{A} = \{S_t\}_{t\in N}$ an admissible family in Y of codimension s with $m - s = q \geqq 0$. Let $\lambda = \{\lambda_t\}_{t\in N}$ be a singular potential for a nonnegative form $\omega \in A^{2k}(N)$, where $k = \dim N$ (see* [5], [7]*). Define $\Omega = h_*\pi^*(\omega)$ and $\Lambda_t = h_*\pi^*(\lambda_t)$ for $t \in N$. Let (g, G, ψ) be a bump in X. Assume $\chi \in A_1^{q,q}(X)$ is closed and strictly nonnegative. Let $K = \bar{G} \cap \operatorname{supp} \chi$. Then $Q = N_K(f, \mathfrak{A})$ is open and for each $t \in Q$,*

$$T_f(G) + D_f(G, t) = N_f(G, t) + m_f(\Gamma, t) - m_f(\gamma, t).$$

Here

$$T_f(G) = \int_G \psi f^*(\Omega) \wedge \chi \geqq 0 \qquad \textit{(characteristic)},$$

$$m_f(\Gamma, t) = \int_\Gamma f^*(\Lambda_t) \wedge d^\perp\psi \wedge \chi \geqq 0 \quad \textit{(exterior proximity)},$$

$$m_f(\gamma, t) = \int_\gamma f^*(\Lambda_t) \wedge d^\perp\psi \wedge \chi \geqq 0 \quad \textit{(interior proximity)},$$

$$N_f(G, t) = \int_{f^{-1}(S_t)\cap G} \nu_f^t \psi \chi \geqq 0 \qquad \textit{(valence)},$$

$$D_f(G, t) = \int_{G-\bar{g}} f^*(\Lambda_t) \wedge d^c d\psi \wedge \chi \qquad \textit{(deficit)},$$

define continuous functions of t in Q; $d^\perp = i(\partial - \bar\partial) = - d^c$.

REMARKS. (1) Strict nonnegativity of χ means that $\chi \wedge \eta \geqq 0$ for all nonnegative $\eta \in A_0^{s,s}(X)$. The F.M.T. remains valid in case f is almost adapted to $\mathfrak{A}$ and $\chi \geqq 0$ outside a thin analytic subset of X.

(2) The F.M.T. implies the Nevanlinna inequality:

$$N_f(G, t) \leqq T_f(G) + D_f(G, t) + m_f(\gamma, t) \qquad (t \in Q).$$

(3) *Equidistribution.* Some general assumptions shall be stated here for later reference. (I) X is a noncompact complex space of dimension m. (II) N is a connected compact complex manifold of dimension k. (III) $\mathfrak{A} = \{S_t\}_{t\in N}$ is an admissible family in a complex space Y of codimension s with $m - s = q \geqq 0$. (IV) $f\colon X \to Y$ is a holomorphic map almost adapted to $\mathfrak{A}$. (V) λ is a singular potential on N for a nonnegative $\omega \in A_\infty^{2k}(N)$ with $\int_N \omega = 1$. Define $\Omega = h_*\pi^*(\omega)$ on Y and $\Lambda_t = h_*\pi^*(\lambda_t)$ on $Y - S_t$ for $t \in N$.

A C^∞-map $\varphi\colon X \to \boldsymbol{R}_+$ is called an exhaustion function iff each sublevel set $X[r] = \{x \in X \mid \varphi(x) \leqq r\}$ is compact. For each $r > 0$, define $X_r = \{x \in X \mid \varphi(x) < r\}$.

THEOREM 3. *Assume* (I) — (V). *Suppose there exists an exhaustion function* φ: $X \to \boldsymbol{R}_+$ *such that*

(i) $X[0]$ *has measure zero*;

(ii) $\omega_1 = (1/2\pi)\, dd^c \log \varphi \geqq 0$ *on* $X - X[0]$ *and* $\omega_1^m = 0$;

(iii) *either* $\chi_q = ((1/4\pi)/ dd^c\varphi^2)^q > 0$ *on an open set*[1] $U \subset X$ *for which* $f(U)$ *meets at least one* $S_t \in \mathfrak{A}$, *or* $f^*(\Omega) \wedge \chi_q \not\equiv 0$. *Then*

(1) *If* codim $\mathfrak{A} = 1$, Im(f) *meets almost all* $S_t \in \mathfrak{A}$.

(2) *If* codim $\mathfrak{A} > 1$ *and if*

$$\liminf_{r\to\infty} \frac{\int_N(\int_{X(\sigma,r)} f^*(\Lambda_t) \wedge \omega_1^{q+1})\omega}{\int_X \psi_{r\sigma} f^*(\Omega) \wedge \chi_q} = 0,$$

the same conclusion holds. Here $0 < \sigma =$ *constant* $< r$ *and*

$$\begin{aligned} \psi_{r\sigma}(x) &= 0 && \textit{on } X - X_r,\\ &= (1/2q)\,(1/\varphi(x)^{2q} - 1/r^{2q}) && \textit{on } X[\sigma, r],\\ &= (1/2q)\,(1/\sigma^{2q} - 1/r^{2q}) && \textit{on } X_\sigma. \end{aligned}$$

[1] This means χ_q has positive extensions into local imbedding spaces of U.

COROLLARY 3.1. *Let V be a complex vector space of dimension $k + 1$ and $p \in Z[0, k - 1]$. For each $t \in G_{k-p-1}(V)$ define $S_t = \{y \in G_p(V) | E(y) \cap E(t) \neq \emptyset\}$. Then $\mathfrak{A} = \{S_t\}$ forms an admissible family in $G_p(V)$ (with a suitable parameter space). Let X, φ and f be as above with $Y = G_p(V)$. Then $\mathrm{Im}(f)$ meets almost all S_t for $t \in G_{k-p-1}(V)$.*

THEOREM 4. *Assume* (I) — (V) *and*

(i) *X is pseudoconcave, i.e., there exists an exhaustion $\varphi: X \to \boldsymbol{R}_+$ with $dd^c\varphi \leqq 0$ on $X - X[\sigma]$ for some σ;*

(ii) *$\chi \in A_1^{q,q}(X)$ is closed in X and nonnegative outside a thin analytic subset of X;*

(iii) *either $\chi > 0$ on an open set $U \subseteq X$ for which $f(U)$ meets at least one $S_t \in \mathfrak{A}$ or $f^*(\Omega) \wedge \chi \not\equiv 0$.*

Then $\mathrm{Im}(f)$ meets almost all $S_t \in \mathfrak{A}$.

THEOREM 5. *Assume* (I) — (IV) *and N is Kähler with fundamental form ω_0 normalized such that $\int_N \omega_0^k = 1$. Assume X is pseudoconvex, i.e., there exists an exhaution $\varphi: X \to \boldsymbol{R}_+$ such that $\chi_1 = dd^c \varphi$ is nonnegative outside a thin analytic subset of X. Define $\Omega_p = h_*\pi^*(\omega_0^{p+k-s})$ on Y, for $p \in \boldsymbol{Z}[1, s]$. Assume*

(i) *either $\chi_q = \chi_1^q > 0$ on an open set $U \subseteq X$ for which $f(U)$ meets at least one $S_t \in \mathfrak{A}$ or $f^*(\Omega_s) \wedge \chi_q \not\equiv 0$;*

(ii) $\liminf_{r\to\infty} (\int_{X_r} f^*(\Omega_{s-1}) \wedge \chi_{q+1})/(\int_{X_r} \psi_r f^*(\Omega_s) \wedge \chi_q) = 0$.

Here $\psi_r = r - \varphi$ on $X[\sigma, r]$ and $\psi_r = r - \sigma$ on X_σ, $0 < \sigma = const < r$. Then $\mathrm{Im}(f)$ meets almost all $S_t \in \mathfrak{A}$.

Let V be a hermitian vector space of dimension $k + 1$ and $p \in \boldsymbol{Z}[0, k - 1]$. Let $\mathfrak{A}_{p,V} = \{E_t\}_{t \in G_p(V)}$ be the family of projective p-planes in $\boldsymbol{P}(V)$. Denote by ω_0 the Kähler form for the Fubini-Study metric on $\boldsymbol{P}(V)$ defined by the hermitian product on V.

COROLLARY 5.1. *Let X be a complex space of dimension m with pseudoconvex exhaustion φ and $f: X \to \boldsymbol{P}(V)$ a holomorphic map almost adapted to $\mathfrak{A}_{p,V}$. Assume*

(i) $q = m - \operatorname{codim} \mathfrak{A}_{p,V} \geqq 0$;

(ii) *$\chi_q = (dd^c\varphi)^q > 0$ on some nonvoid open set in X; and*

(iii) $\liminf_{r\to\infty} (\int_{X_r} f^*(\omega_0^{s-1}) \wedge \chi_{q+1})/(\int_{X_r} \psi_r f^*(\omega_0^s) \wedge \chi_q) = 0$.

Then $\mathrm{Im}(f)$ meets almost all p-planes in $\boldsymbol{P}(V)$.

REFERENCES

1. S. S. Chern, *Holomorphic curves and minimal surfaces*, Proc. Carolina Conf. on Holomorphic Mappings and Minimal Surfaces (Chapel Hill, N.C., 1970), Dept. of Math., Univ. of North Carolina, Chapel Hill, N.C., 1970, pp. 1–28. MR **42** #8400.

2. ———, *The integrated form of the first main theorem for complex analytic mappings in several variables*, Ann. of Math. (2) **71** (1960), 536–551. MR **23** #A3276.

3. M. Cowen, *Value distribution theory*, Trans. Amer. Math. Soc. **180** (1973), 189–228.

4. P. Griffiths and J. King, *Nevanlinna theory and holomorphic mappings between algebraic varieties*, Acta Math. **130** (1973), 145–220.

5. W. Stoll, *Value distribution of holomorphic maps into compact complex manifolds*, Lecture Notes in Math., Vol. 135, Springer-Verlag, Berlin and New York, 1970. MR **42** #2040.

6. ———, *About the value distribution of holomorphic maps into projective space*, Acta Math. **123** (1969), 83–114. MR **41** #3815.

7. C. Tung, *The first main theorem on complex spaces*, Ph.D. Thesis, University of Notre Dame, 1973.

8. H. Wu, *Remarks on the first main theorem of equidistribution theory*. I, II, III, J. Differential Geometry **2** (1968), 197–202; ibid. **3** (1969), 83–94, 369–384. MR **43** #2247a, b, c.

UNIVERSITY OF CHICAGO

COLUMBIA UNIVERSITY (Current address)

SEMINAR SERIES

GROUP REPRESENTATION AND HARMONIC ANALYSIS

Proceedings of Symposia in Pure Mathematics
Volume 30, 1977

THETA FUNCTIONS WITH CHARACTERISTIC, AND DISTINGUISHED SUBSPACES OF THE HEISENBERG MANIFOLD

LOUIS AUSLANDER

Introduction. In this paper we will discuss the deep relation between first order Jacobi's theta functions with characteristics and the distinguished subspaces of the Heisenberg manifold as introduced in [2], [3] and [4]. Full details will be given elsewhere.

Let us begin by recalling the basic facts about first order theta functions with characteristics. If $\tau \in H^+$, ε, ε', $m \in \boldsymbol{Z}$ and $\zeta \in \boldsymbol{C}$, under H^+ is the upper half of the complex plane, then explicit formulas for first order theta functions with characteristic and period τ are given by

$$\Theta\left[\frac{\varepsilon}{m}, \frac{\varepsilon'}{m}\right](\zeta, \tau) = \sum_{\alpha \in \boldsymbol{Z}} \exp \pi i \left\{\tau\left(\alpha + \frac{\varepsilon}{m}\right)^2 + 2\left(\alpha + \frac{\varepsilon}{m}\right)\left(\zeta + \frac{\varepsilon'}{m}\right)\right\}.$$

The above functions satisfy the functional equation

$$(1) \quad \Theta\left[\frac{\varepsilon}{m}, \frac{\varepsilon'}{m}\right](\zeta + 1, \tau) = \exp\left(2\pi i \frac{\varepsilon}{m}\right) \Theta\left[\frac{\varepsilon}{m}, \frac{\varepsilon'}{m}\right](\zeta, \tau),$$

$$(2) \quad \Theta\left[\frac{\varepsilon}{m}, \frac{\varepsilon'}{m}\right](\zeta + \tau, \tau) = \exp -\left(2\pi i \frac{\varepsilon'}{m}\right) \Theta\left[\frac{\varepsilon}{m}, \frac{\varepsilon'}{m}\right](\zeta, \tau) \exp \pi i(-\tau - \zeta).$$

If $n_1, n_2 \in \boldsymbol{Z}$ and $L(\tau)$ denotes the lattice $n_1 + n_2\tau$, $\tau \in H^+$, then the functional equations (1) and (2) show that $[\varepsilon/m, \varepsilon'/m]$ really is a character on $L(\tau)$ given explicitly by

$$\left[\frac{\varepsilon}{m}, \frac{\varepsilon'}{m}\right](n_1 + n_2\tau) = \exp 2\pi i \left(\frac{n_1\varepsilon}{m} - \frac{n_2\varepsilon'}{m}\right)$$

AMS (*MOS*) *subject classifications* (1970). Primary 33A75, 43A65.

whose range is the mth roots of unity. If $[\varepsilon/m, \varepsilon'/m]$ is a surjection it will be called an m-nondegenerate character. The fundamental uniqueness theorem is that if $\psi(\zeta, \tau)$ is any entire function of ζ, satisfying the functional equations (1) and (2) for all $\tau \in H^+$ and $[\varepsilon/m, \varepsilon'/m]$ then $\psi(\zeta, \tau) = C\Theta[\varepsilon/m, \varepsilon'/m]\,(\zeta, \tau)$ where C is a constant.

We will introduce the notation $\Theta[\varepsilon/m, \varepsilon'/m]\,(0, \tau) = \Theta[\varepsilon/m, \varepsilon'/m]$. (Warning: In many modern texts $\Theta[\varepsilon/m, \varepsilon'/m]$ is what is called a theta function.)

1. The Heisenberg group. The Heisenberg group is written in different forms in the literature. We will call these different ways of writing the Heisenberg group a presentation and briefly indicate that they reflect a finer algebraic object than the Heisenberg group.

If N_3 is the Heisenberg group and x_1, y, $t \in \boldsymbol{R}$ and $\zeta \in \boldsymbol{C}$, $\zeta = x + iy$, then we may write N_3 in any of the ways listed below.

$$(x_1, y_1, t_1)(x_2, y_2, t_2) = (x_1 + x_2, y_1 + y_2, t_1 + t_2 + y_1x_2), \tag{3}$$

$$(x_1, y_1, t_1)(x_2, y_2, t_2) = (x_1+x_2, y_1+y_2, t_1+t_2+(y_1x_2)/k), \quad k \neq 0, k \in \boldsymbol{R}, \tag{4}$$

$$(\zeta_1, t_1)(\zeta_2, t_2) = (\zeta_1 + \zeta_2, t_1 + t_2 + \operatorname{Im}(z_1\bar{z}_2)), \tag{5a}$$

$$(x_1, y_1, t_1)(x_2, y_2, t_2) = (x_1 + x_2, y_1 + y_2, t_1 + t_2 + (x_2y_1 - x_1y_2)), \tag{5b}$$

where (5a) and (5b) are clearly the same way of writing N_3 except for a notational change. Let $\mathfrak{A}$ be a finite dimension nilpotent associative algebra. Then, if $I + \mathfrak{A}$ denotes the algebra obtained by adjoining an identity to $\mathfrak{A}$, where I will denote the identity, the following are easily verified and the details are in **[6]**:

(6) $N = \{(1 + a) \in I + \mathfrak{A}\}$ is a nilpotent Lie group.

(7) If $\mathfrak{A}_L$ denotes the Lie algebra of $\mathfrak{A}$ given by $[a, b] = ab - ba$, $a, b \in \mathfrak{A}$, we have $\mathfrak{A}_L$ is the Lie algebra of N.

Now we may write (3) and (5b) as the matrix groups

$$\begin{pmatrix} 1 & y & t \\ 0 & 1 & x \\ 0 & 0 & 1 \end{pmatrix}, \qquad \begin{pmatrix} 1 & -x & y & t \\ 0 & 1 & 0 & y \\ 0 & 0 & 1 & x \\ 0 & 0 & 0 & 1 \end{pmatrix}.$$

If we subtract the identity matrix we get two nilpotent associative alegbras of matrices which we will denote by $\mathfrak{A}_1$ and $\mathfrak{A}_2$ respectively. It is important to note that $\mathfrak{A}_1$ and $\mathfrak{A}_2$ are not isomorphic as associative algebras because the left annihilator of $\mathfrak{A}_1$ has dimension two and the left annihilator of $\mathfrak{A}_2$ has dimension one.

Thus we may call $I + \mathfrak{A}$ a presentation of the Heisenberg group N_3 and we will say two presentations $I + \mathfrak{A}_1$, $I + \mathfrak{A}_2$ are different if $\mathfrak{A}_1$ and $\mathfrak{A}_2$ are not alegbra isomorphic.

In this paper we will work with the presentation of N_3 given by formula (3).

2. The Heisenberg manifold. In N_3 consider the subgroup

$$\Gamma_0 = \{(n_1, n_2, n_3) \in N_3 \mid n_i \in Z, i = 1, 2, 3\}.$$

Then Γ_0 is a discrete subgroup of N_3 and $\Gamma_0\backslash N_3$ is compact. In N_3 consider the set $\mathscr{D}$ of all subgroups isomorphic to Γ_0. Then by the results in **[1]** it follows that if $\Gamma \in \mathscr{D}$ then Γ is a discrete subgroup of N_3 and $\Gamma\backslash N_3$ is compact. Indeed, any nonabelian subgroup Γ of N_3 generated by two elements is an element of $\mathscr{D}$. Again this may be seen from the results in **[1]**. The facts are well known and we will review them briefly. **[1]** and **[3]** will serve as general references for the statements contained in the rest of this section.

(1) Every isomorphism $\alpha\colon \Gamma_1 \to \Gamma_2, \Gamma_i \in \mathscr{D}, i = 1, 2,$ is uniquely extendable to an isomorphism $\alpha_1\colon N_3 \to N_3$ and so induces a diffeomorphism $\alpha_2\colon \Gamma_1\backslash N_3 \to \Gamma_2\backslash N_3$.

(2) The 3-form $\omega = dx \wedge dy \wedge dt$ determines a two-sided invariant measure on N_3.

(3) We define N_3 acting on $\Gamma\backslash N_3$ by $\Gamma n \to \Gamma ng$, $g, n \in N_3$, and if F is a function on $\Gamma\backslash N_3$ we define $(R(g)F)(\Gamma n) = F(\Gamma ng)$.

(4) ω induces a measure on $\Gamma\backslash N_3$ that is invariant under the action of N_3 on $\Gamma\backslash N_3$. We will let $L^2(\Gamma\backslash N_3)$ denote the L^2 spaces in $\Gamma\backslash N_3$ with respect to the probability measure of the form $C\omega$, $C > 0$.

(5) Let $F \in L^2(\Gamma\backslash N_3)$. Then $F \to R(g)F$ is a unitary operator and determines a unitary representation R_Γ of N_3.

(6) If $\alpha\colon \Gamma_1 \to \Gamma_2, \Gamma_i \in \mathscr{D}$ is an isomorphism then α_2 determines an intertwining operator

$$\alpha^*\colon L^2(\Gamma_2\backslash N_3) \to L^2(\Gamma_1\backslash N_3)$$

for the representations R_{Γ_2} and R_{Γ_1}.

(7) Let $z(\ )$ denote the center of the group in the bracket. Then $\Gamma\backslash N_3$ is a principal fiber bundle over $\Gamma_z(N_3)\backslash N_3 = \Pi^2$, the two-dimensional torus, with $z(N_3)/z(\Gamma)$ as group and fiber.

(8) Let $a(\Gamma)$ be the positive generator of the cyclic group $\Gamma \cap z(N_3) \subset (0, 0, t)$, $t \in R$. Then

$$L^2(\Gamma\backslash N_3) = \bigoplus \sum_{m \in \mathbf{Z}} H_m(\Gamma)$$

where the sum is the orthogonal direct sum, and $F \in H_m(\Gamma)$ if and only if $F \in L^2(\Gamma\backslash N_3)$ and

$$F(x, y, z + t) = \exp\{2\pi i(m/a(\Gamma))z\}\, F(x, y, t).$$

3. $(m_1\Gamma_0)$-equivalence. Let $\Gamma \in \mathscr{D}$. We define $\Gamma(m)$ as the subgroup of N_3 generated by Γ and the element $m^{-1}\, a(\Gamma)$, $m \in Z - \{0\}$. The important property of $\Gamma(m)$ is that if $F \in H_m(\Gamma)$ then F is $\Gamma(m)$ invariant. Notice that for $m \neq \pm 1$, $\Gamma(m) \notin \mathscr{D}$.

DEFINITIONS. (1) Let $\Gamma \subset \Gamma_0$, $\Gamma \in \mathscr{D}$ and let $a(\Gamma) = m$. We will call Γ (m, Γ_0)-admissible.

(2) We will call Γ_1 and Γ_2 (m, Γ_0)-equivalent if Γ_1 and Γ_2 are (m, Γ_0)-admissible and $\Gamma_1(m) = \Gamma_2(m)$.

The following characterization of the concept of (m, Γ_0)-equivalence should help the reader to understand this concept.

Let Γ_1 and Γ_2 be (m, Γ_0)-admissible. Then Γ_1 and Γ_2 are (m, Γ_0)-equivalent if and only if the images of Γ_1 and Γ_2 in $N_3/z(N_3)$ coincide.

The following result is the main reason for introducing the concept of (m, Γ_0)-equivalence.

THEOREM 1. *Let Γ_1 and Γ_2 be (m, Γ_0)-admissible. Then Γ_1 and Γ_2 are (m, Γ_0)-equivalent if and only if $H_m(\Gamma_1) = H_m(\Gamma_2)$.*

4. From theta functions to $H_m(\Gamma)$. Let $[\varepsilon/m, \varepsilon'/m]$ be an m-nondegenerate character on $L(\tau)$. Let β be the homomorphism of Γ to $L(\tau)$ defined by

$$\beta(n_1, n_2, n_3) = n_1 + n_2\tau, \qquad \tau \in H^+.$$

Let K^* be kernel of the character $[\varepsilon/m, \varepsilon'/m] \circ \beta$. We define $\Gamma_\alpha[\varepsilon/m, \varepsilon'/m]$ to be the (m, Γ_0)-admissible groups such that $\Gamma_\alpha[\varepsilon/m, \varepsilon'/m](m) = K^*$, $\alpha = 1, \cdots, N$.

THEOREM 2. *Let all notation be as above. The groups $\Gamma_\alpha[\varepsilon/m, \varepsilon'/m]$, $\alpha = 1, \cdots, N$, are exactly the elements in an (m, Γ_0)-equivalence class. Conversely, let $\{\Gamma\}$ be an (m, Γ_0)-equivalence class; then there exists an m-nondegenerate character $[\varepsilon/m, \varepsilon'/m]$ such that the groups $\Gamma_\alpha[\varepsilon/m, \varepsilon'/m]$, $\alpha = 1, \cdots, N$, are the elements of $\{\Gamma\}$.*

Since $H_m(\Gamma_\alpha[\varepsilon/m, \varepsilon'/m])$ are the same for all α, we will denote this space simply by $H_m(\Gamma[\varepsilon/m, \varepsilon'/m])$.

THEOREM 3. *Let K be the kernel of an m-nondegenerate character $[\varepsilon/m, \varepsilon'/m]$ and let $[\varepsilon_i/m, \varepsilon_i'/m]$, $i = 0, \cdots, m-1$, be the m-distinct characters on $L(\tau)/K$. Then the following are true.*

(1) $e^{2\pi i t}e^{\pi i \tau y^2}\, \Theta[\varepsilon_i/m, \varepsilon_i'/m](x + zy, \tau) \in H_m(\Gamma[\varepsilon/m, \varepsilon'/m])$, $i = 0, \cdots, m-1$.

(2) *If $\Theta_i(m, [\varepsilon/m, \varepsilon'/m], \Gamma_0)$ is the irreducible subspace under R of $H_m(\Gamma[\varepsilon/m, \varepsilon'/m])$ that contains the function in* (1) *above, then*

$$H_m\left(\Gamma\left[\frac{\varepsilon}{m}, \frac{\varepsilon'}{m}\right]\right) = \oplus \sum_{i=0}^{m} \Theta_i\left(m, \left[\frac{\varepsilon}{m}, \frac{\varepsilon'}{m}\right], \Gamma_0\right).$$

(3) *Let $B\colon \Gamma_0 \to \Gamma_\alpha[\varepsilon/m, \varepsilon'/m]$ be an isomorphism. Then*

$$H_m(\Gamma_0) = \oplus \sum B^* \Theta_i\left(m, \left[\frac{\varepsilon}{m}, \frac{\varepsilon'}{m}\right], \Gamma_0\right).$$

(4) *As $[\varepsilon/m, \varepsilon'/m]$ varies over all m-nondegenerate characters the subspaces $B^*\Theta_i(m, [\varepsilon/m, \varepsilon'/m], \Gamma_0)$, $i = 0, \cdots, m-1$, are exactly the set of distinguished subspaces of Auslander and Brezin* [**2**], [**3**], [**4**].

REFERENCES

1. L. Auslander, *An exposition of the structure of solvmanifolds.* I, *Algebraic theory*, Bull. Amer. Math. Soc. **79** (1973), 227–261.

2. L. Auslander and J. Brezin, *Invariant subspaces theory for three-dimensional nilmanifolds*, Bull. Amer. Math. Soc. **78** (1972), 255–258. MR **44** #6903.

3. ———, *Translation-invariant subspaces on L of a compact nilmanifold.* I, Invent Math. **20** (1973), 1–14. MR **48** #464.

4. ———, *Fiber bundle structure and harmonic analysis of compact Heisenberg manifolds*, Lecture Notes in Math., vol. 266, Springer-Verlag, Berlin and New York, 1971, pp. 17–22.

5. L. Auslander, R. Tolimieri (with the assistance of H. E. Rauch), *Abelian harmonic analysis, theta functions and function algebras on a nilmanifold*, Lecture Notes in Math., vol. 436, Springer-Verlag, Berlin and New York, 1975.

6. L. Auslander, *Simply transitive groups of affine motions* (to appear).

7. H. E. Rauch and A. Lebowitz, *Elliptic functions, theta functions and Riemann surfaces*, Williams and Wilkins, Baltimore, Md., 1973. MR **50** #2486.

CITY UNIVERSITY OF NEW YORK

Proceedings of Symposia in Pure Mathematics
Volume 30, 1977

THE COMPLEX-ANALYTIC EXTENSION OF THE FOURIER SERIES ON LIE GROUPS*

LUIS A. FROTA-MATTOS

1. Introduction. This paper primarily studies the C^∞ functions on a connected compact real Lie group K which have an analytic continuation into a neighbourhood of K in the complexification G of K. We obtain an equivalent condition through the Fourier coefficients of such a function showing that they have to decrease exponentially. The rate of decay is analyzed and we achieve the optimum rate.

The region of analyticity naturally arising from the analytic continuation of the Peter-Weyl expansion is shown to be a domain of holomorphy, having some analogy with classical Reinhardt and tube domains.

As an application we obtain the well-known result in classical analysis that any analytic function in the interior of an ellipse with foci ± 1 can be represented there by a convergent series of Gegenbauer polynomials [**1**]. Actually, we achieve a result which proves more than this classical one. We obtain uniform absolute convergence of the series on the compact subsets of the ellipse while the classical result asserts uniform convergence on the compact subsets of the ellipse [**2**].

2. Analytic continuation. Let K be a connected compact real Lie group with Lie algebra $\mathfrak{K}$. Let G be the connected complex Lie group, with Lie algebra $\mathfrak{G} = \mathfrak{K} + i\mathfrak{K}$, $\sqrt{-1} = i$, containing K. Call G the complexification of K.

Let $\mathfrak{H}$ be a maximal abelian subalgebra of $\mathfrak{K}$. Set $\mathfrak{A} = i\mathfrak{H}$ and $A = \exp \mathfrak{A}$. It is true that G has the following decomposition, $G = KAK$ [**3**]. Let $\|\cdot\|$ be the norm defined in terms of the Killing form.

DEFINITION 1. If $\varepsilon > 0$ then define $\mathfrak{A}_\varepsilon = \{X \in \mathfrak{A} \mid \|X\| < \varepsilon\}$. If $\exp \mathfrak{A}_\varepsilon = A_\varepsilon$ then set $G_\varepsilon = KA_\varepsilon K$.

AMS (MOS) subject classifications (1970). Primary 43A75, 22E30, 32D05, 32A07, 33A75.

*This work reflects the author's 1975 Ph.D. thesis, written under the direction of Professor Roe Goodman at Rutgers University [**3**].

A representation π of K acting on a complex Hilbert space of finite dimension d_π can be taken unitary on K. Let $\hat{K}$ be the set of equivalence classes of finite-dimensional irreducible unitary representations of K. For each $\pi \in \hat{K}$ there exists one and only one corresponding highest weight λ_π.

By the Peter-Weyl theorem, for a given function f in $L^2(K)$, the series

$$f(k) \simeq \sum_{\pi \in \hat{K}} d_\pi \operatorname{tr}(\pi(k)\hat{f}(\pi))$$

converges in $L^2(K)$, where $\hat{f}(\pi) = \int_K f(k)\overline{\pi(k)}\, dk$.

For a detailed discussion on this subject see N. Wallach's book **[4]**.

THEOREM 1. *Let f be a complex-valued C^∞ function defined on K. Then f has a complex-analytic extension into G_ε if and only if* $(\forall\, \pi \in \hat{K})$ $(\forall\, \delta, 0 < \delta < \varepsilon)$ $\|\hat{f}(\pi)\|_\infty < C_\delta \exp(-\delta \|\lambda_\pi\|)$, *where C_δ is a constant.*

Here $\| \ \|_\infty$ denotes the sup norm **[9]**.

We mainly use the fact that λ_π has highest length to prove the sufficiency.

The necessity follows with the help of the operator $\theta(t)$ defined as follows. Let L be the left regular representation of K and Ad the adjoint representation of K. For fixed $X \in \mathfrak{A}$ and t real such that $\|tX\| \leqq \delta < \varepsilon$, define

$$\theta(t) = \int_K L(\exp(\operatorname{Ad}(k)(-tX)))\, dk.$$

The operator $\theta(t)$ is evaluated on the complex-analytic functions from G_ε into the complex plane $\boldsymbol{C}$. We get, by Schur's lemma, that $\theta(t)\pi_{ij} = m_\pi(t)\pi_{ij}$, where

$$m_\pi(t) = (1/d_\pi) \sum_{p=1}^{d_\pi} \pi_{pp}(\exp(tX)),$$

if π_{ij} denotes an entry of the representation $\pi \in \hat{K}$ **[3]**. The operator $\theta(t)$ was already used by Beers and Dragt in their paper **[6]** for the special case of $K = SU(2)$.

This result includes a consequence of R.T. Seeley's theorem **[7]** that f is real-analytic on K if its Fourier coefficients have an exponential decay.

A set G_ε into which f has a complex-analytic extension can be determined by techniques similar to those for Taylor series. We refer to Beers and Dragt for the next definition and the next two theorems for the case $K = SU(2)$.

DEFINITION 2. Let f be an $L^2(K)$ function. Let $\delta \geqq 0$. Consider the sequence of numbers $a_\pi(\delta) = \|\hat{f}(\pi)\|_\infty \exp(\|\lambda_\pi\|\delta)$ where π ranges over $\hat{K}$. Define

$$\rho = \sup\{\delta \geqq 0 \,|\, \{a_\pi(\delta)\}_{\pi \in \hat{K}} \text{ is bounded}\}.$$

THEOREM 2. *The following formula holds:*

$$\rho = -\log\left[\limsup \|\hat{f}(\pi)\|_\infty^{1/\|\lambda_\pi\|}\right],$$

where the limit superior is taken over all $\pi \in \hat{K}$ as $\|\lambda_\pi\| \to \infty$.

THEOREM 3. *Let f be a C^∞ function on K. Let ρ be as in Theorem 2. Then f has a complex-analytic extension into G_ρ.*

3. Domains of holomorphy. In general, to decide if an open connected set $U \subset \boldsymbol{C}^n$ is a domain of holomorphy is a rather complicated business. One process used in classical analysis for this purpose deals with the analytic continuation of the Taylor series. Here we make use of the anlaytic extension of the Peter-Weyl series as a tool to decide if the set G_ε is a domain of holomorphy of the complex Lie group G.

THEOREM 4. *The set G_ε is a domain of holomorphy.*

This theorem induces us to generalize the concept of Reinhardt domain and to have a concept of log-convex set **[8]** for a complexification G of the real connected compact Lie group K.

DEFINITION 3. A Reinhardt domain Ω is an open connected subset of G such that if $g \in \Omega$ then $k_1 g k_2 \in \Omega$ for all $k_1, k_2 \in K$, that is, if $K\Omega K \subset \Omega$.

DEFINITION 4. Let $\Omega \subset G$ be a Reinhardt domain. Its real representation in A, denoted $|\Omega|$, is $\{a \in A | \ k_1, k_2 \in K \ni k_1 a k_2 \in \Omega\}$.

DEFINITION 5. A subset $\Omega \subset A$ is logarithmically convex (log-convex) if $L(\Omega) = \{\log a \,|\, a \in \Omega\}$ is a bounded open convex subset of $\mathfrak{A}$, invariant under the action of the Weyl group.

DEFINITION 6. A Reinhardt domain Ω is said to be log-convex if its real representation $|\Omega|$ is log-convex.

We believe that any log-convex Reinhardt domain $\Omega \subset G$ can be shown to be a domain of holomorphy, following the same procedure as in Theorem 4.

4. Application to a Gegenbauer polynomial expansion of an analytic function.

THEOREM 5. *Any complex-analytic function in the interior of an ellipse with foci ± 1 can be represented there by an absolutely convergent series of Gegenbauer polynomials.*

We make use of the special orthogonal group $SO(n)$ to obtain that result. We consider the C^∞ functions on $SO(n)$ which are $SO(n-1)$-bi-invariant and the fact that the Peter-Weyl series for such functions reduces to a sum over the class 1 representations of $SO(n)$ **[5]**.

Also the following function p takes an important role in the proof. Let p_0 be the column vector $(0, \cdots, 0, 1)^t \in C^n$. Denote by $(\ , \ )$ the nondegenerate symmetric bilinear form of rank n given by $((z_1, \cdots, z_n), (w_1, \cdots, w_n)) = \sum z_i w_i$. Define p: $SO(n, C) \to C$ by the expression $p(g) = (g \cdot p_0, p_0)$, for all g in $G = SO(n, C)$, where $g \cdot p_0$ denotes matrix multiplication. Now, if E_ε denotes the set $\{\cos\phi \,|\, |\cos\phi|^2 + |\sin\phi|^2 < \alpha(\varepsilon), \phi \in C\}$ where $\alpha(\varepsilon) = \cosh(2\varepsilon/(2n-4)^{1/2})$, and $n > 2$, then E_ε is an ellipse with foci ± 1 and $p(G_\varepsilon) = E_\varepsilon$. Hence given any compact subset S of E_ε there exists δ, $0 < \delta < \varepsilon$, so that $p(\bar{G}_\delta) \supset S$. Thus the uniform absolute convergence follows.

REFERENCES

1. R. P. Boas, Jr. and R. C. Buck, *Polynomial expansions of analytic functions*, Ergebnisse der Mathematik und ihrer Grenzgebiete, N. F., Heft 19, Springer-Verlag, Berlin, 1958. MR **20** #984.
2. E. T. Whittaker and G. N. Watson, *A course of modern analysis*, 4th. ed., Cambridge Univ. Press, New York, 1927.
3. L. A. Frota-Mattos, *Analytic continuation of the Fourier series on connected compact Lie groups*, Thesis, Rutgers University, 1975.
4. N. Wallach, *Harmonic analysis on homogeneous spaces*, Dekker, New York, 1973.

5. N. Wallach, *Minimal immersions of symmetric spaces into spheres: Symmetric spaces*, Dekker, New York, 1972.

6. B. L. Beers and A. J. Dragt, *New theorems about spherical harmonic expansions on SU*(2), J. Mathematical Phys. **11** (1970), 2313–2328. MR **42** #8788.

7. R. T. Seeley, *Eigenfunction expansions of analytic functions*, Proc. Amer. Math. Soc. **21** (1969), 734–738. MR **39** #2180.

8. L. Nachbin, *Holomorphic functions, domains of holomorphy and local properties*, North-Holland Math. Studies, 1, North-Holland, Amsterdam; American Elsevier, New York, 1970. MR **43** #558.

9. R. E. Edwards, *Integration and harmonic analysis on compact groups*, Cambridge Univ. Press, New York, 1972.

FEDERAL UNIVERSITY OF RIO DE JANEIRO

Proceedings of Symposia in Pure Mathematics
Volume 30, 1977

ANALYSIS ON MATRIX SPACE AND CERTAIN SIEGEL DOMAINS

KENNETH I. GROSS* AND RAY A. KUNZE*

This note concerns interrelationships between Bessel functions and Fourier analysis on matrix space, and analysis and representation theory associated with certain classical Siegel domains of tube type. Details will appear in [5], [6], [7].

1. Classical results. To help motivate the general theory to follow, we present the circle of ideas relating spaces of holomorphic functions on the ordinary upper half-plane H to Fourier analysis on real Euclidean space.

Let $G = \mathrm{SL}(2, \boldsymbol{R})$, acting on H by linear fractional transformations $z \cdot g = (zg_{12} + g_{22})^{-1}(zg_{11} + g_{21})$; equip H with the G-invariant measure $d_*z = y^{-2}\,dx\,dy$; and let $P = (0, \infty)$ be endowed with multiplication-invariant measure $d_*r = r^{-1}dr$. Then for each real number $l > 1$, *the l-adapted Laplace transform* $\mathscr{T}_l\colon \phi \to F$ given by

$$F(z) = (2\pi)^{-1/2} \int_0^\infty e^{izr} r^{l/2} \phi(r)\, d_*r \tag{1.1}$$

exists for each $\phi \in L^2(P)$ and defines a holomorphic function F on H. Moreover, by the Plancherel theorem on $\boldsymbol{R}$,

$$\int_0^\infty |\phi(r)|^2\, d_*r = \frac{2^{l-1}}{\Gamma(l-1)} \int_H |F(x+iy)|^2 y^l\, d_*z, \tag{1.2}$$

so $\mathscr{T}_l$ is a unitary operator from $L^2(P)$ onto a Hilbert space $\mathfrak{M}_l$ of holomorphic functions on H. In fact, $\mathfrak{M}_l$ is precisely the space of holomorphic functions on H for which the right side of (1.2) is finite.

AMS (MOS) subject classifications (1970). Primary 22E30, 22E45, 33A75, 32M15, 46E20; Secondary 33A40, 43A75, 43A80, 43A85, 57E25, 81A78.

*This research was partially supported by National Science Foundation grants MPS73–08754 and GP-36565X.

When l is an integer, $l > 1$, these spaces arise in the representation theory of the group G. Indeed, the formula

$$(T(g, l)F)(z) = (zg_{12} + g_{22})^{-l}F(z \circ g) \tag{1.3}$$

defines an irreducible unitary representation $T(\cdot, l)$ of G on the space $\mathfrak{M}_l$ which is *square-integrable* in the sense that the matrix entries $(T(\cdot, l)F_1 | F_2)$ are in $L^2(G)$. These representations $T(\cdot, l)$, $l = 2, 3, \cdots$, constitute the *holomorphic discrete series* for G. When $l = 1$ equation (1.1) still defines a bijection $\mathscr{T}_1$ of $L^2(P)$ with a space $\mathfrak{M}_1$ of holomorphic functions on H, although (1.2) is no longer valid. As a Hilbert space (in the inner product inherited from that of $L^2(P)$) $\mathfrak{M}_1$ is the classical Hardy space, H^2 of the upper half-plane. Then for $l = 1$, formula (1.3) defines another irreducible unitary representation $T(\cdot, 1)$ of G, no longer square-integrable, which is said to be a *limit of holomorphic discrete series.* Finally, we note that entirely analogous *conjugate-holomorphic* versions of the above constructions result from the substitution $z \to -\bar{z}$ in (1.1). For further details on these spaces and representations we refer to **[2]**, **[10]**, and **[12]**.

These same Hilbert spaces also relate to Fourier analysis on Euclidean space $\boldsymbol{R}^k$, $k \geqq 2$, under the action of the rotation group $SO(k)$; in particular, to the well-known decomposition of the Fourier transform $\mathscr{F}: f \to \hat{f}$ on $L^2(\boldsymbol{R}^k)$,

$$\hat{f}(x) = \pi^{-k/2} \int_{\boldsymbol{R}^k} e^{2i(x|y)} f(y)\, dy, \tag{1.4}$$

into *Hankel transforms* $\mathscr{J}_l^{(k)}$, $l = 0, 1, 2, \cdots$, on $L^2(P)$ given by

$$(\mathscr{J}_l^{(k)}\phi)(r) = i^{l+1} \int_0^\infty J_{l-1+k/2}(2r^{1/2}s^{1/2}) r^{1/2}s^{1/2}\phi(s)\, d^*s. \tag{1.5}$$

Here, $J_{l-1+k/2}$ is the classical Bessel function of the indicated order. More specifically, relative to the action of $SO(k)$ and a variant of polar coordinates,

$$L^2(\boldsymbol{R}^k) \cong \sum_{l=0}^{\infty} \oplus (\mathscr{V}_l^* \otimes L^2(P)) \tag{1.6}$$

where $\mathscr{V}_l^*$ is the space of spherical harmonics of degree l. Then since $\mathscr{F}$ commutes with rotations, it splits relative to (1.6); in fact, as

$$\mathscr{F} \cong \sum_{l=0}^{\infty} \oplus (I_l \otimes \mathscr{J}_l^{(k)}) \tag{1.7}$$

where I_l is the identity on $\mathscr{V}_l^*$. We refer to **[15]** for a classical treatment of this decomposition.

The preceding analysis has implications for the representation theory of G. Indeed, for each integer $k \geqq 2$ there exists a so-called *metaplectic representation* $\mathscr{R}^{(k)}$ of G in the space $L^2(\boldsymbol{R}^k)$, with the characteristic property that for the element $p = \left(\begin{smallmatrix} 0 & -1 \\ 1 & 0 \end{smallmatrix}\right) \in G$, $\mathscr{R}^{(k)}(p)$ is essentially the Fourier transform. Such representations arise in physics **[13]**, **[14]** and number theory **[17]**, and can be described naturally in terms of the action of G as automorphisms of a Heisenberg group. When k is odd, $\mathscr{R}^{(k)}$ is double-valued; or alternatively, $\mathscr{R}^{(k)}$ actually represents a two-fold covering of G. Now, by a classical Bessel function formula (cf. (4.9) below),

$$(\mathscr{T}_{l+k/2}\mathscr{J}_l^{(k)}\mathscr{T}_{l+k/2}^{-1}F)(z) = i^{k/2}(-z)^{-(l+k/2)}F(-1/z) \tag{1.8}$$

for $F \in \mathfrak{M}_{l+k/2}$; or in terms of the representations (1.3), $i^{-k/2}\mathscr{I}_l^{(k)}$ is unitarily equivalent to the operator $T(p, l + k/2)$. Thus, (1.7) implies the primary decomposition

$$\mathscr{R}^{(k)} \cong \sum_{l=0}^{\infty} \oplus \, (I_l \otimes T(\cdot, l + k/2)). \tag{1.9}$$

In particular, when $k = 2$ the entire holomorphic discrete series of G and the limit representation appear in the metaplectic representation.

We shall now outline the generalization of these results from $\boldsymbol{R}^k$ to matrix space over the real, complex, or quaternionic fields, and from H to the (generalized) Siegel upper half-planes in the complexification of matrix algebras over these fields.

2. Peter-Weyl transforms. Suppose (U, X) is a topological transformation group in which the group U is compact and X is a locally compact Hausdorff space with a U-invariant Baire measure dx which is positive on open sets. Then the equation $(L(u)f)(x) = f(u^{-1}x)$ defines a unitary representation L of U on $L^2(X)$. We describe the extension of the Peter-Weyl theory [3] which provides an explicit decomposition of L into its primary constituents. Fix representations λ for the unitary dual $\tilde{U}$, let $\mathscr{V}_\lambda$ be the space of λ, let $\mathscr{L}_\lambda \cong \mathscr{V}_\lambda^* \otimes \mathscr{V}_\lambda$ be the space of linear transformations on $\mathscr{V}_\lambda$, define $L^2_\lambda(X, \mathscr{L}_\lambda)$ to be the space of all square-integrable Baire functions $f: X \to \mathscr{L}_\lambda$ which are λ-covariant in the sense that $f(ux) = \lambda(u)f(x)$, and set

$$L^2(X)^\sim = \sum_{\lambda \in \tilde{U}} \oplus \, L^2_\lambda(X, \mathscr{L}_\lambda). \tag{2.1}$$

Then there exists a unitary mapping $\Phi : f \to \tilde{f}$, called the *Peter-Weyl transform for* (U, X), from $L^2(X)$ to $L^2(X)^\sim$ defined by $\tilde{f}(\lambda)(x) = \int_U f(u^{-1}x)\lambda(u)\,du$, such that the subspace $\Phi^{-1}(L^2_\lambda(X, \mathscr{L}_\lambda)) = L^2_\lambda(X)$ of $L^2(X)$ is L-invariant, and on this subspace L transforms according to the irreducible representation $\bar{\lambda}$ conjugate to λ.

In the case in which U acts transitively on X, we can replace X by $\Sigma = U/U_0$ for some closed subgroup U_0 of U. In this well-known classical context, the equivalent spaces $L^2_\lambda(\Sigma)$ and $L^2_\lambda(\Sigma, \mathscr{L}_\lambda)$ are finite-dimensional and each consists of continuous functions ("Σ-harmonics") on Σ. Indeed, let $\mathscr{W}_\lambda = \{v \in \mathscr{V}_\lambda\colon \lambda(u_0)v = v$ for all $u_0 \in U_0\}$, the space of U_0-*invariant vectors*, and set $\tilde{U}_\Sigma = \{\lambda \in \tilde{U}\colon \mathscr{W}_\lambda \neq 0\}$. Then $L^2_\lambda(\Sigma, \mathscr{L}_\lambda) \cong \mathscr{V}_\lambda^* \otimes \mathscr{W}_\lambda$,

$$L^2(\Sigma) \cong \sum_{\lambda \in \tilde{U}_\Sigma} \oplus \, (\mathscr{V}_\lambda^* \otimes \mathscr{W}_\lambda), \tag{2.2}$$

and

$$L \cong \sum_{\lambda \in \tilde{U}_\Sigma} \oplus \, (\bar{\lambda} \otimes I_{\mathscr{W}_\lambda}) \tag{2.3}$$

is the primary decomposition of L. In particular, $\bar{\lambda}$ appears in L with multiplicity equal to $\dim \mathscr{W}_\lambda$.

3. Generalized Bessel functions. Now suppose X is a real finite-dimensional inner product space on which the compact group U acts orthogonally. Then the Fourier transform $\mathscr{F}$ on $L^2(X)$, given by (1.4) with $k = \dim X$, commutes with L, and consequently the equivalent operator $\Phi\mathscr{F}\Phi^{-1}$ on $L^2(X)^\sim$ splits according to (2.1) as a direct sum.

In fact,

(3.1)
$$\mathscr{F} \cong \sum_{\lambda \in \hat{U}} \oplus \mathscr{J}_\lambda$$

where $\mathscr{J}_\lambda$ is the unitary operator on $L^2_\lambda(X, \mathscr{L}_\lambda)$ of the form

(3.2)
$$(\mathscr{J}_\lambda f)(x) = \pi^{-k/2} \int_X J_\lambda(2x, y) f(y)\, dy$$

and $J_\lambda \colon X \times X \to \mathscr{L}_\lambda$ is defined by

(3.3)
$$J_\lambda(x, y) = \int_U e^{i(x|uy)} \lambda(u)\, du.$$

The mapping J_λ is called the *λth order (generalized) Bessel function* for (U, X) and $\mathscr{J}_\lambda$ is the associated *Bessel*, or *Hankel, transform*. Of course, $L^2_\lambda(X, \mathscr{L}_\lambda)$ may be 0, in which case $J_\lambda = 0$.

In the presence of polar coordinates $\Sigma \times P$ for the transformation group (U, X), the decomposition (3.1) can be determined more precisely. Thus, we suppose that $\Sigma = U/U_0$ where U_0 is a fixed closed subgroup of U; P is a regular convex cone in the subspace of X of points fixed under U_0, such that each orbit Ur for $r \in P$ is equivalent to Σ; and the mapping $(uU_0, r) \to x = ur$ is a bijection of $\Sigma \times P$ with an open dense subset of X, the complement of which has Lebesgue measure 0. Under these assumptions there exists a nonnegative Baire function ω on P relative to which Lebesgue measure decomposes as $dx = du \times \omega(r)dr$ for $x = ur$, where du is normalized Haar measure on U and dr is Lebesgue measure on P. Then $L^2(X) \cong L^2(\Sigma) \otimes L^2(P, \omega(r)dr)$, and with notation as in §2, $L^2_\lambda(X, \mathscr{L}_\lambda) \cong (\mathscr{V}^*_\lambda \otimes \mathscr{W}_\lambda) \otimes L^2(P, \omega(r)dr) \cong \mathscr{V}^*_\lambda \otimes L^2(P, \mathscr{W}_\lambda, \omega(r)dr)$, and

(3.4)
$$L^2(X) \cong \sum_{\lambda \in \hat{U}_\Sigma} \oplus (\mathscr{V}^*_\lambda \otimes L^2(P, \mathscr{W}_\lambda, \omega(r)dr)).$$

In terms of the decomposition (3.4), formula (3.1) has the refinement

(3.5)
$$\mathscr{F} \cong \sum_{\lambda \in \hat{U}_\Sigma} \oplus (I_\lambda \otimes \mathscr{J}_\lambda)$$

where $\tilde{\mathscr{J}}_\lambda$ acts on $L^2(P, \mathscr{W}_\lambda, \omega(r)dr)$ by the formula

(3.6)
$$(\tilde{\mathscr{J}}_\lambda \phi)(r) = \pi^{-k/2} \int_P J_\lambda(2r, s)\, \phi(s) \omega(s)\, ds.$$

4. Decomposition of matrix space. Let $\boldsymbol{F}$ be a real finite-dimensional division algebra equipped with its usual conjugation. We consider the case of the transformation group (U, X) in which $X = \boldsymbol{F}^{k\times n}$, $k \geqq 2n$, with inner product $(x|y) = \operatorname{Re} \operatorname{tr}(y'x)$, $y' = \bar{y}^t$, and U is the component of 1 in the "orthogonal" group $\mathscr{O}(k, \boldsymbol{F}) = \{u \in \boldsymbol{F}^{k\times k} \colon uu' = 1_k\}$. In particular, U is $SO(k)$, $U(k)$, or $Sp(k)$ as $\boldsymbol{F}$ is real, complex, or quaternionic. Here, U_0 is the stability group of $\mathbf{1} = \binom{1_n}{0}$, $\Sigma = U/U_0$ is isomorphic to the *Stiefel manifold* $\{\sigma \in X \colon \sigma'\sigma = 1_n\}$, and P is the cone of positive-definite selfadjoint matrices in $\boldsymbol{F}^{n\times n}$.

The idea in what follows is to transfer problems of representation theory and analysis for the rectangular $k \times n$ matrix space to the context of square $n \times n$ matrix space. That is to say, there exists a reduction from (U, X) to $(\mathscr{O}_0, A_0)$ where A_0 is the component of 1 in $\mathrm{GL}(n, \boldsymbol{F})$ and $\mathscr{O}_0$ is the subgroup $A_0 \cap \mathscr{O}(n, \boldsymbol{F})$. Crucial to this procedure is the invariant-theoretic theorem below. Let $A = \mathrm{GL}(n, \boldsymbol{F}^{C})$ be the group of invertible elements of the complexified algebra $(F^C)^{n\times n}$, and denote by $\tilde{A}_p$ a complete set of representatives for the irreducible finite-dimensional polynomial representations π of A. For $\pi \in \tilde{A}_p$, the inner product on the space $\mathscr{H}_\pi$ is chosen so that the restriction of π to the unitary subgroup of A is a unitary representation. Note that the group A is isomorphic to GL(n, C), GL(n, C) $\times$ GL(n, C), or GL($2n$, C) as F is real, complex, or quaternionic.

THEOREM 1. (1) *There exists a one-to-one mapping* $\pi \to \lambda = \lambda(\cdot, \pi)$ *from* $\tilde{A}_p$ *to* $\tilde{U}_\Sigma$. *When* $F = C$ *or* H, *or when* $F = R$ *and* $k > 2n$, *this mapping is a bijection onto* $\tilde{U}_\Sigma$.

(2) *For each* $\pi \in \tilde{A}_p$ *there exists a linear transformation* $T = T_\pi : \mathscr{V}_\lambda \to \mathscr{H}_\pi$ *such that its restriction* T_0 *to the subspace* $\mathscr{W}_\lambda$ *is an isomorphism of* $\mathscr{W}_\lambda$ *with* $\mathscr{H}_\pi$. *Moreover, for* $u \in U$,

$$T\lambda(u)T_0^{-1} = \pi(u_{11} + iu_{21}), \tag{4.1}$$

where $u = (u_{ij})$, $1 \leqq i, j \leqq 3$, *relative to the* $(n, n, k - 2n)$ *block decompostion of a* $k \times k$ *matrix.*

(3) *Let* $\rho(\mathcal{O}_0)$ *be the subgroup of* U, *isomorphic to* $\mathcal{O}_0$, *of matrices* $\rho(\tau) =$ diag$(\tau, \tau, 1_{k-2n})$ *with* $\tau \in \mathcal{O}_0$. *Then the restriction of* λ *to* $\rho(\mathcal{O}_0)$ *leaves* $\mathscr{W}_\lambda$ *invariant and*

$$T_0\lambda(\rho(\tau))T_0^{-1} = \pi(\tau), \qquad \tau \in \mathcal{O}_0. \tag{4.2}$$

The proof of Theorem 1 depends upon the explicit construction of λ as an "induced" representation. Then λ acts by right translation in a space $\mathscr{V}_\lambda$ of holomorphic functions $f : U^C \to \mathscr{H}_\pi$, the mapping T of the theorem is point evaluation at 1, and the U_0-invariant vectors are of the form $f(g) = \pi(g_{11} + ig_{21})v$ for $v \in \mathscr{H}_\pi$, $g \in U^C$. This accounts for (4.1), and (4.2) is a special case of (4.1). We remark that the case $R^{2n\times n}$ is exceptional in that the mapping $\pi \to \lambda(\cdot, \pi)$ only determines half of $\tilde{U}_\Sigma$; however, the remaining half is defined by composition of the representations $\lambda(\cdot, \pi)$ with a fixed automorphism of $SO(2n)$. The coincidence of the numbers dim $\mathscr{W}_\lambda$ and deg π—implied by (2)—was first observed by S. Gelbart [**1**], who, by methods different from ours, treated the real case.

We apply Theorem 1. For $\lambda = \lambda(\cdot, \pi) \in \tilde{U}_\Sigma$ the *reduced Bessel function* $K_\pi : A_0 \to \mathscr{L}_\pi$ is defined by

$$K_\pi(a) = \int_U e^{i \operatorname{Re} \operatorname{tr}(au_{11}')} \pi(u_{11} + iu_{21})\, du. \tag{4.3}$$

Then from (4.1)

$$T_0 J_\lambda(a) T_0^{-1} = K_\pi(a), \qquad a \in A_0, \tag{4.4}$$

and J_λ is completely determined from K_π by λ-covariance. Let $L_\pi^2(A_0)$ consist of all Baire functions $\psi : A_0 \to \mathscr{H}_\pi$ which are square-integrable with respect to Haar measure d_*a on A_0 and satisfy the π-covariance $\psi(\tau a) = \pi(\tau)\psi(a)$ for $(\tau, a) \in \mathcal{O}_0 \times A_0$. Then the *reduced Bessel transform* $\mathscr{K}_\pi$ is defined on $L_\pi^2(A_0)$ by the formula

$$(\mathscr{K}_\pi\psi)(a) = \pi^{-k\nu/2}\beta \int_{A_0} K_\pi(2ab')|\Delta(ab')|^{k\nu/2}\psi(b)d_*b \tag{4.5}$$

where $\nu = \dim F$, β is a constant, and Δ is a character of A_0 described as follows:

$\Delta(a) = (\det a)^j$ where $j = 1$ for $\boldsymbol{F} = \boldsymbol{R}$ or $\boldsymbol{C}$ and $j = \frac{1}{2}$ for $\boldsymbol{F} = \boldsymbol{H}$, "det" referring to the reduced norm on $\boldsymbol{H}^{n\times n}$.

THEOREM 2. *Let $\pi \in \tilde{A}_p$ and $\lambda = \lambda(\cdot, \pi)$. Then relative to the norm $\|\psi\|^2_{T_0} = \int_{A_0} \|T_0^{-1}\psi(a)\|^2 \, d_*a$ on $L^2_\pi(A_0)$, $\mathscr{K}_\pi$ is unitary. Moreover, there exists a unitary operator from $L^2(P, \mathscr{W}_\lambda, \omega(r)dr)$ to $L^2_\pi(A_0)$ which intertwines $\mathscr{J}_\lambda$ with $\mathscr{K}_\pi$. In terms of the reduced Bessel transforms,* (3.5) *becomes*

$$\mathscr{F} \cong \sum_{\pi\in\tilde{A}_p} \oplus (I_\lambda \otimes \mathscr{K}_\pi) \tag{4.6}$$

for $\boldsymbol{F} = \boldsymbol{C}$ or $\boldsymbol{H}$, or $\boldsymbol{F} = \boldsymbol{R}$ and $k > 2n$, and

$$\mathscr{F} \cong \sum_{\pi\in\tilde{A}_p} \oplus 2(I_\lambda \otimes \mathscr{K}_\pi) \tag{4.7}$$

for $\boldsymbol{F} = \boldsymbol{R}$ and $k = 2n$.

The reduced Bessel functions can alternatively be realized as inverse Laplace transforms with respect to the *Siegel upper half-plane* $H = \{z = x + iy : x = x', y = y' \in \boldsymbol{F}^{n\times n},\ y > 0\}$ in $(\boldsymbol{F}^{\boldsymbol{C}})^{n\times n}$. In fact, it follows from a "Bochner-type theorem" for the Fourier transform of the product of the Gaussian with a certain covariant harmonic polynomial, that

$$\begin{aligned}\int_{A_0} e^{i\,\mathrm{tr}\,(zb'b)} K_\pi(2ab')\pi(b)|\Delta(b)|^{k\nu}\, d_*b \\ = \beta^{-1}(\pi i)^{k\nu/2} e^{-i\,\mathrm{tr}\,(za'a)} \Delta(-z^{-1})^{k\nu/2}\pi(-az^{-1})\end{aligned} \tag{4.8}$$

for all $z \in H$. This formula is crucial to the analysis that follows.

From (4.8) it can be observed that when $n = 1$, so k-dimensional Euclidean space over $\boldsymbol{F}$ is under discussion, the reduced Bessel functions are classical. For in that case H is the usual upper half-plane in $\boldsymbol{C}$, the restriction of π to H is scalar, and there exists a positive integer l such that

$$K_\pi(2r^{1/2}s^{1/2}) = \beta^{-1}\pi^{k\nu/2} i^{l+1}(r^{1/2}s^{1/2})^{1-k\nu/2} J_{l-1+k\nu/2}(2r^{1/2}s^{1/2}). \tag{4.9}$$

In particular, when $\boldsymbol{F} = \boldsymbol{R}$ we recapture the classical theory of §1.

5. Analysis and representation theory on H. In analogy to the theory described in §1, one constructs Hilbert spaces of holomorphic functions on the Siegel upper half-plane H in $(\boldsymbol{F}^{\boldsymbol{C}})^{n\times n}$ on which representations are realized of the group $G = G(n, \boldsymbol{F})$ of biholomorphic automorphisms of H. Now, H is an unbounded version of the Hermitian symmetric space G/K, where K, the stability group of i, is the maximal compact subgroup of G. In the present context G consists of all 2×2 matrices over $\boldsymbol{F}^{n\times n}$ such that $gpg' = p$, where

$$p = \begin{pmatrix} 0 & -1_n \\ 1_n & 0 \end{pmatrix},$$

and K is isomorphic to the unitary subgroup of A. The action of G on H is given by linear fractional transformation $z \circ g = (zg_{12} + g_{22})^{-1}(zg_{11} + g_{21})$, and the measure $d_*z = \Delta(y)^{-m}dxdy$, $m = 1 + (n - 1)\nu/2$, on H is G-invariant. In customary notation, G is $Sp(n, \boldsymbol{R})$, $U(n, n)$, or $\mathcal{O}_*(4n)$ according as $\boldsymbol{F}$ is $\boldsymbol{R}$, $\boldsymbol{C}$, or $\boldsymbol{H}$.

Let $\tilde{A}_h$ be a complete set of representatives π for the irreducible *holomorphic* finite-dimensional representations of A. For $\pi \in \tilde{A}_h$, let $\mathfrak{M}_\pi$ consist of all holomorphic functions F from H to the space $\mathscr{H}_\pi$ of π such that

$$\|F\|_\pi^2 = \int_H \|\pi(y)^{1/2}F(x+iy)\|^2\, d_*z < \infty. \tag{5.1}$$

In general, $\mathfrak{M}_\pi$ may be 0. However, whenever $\mathfrak{M}_\pi \neq 0$ the equation

$$(T(g,\pi)F)(z) = \pi(g'_{12}z + g'_{22})^{-1}F(z \circ g) \tag{5.2}$$

defines an irreducible, unitary, square-integrable representation $T(\cdot,\pi)$ of G on $\mathfrak{M}_\pi$, and these representations form the holomorphic discrete series for G **[8]**.

The space $\mathfrak{M}_\pi$ can be described by means of "π-adapted" Laplace transforms, a procedure which shows that the nonvanishing of $\mathfrak{M}_\pi$ is equivalent to the absolute convergence of the *generalized gamma integral*

$$\gamma(\pi) = \int_P e^{-2\,\mathrm{tr}\, r}\pi(r)\Delta(r)^{-2m}\, dr. \tag{5.3}$$

Thus, if $\pi \in \tilde{A}_h$ is such that (5.3) converges absolutely, then the *Laplace transform* $\mathscr{T}_\pi : \psi \to F$, given by

$$F(z) = (2\pi)^{-nm}\int_{A_0} e^{i\,\mathrm{tr}\, za'a}\pi(a')f(a)\, d_*a \tag{5.4}$$

is a unitary map from $L_\pi^2(A_0)$ to $\mathfrak{M}_\pi$, where the appropriate norm on $L_\pi^2(A_0)$ is

$$\|\psi\|_\pi^2 = \int_{A_0} \|\gamma(\pi)^{1/2}\psi(a)\|^2\, d_*a. \tag{5.5}$$

The representations $R(\cdot,\pi) = \mathscr{T}_\pi^{-1}T(\cdot,\pi)\mathscr{T}_\pi$ on $L_\pi^2(A_0)$ give an alternate realization of the holomorphic discrete series. We remark that the Laplace transform exists more generally for those $\pi \in \tilde{A}_h$ such that

$$\gamma_0(\pi) = \int_P e^{-2\,\mathrm{tr}\, r}\pi(r)\Delta(r)^{-m}\, dr \tag{5.6}$$

is absolutely convergent, and the above construction leads in principle to certain "limits" of holomorphic discrete series **[4]**, **[9]**, **[11]**, **[16]**.

Now, as in the classical case, for each k there exists a metaplectic representation $\mathscr{R}^{(k)}$ of G in the space $L^2(F^{k\times n})$ for which $\mathscr{R}^{(k)}(p) = i^{-k\nu/2}\mathscr{F}$. From (4.9),

$$\mathscr{T}_{\pi_k}^{-1}T(p,\pi^{(k)})\mathscr{T}_{\pi_k} = i^{-k\nu/2}\,\mathscr{K}_\pi \tag{5.7}$$

(relative to (5.5) as norm) where $\pi_k = \delta_k \otimes \pi$ with π *polynomial* and δ_k a certain fixed holomorphic character of A depending upon k. (When $\boldsymbol{F} = \boldsymbol{R}$ and k is odd, δ_k is double-valued.) Then the decompostions (4.7) and (4.8) are special cases of the primary decompositions

$$\mathscr{R}^{(k)} \cong \sum_{\pi \in \tilde{A}_p} \oplus\, (I_\lambda \otimes T(\cdot,\pi_k)) \tag{5.8}$$

when $\boldsymbol{F} = \boldsymbol{C}$ or $\boldsymbol{H}$, or $\boldsymbol{F} = \boldsymbol{R}$ and $k > 2n$, and

$$R^{(2n)} \cong \sum_{\pi \in \tilde{A}_p} \oplus 2(I_\lambda \otimes T(\cdot,\pi_{2n})) \tag{5.9}$$

for $\boldsymbol{F} = \boldsymbol{R}$ and $k = 2n$. In (5.8) only holomorphic discrete series appear. But in (5.9) not only do holomorphic discrete series of $Sp(n, \boldsymbol{R})$ appear (in fact, the full holomorphic discrete series appears) but certain limits enter as well.

REFERENCES

1. S. Gelbart, *A theory of Stiefel harmonics*, Trans. Amer. Math. Soc. **192** (1974), 29–50.

1a. ———, *Holomorphic discrete series for the real symplectic group*, Invent. Math. **19** (1973), 49–58. MR **47** #8770.

2. I. M. Gel'fand and M. I. Graev, *Unitary representations of the real unimodular group (principal nondegenerate series)*, Izv. Akad. Nauk SSSR Ser. Mat. **17** (1953), 189–248; English transl., Amer. Math. Soc. Transl. (2) **2** (1956), 147–205. MR **15**, 199; **17**, 876.

3. K. I. Gross and R. A. Kunze, *Fourier decompositions of certain representations*, Symmetric Spaces, Dekker, New York, 1972.

4. ———, *Fourier-Bessel transforms and holomorphic discrete series*, Conference on Harmonic Analysis, Lecture Notes in Math., vol. 266, Springer-Verlag, Berlin and New York, 1972.

5. ———, *Bessel functions and representation theory*. I, J. Functional Analysis **22** (1976), 73–105.

6. ———, *Bessel functions and representation theory*. II, J. Functional Analysis (to appear).

7. ———, *Finite-dimensional induction and invariants for groups of forms* (in preparation).

8. Harish-Chandra, *Representations of semisimple Lie groups*. V, Amer. J. Math. **78** (1956), 1–41. MR **18**, 490.

9. A. Knapp and K. Okamoto, *Limits of holomorphic discrete series*, J. Functional Analysis **9** (1972), 375–409. MR **45** #8774.

10. S. Lang, $SL_2(R)$, Addison-Wesley, Reading, Mass., 1974.

11. H. Rossi and M. Vergne, *Analytic continuation of the holomorphic discrete series*, Acta Math. **136** (1976), 1–59.

12. P. J. Sally, Jr., *Analytic continuation of the irreducible unitary representations of the universal covering group of* $SL(2, R)$, Mem. Amer. Math. Soc. No. 69 (1967). MR **38** #3380.

13. I. E. Segal, *Transforms for operators and symplectic automorphisms over a locally compact abelian group*, Math. Scand. **13** (1963), 31–43. MR **29** #486.

14. D. Shale, *Linear symmetries of free Boson fields*, Trans. Amer. Math. Soc. **103** (1962), 149–167. MR **25** #956.

15. E. M. Stein and G. Weiss, *Introduction to Fourier analysis on Euclidean space*, Princeton Math. Ser., no. 32, Princeton Univ. Press, Princeton, N. J., 1971. MR **46** #4102.

16. N. Wallach, *Analytic continuation of the discrete series*. I, II, Trans. Amer. Math. Soc. (to appear).

17. A. Weil, *Sur certain groups d'opérateurs unitaires*, Acta Math. **111** (1964), 143–211. MR **29** #2324.

UNIVERSITY OF NORTH CAROLINA AT CHAPEL HILL
UNIVERSITY OF CALIFORNIA AT IRVINE

Proceedings of Symposia in Pure Mathematics
Volume 30, 1977

BOUNDARY BEHAVIOUR OF HOLOMORPHIC FUNCTIONS ON BOUNDED SYMMETRIC DOMAINS

ADAM KORÁNYI

The first part will contain a brief description without proofs of some results I have recently obtained about the boundary behaviour of Poisson integrals on Riemannian symmetric spaces [2]. These results apply immediately to bounded holomorphic functions on symmetric domains in $\boldsymbol{C}^n$. In the second part I will show, with sketches of proofs, how the general convergence results can be put into an explicit geometric form in the case of the symmetric domains.

1. Let $X = G/K$ be a Riemannian symmetric space of noncompact type. Let $\mathfrak{g} = \mathfrak{k} + \mathfrak{p}$ be the Cartan decomposition of the Lie algebra of G, let $\mathfrak{a}$ be maximal Abelian in $\mathfrak{p}$ and let Π be a simple system of restricted roots. For any subset $E \subset \Pi$ let $\mathfrak{a}(E)$ be the subspace of $\mathfrak{a}$ annihilated by all $\lambda \in E$; let $\mathfrak{n}(E)$ (resp. $\bar{\mathfrak{n}}(E)$) be the sum of all root spaces corresponding to the positive (resp. negative) restricted roots that do not vanish on $\mathfrak{a}(E)$; let $A(E)$, $N(E)$, $\bar{N}(E)$ be the corresponding analytic subgroups. The normalizer of $\mathfrak{n}(E)$ is a parabolic subgroup $B(E)$ with Langlands decomposition $B(E) = M(E)A(E)N(E)$. The group G_n^E, corresponding to the algebra $\mathfrak{g}_n^E$ generated by the root spaces corresponding to the linear span of E, is a normal subgroup of $M(E)$. Writing o for the identity coset in G/K, the orbit $X^E = M(E)\cdot o = G_n^E \cdot o$ is a symmetric subspace of X.

Given $E \subset \Pi$, X^E splits into a product of irreducible spaces X^{E_j} (and $\mathfrak{g}_n^E$ into a direct sum of ideals $\mathfrak{g}_n^{E_j}$), with E the disjoint union of the E_j's which are called the *components* of E. Given $E_0 \subset \Pi$, a set E is called *E_0-connected* if no component of E is contained in E_0.

Let E_0 be such that Π is E_0-connected. Every such E_0 determines a nondegenerate Satake-Furstenberg compactification $\bar{X}(E_0)$ of X [3], [4]. For any E_0-connected E,

AMS (MOS) subject classifications (1970). Primary 32M15.

let E'' be the set of all $\lambda \in \Pi - E$ such that $E \cup \{\lambda\}$ is not E_0-connected, and let $E' = E \cup E''$. Then $X^{E'} = X^E \times X^{E''}$, the action of $M(E')$ projects to X^E and then extends trivially to an action of $B(E')$ on X^E. Now $\bar{X}(E_0)$ is a G-space in which X is dense; for each E_0-connected E there is an imbedding $\iota_E: X^E \to \bar{X}(E_0)$ equivariant for the $B(E')$-action just described, and $\bar{X}(E_0)$ is the disjoint union of the orbits $G \cdot \iota_E(X^E)$, the isotropy group of the set $\iota_E(X^E)$ being $B(E')$. These properties, together with a direct description of convergence in the closure of $(\exp \mathfrak{a}) \cdot o$, determine $\bar{X}(E_0)$ completely [4].

It is easy to see that $G \cdot \iota_E(X^E) = G \cdot \iota_E(o) = K \cdot \iota_E(X^E)$. Hence $\bar{X}(E_0)$ has exactly one compact G-orbit, namely $G \cdot \iota_\phi(o) = K \cdot \iota_\phi(o)$; it is called the *distinguished boundary* $S(E_0)$ of X in $\bar{X}(E_0)$. For any integrable function f on $S(E_0)$, the *Poisson integral* of f is defined by $F(g \cdot o) = \int f(g \cdot u)\, d\mu(u)$ where μ is the normalized K-invariant measure.

The behaviour of the Poisson integral near the distinguished boundary has been studied in detail. It is known that if f is bounded then, at a.a. $u \in S(E_0)$, F tends to $f(u)$ in a sense that generalizes nontangential convergence in the ordinary unit disc [1]. There are also similar results about not necessarily bounded f (the most important one of these concerns the symmetric domains with their Bergman-Šilov boundary [5], [1]). The new results in [2], however, concern the behaviour of F on the *entire* boundary of X in $\bar{X}(E_0)$. To describe them we have to define the corresponding generalizations of nontangential domains.

For $T \in \mathfrak{a}(E')$ let $A(E')^T$ be the set of exponentials of all $H \in \mathfrak{a}(E')$ such that $\lambda(H) \geqq \lambda(T)$ for all $\lambda \in E'$. Let V be a (small) compact neighborhood of e in G_n^E, and let U, C be (fixed) compact neighborhoods of e in $\bar{N}(E')$, resp. $G_n^{E''}$. Now we set

$$\Gamma_{U,C}^{T,V} = A(E')^T\, UCV \cdot o.$$

We say that F *converges admissibly* to the number c at $g \cdot \iota_E(o)$ if for every C, U and $\varepsilon > 0$, there exist T, V such that $x \in g\Gamma_{U,C}^{T,V}$ implies $|F(x) - c| < \varepsilon$. We say that F converges admissibly at the boundary component $g \cdot \iota_E(X^E)$ if it converges admissibly at every point of this set (the convergence is then automatically uniform on compact subsets, provided F is continuous).

At the distinguished boundary this notion specializes to the "semirestricted admissible convergence" of [1]. We drop the adjective "semirestricted", since this seems to be the most natural convergence notion of all. In fact, examples show [5], that what is called "unrestricted admissible convergence" in [5], [1], fails when it differs from "semirestricted". In the case of the polydisc the two notions coincide; in this case our notion of admissible convergence at general boundary points also coincides with the one used in [7, Chapter XVII].

Now we can state the main result of [2]. If F is the Poisson integral of a bounded function f on $S(E_0)$, then F converges admissibly at a.a. boundary components in $\bar{X}(E_0)$; the limit function on a component is equal to the Poisson integral with respect to this component of an appropriate restriction of f. The proof proceeds by reducing the general case to the case $E_0 = \varnothing$, and then applying a differentiation theorem similarly as in the special case treated in [1].

2. In talking about bounded symmetric domains we restrict ourselves to the irreducible case, the general case being a trivial extension. It is known then that the elements of Π are $\lambda_j = \frac{1}{2}(\gamma_j - \gamma_{j+1})$ $(1 \leqq j \leqq l-1)$ and $\lambda_l = \gamma_l$ or $\frac{1}{2}\gamma_l$, where the γ_j are certain orthogonal coordinates in $\mathfrak{a}$ [3]. The closure of X in the canonical Harish-Chandra realization as a bounded domain is isomorphic with $\bar{X}(E_0)$ where $E_0 = \{\lambda_1, \cdots, \lambda_{l-1}\}$. If E is E_0-connected, then it is of the form $\{\lambda_{k+1}, \cdots, \lambda_l\}$. We fix such an E and proceed to describe admissible convergence at the boundary components $g \cdot \iota_E(X^E)$.

We will first consider the realization of X as a Siegel domain of type III with base $\iota_E(X^E)$ and get an explicit description of the sets $\Gamma_{U;C}^{T;V}$ corresponding to E. Then we will reformulate this in geometric terms that are invariant under G and under the Cayley transformation which maps the Siegel domain onto the canonical bounded realization.

For brevity we write the splitting described in [6, §7] as $\mathfrak{p}^- = V_1 \oplus V_2 \oplus V_3$. G acts via the Harish-Chandra imbedding followed by the Cayley transform, o is a point in V_1, $\iota_E(o)$ is the origin 0. $A(E')$ is one-dimensional and acts by $(z_1, z_2, z_3) \to (\xi^2 z_1, \xi z_2, z_3)$ $(\xi > 0)$; $A(E')^T$ is the set of elements such that $0 < \xi < \delta$ (with some δ depending on T). $A(E')G_n^{E''}$ is $K_{\Delta-\Gamma,1}^*$ in [6], the orbit of o under it is a self-dual cone Ω in a real form of V_1 ($\mathfrak{c}^\Gamma$ in [6]). $\bar{N}(E')$ is $N^{\Gamma-}$ in [6]. $\Lambda_0 : V_2 \times V_2 \to V_1$ is a Hermitian bilinear form, "Ω-positive" in the sense that $\Lambda_0(u, u) \in \bar{\Omega}$ for all u. For $z_3 \in V_3$, $\mu(z_3) : V_2 \to V_2$ is a complex antilinear map, depending linearly on z_3. For $z_3 \in \iota_E(X^E)$, $\Lambda_{z_3}(u, v) = \Lambda_0(u, (I + \mu(z_3))^{-1}v)$ is the sum of an Ω-positive form $\Lambda_{z_3}^{(1)}$ and of the symmetric bilinear form $\Lambda_{z_3}^{(2)}(u, v) = \Lambda_0(u, (I + \mu(z_3))^{-1}\mu(z_3)v)$. Writing $t(z) = \operatorname{Im} z_1 - \operatorname{Re} \Lambda_{z_3}(z_2, z_2)$, X is the set of all $z = (z_1, z_2, z_3)$ such that $t(z) \in \Omega$.

It is now easy to describe $\Gamma_{U;C}^{T;V}$. $C \cdot o$ is a compact subset of $i\Omega$ (the cones spanned by such subsets are all the cones $i\omega$ such that $\bar{\omega} \subset \Omega \cup \{0\}$). V and C commute, so $CV \cdot o$ is the set of $(iy_1, 0, z_3)$ with $iy_1 \in C \cdot o$ and z_3 in a neighborhood V' of 0 in V_3. By [6, Proposition 7.5], $UCV \cdot o$ is the set of $(x_1 + iy_1 + i\Lambda_{z_3}(f_v(z_3), f_v(z_3)), f_v(z_3), z_3)$ where $f_v(z_3) = v + \mu(z_3)v$; y_1, z_3 are as before, and (x_1, v) is in a compact subset of $(\operatorname{Re} V_1) \times V_2$. $\Gamma_{U;C}^{T;V}$ is obtained by applying to this $A(E')^T$, whose action has been described above.

Now we start replacing the family of sets $\Gamma_{U;C}^{T;V}$, by other families which still give the same notion of convergence at $\iota_E(o)$. An obvious change of this kind is to replace V' by the δ-neighborhood of 0 in V_3 (the same δ as the one determined by T), and to restrict U to the more special sets $\{|x_1|, |v| \leqq M\}$ $(M > 0)$.

Furthermore, for small $|z_3|$, $f_v(z_3)$ is comparable to and can be replaced by v. Taking into account that by the reality of $\Lambda^{(1)}$ and the explicit formula for $\Lambda^{(2)}$ we have $|\operatorname{Im} \Lambda_{z_3}(z_2, z_2)| \leqq \frac{1}{2}|z_2|^2$ (for small $|z_3|$), and that $iy_1 \in C \cdot o$ implies that $|y_1|$ is between positive bounds, we see that the $\Gamma_{U;C}^{T;V}$ can be replaced by the sets

$$\Gamma_{M,\omega}^{\delta} = \{z \in X \mid |z| \leqq \delta;\, t(z) \in \omega;\, |\operatorname{Re} z_1|, |z_2|^2 \leqq M|t(z)|\}$$

(F converges admissibly at $\iota_E(o)$ if and only if for all M, ω and all $\varepsilon > 0$ there exists $\delta > 0 \cdots$).

Next we prove that the $\Gamma_{M,\omega}^{\delta}$ can be replaced by the sets

$$\tilde{\Gamma}_M^{\delta} = \{z \in X \mid |z| \leqq \delta;\, |z_1|, |z_2|^2 \leqq Md(z, \partial X)\}$$

where $d(z, \partial X)$ is the distance of z to the boundary of X (in the Euclidean metric of the ambient vector space). We first show that, $\tilde{\Gamma}_M^{\delta} \subset \Gamma_{M,\omega}^{\delta}$ for small δ, with some ω

depending only on M. By looking (for given z) at the boundary points of the form $(\operatorname{Re} z_1 + iu_1, z_2, z_3)$ $(u_1 \in \operatorname{Re} V_1)$ we see that

$$d(z, \partial X) \leqq d(t(z), \partial\Omega) \leqq |t(z)|.$$

Hence $z \in \tilde{\Gamma}^\delta_M$ implies $|\operatorname{Re} z_1|, |z_2|^2 \leqq M|t(z)|$. To see that $t(z) \in \omega$, we write $y_1 = \operatorname{Im} z_1$, $v = \Lambda^{(1)}_{z_3}(z_2, z_2)$, $u = \operatorname{Re} \Lambda^{(2)}_{z_3}(z_2, z_2)$. Since $t(z)$, $v \in \Omega$ and Ω is self-dual, $(t(z)|v) \geqq 0$, and so $|y_1 - u| \geqq |t(z)|$. By the formula for $\Lambda^{(2)}$ also $|u| \leqq \frac{1}{2}|t(z)|$ when δ is small, so it follows that $|t(z)| \leqq 2|y_1|$. But then $|t(z)| \leqq 2|z_1| \leqq 2Md(z, \partial X) \leqq 2Md(t(z), \partial\Omega)$, which implies that $t(z)$ stays in a certain ω. Next we have to show that $\Gamma^\delta_{M,\omega} \subset \Gamma^\delta_{M'}$ with some M' depending only on M and ω. We observe first that there exists $r > 0$ such that $|w_j| < r|t(z)|$ $(1 \leqq j \leqq 3)$ implies $z + w \in X$; in fact, when $|t(z)| = 1$ this follows by compactness, and hence the case $|t(z)| < 1$ follows by applying an appropriate element of $A(E')$. From this observation it follows that (for small δ) there exists $k > 0$, depending on M and ω only, such that $z \in \Gamma^\delta_{\omega,M}$ implies $|t(z)| \leqq kd(z, \partial X)$. But then we also have $|z_2|^2 \leqq Mkd(z, \partial X)$, and writing $z_1 = \operatorname{Re} z_1 + it(z) + i \operatorname{Re} \Lambda_{z_3}(z_2, z_2)$ we obtain a similar estimate for $|z_1|$, finishing the proof.

For the last reformulation we note that the condition $|z_2|^2 < Md(z, \partial X)$ in the definition of $\tilde{\Gamma}^\delta_M$ can be replaced by $|z_1|^2 + |z_2|^2 < Md(z, \partial X)$ without changing anything. Now $|z_1|^2$ and $|z_1|^2 + |z_2|^2$ can be written as $|\pi(z)|^2$ with certain projections π. In such an expression only the kernel of the projection matters, since if π and π' have the same kernel, $\pi(z) \to \pi'(z)$ is an isomorphism of the ranges, and hence $|\pi(z)|$ and $|\pi'(z)|$ are comparable. Now we have to remark only that the Cayley transformation of **[6]** is a map between convex domains which extends holomorphically to their closure, that the same is true for the elements of G, and we have the following final result.

Let D be a symmetric domain in the canonical bounded realization in $\mathbf{C}^n$. For $\zeta \in \partial D$, let O_ζ be the orbit of ζ under the holomorphic automorphisms of D; let H be the holomorphic tangent space of O_ζ at ζ (i.e., the largest complex subspace of the tangent space); let C $(\subset H)$ be the tangent space at ζ of the boundary component of D which contains ζ. Let π_H, π_C be projections of $\mathbf{C}^n$ with kernels H, C (identified in the obvious way with subspaces of $\mathbf{C}^n$). For $\delta > 0$, $M > 0$, let

$$\Gamma^\delta_M(\zeta) = \{z \in D \,|\, |z - \zeta| < \delta;\, |\pi_C(z - \zeta)|^2, |\pi_H(z - \zeta)| \leqq Md(z, \partial D)\}.$$

Now a function F on D converges admissibly to the number c at ζ if and only if for every $M > 0$ and $\varepsilon > 0$ there exists $\delta > 0$ such that $z \in \Gamma^\delta_M(\zeta)$ implies $|F(z) - c| < \delta$.

References

1. A. Korányi, *Harmonic functions on symmetric spaces* (Boothby and Weiss, eds.), Symmetric spaces, M. Dekker, New York, 1972.

2. ———, *Poisson integrals and boundary components of symmetric spaces* Invent. Math. **34** (1976), 19–35.

3. C. C. Moore, *Compactifications of symmetric spaces*. I, II, Amer. J. Math. **86** (1964), 201–218, 358–378. MR **28** #5146; #5147.

4. I. Satake, *On representations and compactifications of symmetric Riemannian spaces*, Ann. of Math. (2) **71** (1960), 77–110. MR **22** #9546.

5. E. M. Stein and N. J. Weiss, *On the convergence of Poisson integrals*, Trans. Amer. Math. Soc. **140** (1969), 35–54. MR **39** #3024.

6. J. A. Wolf and A. Korányi, *Generalized Cayley transformations of bounded symmetric domains*, Amer. J. Math. **87** (1965), 899–939. MR **33** #229.

7. A. Zygmund, *Trigonometric series*. Vols. I, II, 2nd rev. ed., Cambridge Univ. Press, New York, 1959. MR **21** #6498.

YESHIVA UNIVERSITY
UNIVERSITÉ DE STRASBOURG

Proceedings of Symposia in Pure Mathematics
Volume 30, 1977

RANKIN-SELBERG METHOD IN THE THEORY OF AUTOMORPHIC FORMS

I. I. PYATETSKII-SHAPIRO AND M. E. NOVODVORSKY

One of the most interesting topics in the theory of automorphic forms is, we think, their connection with zeta-functions. The famous work of Hecke **[1]** about automorphic forms on the upper complex half-plane H^1 may be considered as the origin and, perhaps, the pattern in the area. We will briefly describe Hecke's results for cusp forms of the modular group on H^1.

The Mellin transform

$$f \to D_f(s) = (2\pi)^s/\Gamma(s) \int_0^\infty f(iy)y^{s-1}\, dy \tag{1}$$

attaches a function D_f analytic in a half-plane Re $s > k/2 + 1$ (where the integral is absolutely convergent) to any such form f of weight k. Substitution of Fourier decomposition of the form f into this integral gives the expression of D_f as a Dirichlet series

$$f(z) = \sum_{n=1}^{\infty} a_n e^{2\pi i n z} \Rightarrow D_f(s) = \sum_{n=1}^{\infty} a_n n^{-s}. \tag{2}$$

Hecke introduced a family of operators T_n now bearing his name and proved: (i) If f is an eigenform of all these operators, then D_f is decomposable into the Euler product

$$D_f(s) = \prod_p [Q_p(p^{-s})], \qquad Q_p(x) = 1 - \lambda_p x - p^{k-1}x^2, \text{Re } s > k/2 + 1; \tag{3}$$

here p runs over all prime numbers, λ_p is the eigenvalue of the operator T_p on the form f, and the infinite product is absolutely convergent. (ii) For any cusp form f of the modular group the function D_f admits analytic continuation on the whole

AMS (MOS) subject classifications (1970). Primary 12A85, 32N10.

complex plane; this continuation is an entire function satisfying a certain functional equation. (iii) The converse theorem: If a Dirichlet series $\sum_{n=1}^{\infty} b_n n^{-s}$ with power growth of coefficients b_n admits analytic continuation which is an entire function satisfying that functional equation, then this series is the Mellin transform of a certain cusp form of the modular group on the upper half-plane H^1.

Siegel invented and investigated automorphic forms on the spaces H^n now called Siegel half-planes (n is the rank of the group of automorphisms of the space H^n) [2]—[4]. Maass continued these investigations [5], [6]. A natural generalization of Mellin transform attaches a Dirichlet series to a cusp form of a Siegel modular group on H^n:

$$(4) \qquad D_f(s) = \frac{\pi^{1/2}(2\pi)^{2s}}{(\Gamma(s)\cdot\Gamma(s-\frac{1}{2}))} \int_{\Omega} f(iY)(\det\, Y)^{s-(n+1)/2}\, dY = \sum_{\{N\}} \frac{a_N}{\varepsilon(N)(\det N)^s};$$

here Ω is the fundamental domain of the group SL(n, Z) in the space of all positive symmetric $n \times n$ matrices, N runs over all classes of equivalence of such matrices with half-integer coefficients under SL$(n, \boldsymbol{Z})$, $\boldsymbol{Z}$ integers, and $\varepsilon(N)$ is the order of the stabilizer of the matrix N in this group. The integral and the series (4) are absolutely convergent in a half-plane Re $s \gg$ and define D_f as an analytic function in this half-plane; this function admits meromorphic continuation satisfying a certain functional equation (cf. [6, §15]).

In 1967 A. Weil [7] reformulated Hecke's results and extended them to the forms on H^1 automorphic under congruence subgroups. This enabled Jacquet and Langlands [8] to interpret (this completed by Weil) Hecke theory in the terms of the representations of adele groups.

Perhaps the first work considering both Euler decomposition and meromorphic continuation of Dirichlet series associated to automorphic forms of several variables is due to Andrianov [9]. He considered automorphic forms of the Siegel modular group on H^2. Instead of integrals (4) he integrated such forms over a 3-dimensional Lobachevsky subspace in H^2. Based on some explicit relations between Fourier coefficients of Siegel automorphic forms and eigenvalues of Hecke operators, he proved that the Dirichlet series obtained in this way admit Euler decompositions and meromorphic continuations satisfying certain functional equations. The proof contains rather bulky computations; it seems difficult to extend it even to forms automorphic under congruence subgroups, to say nothing about forms of more variables.

However, the difficulties were overcome with the help of the language of the representation theory which allowed us to obtain Euler decompositions and meromorphic continuations for zeta-functions associated to cusp forms on the symplectic group of rank 2 [15], [16], [18] and on orthogonal groups [17], [19] over any global field. It includes, particularly, Hilbert-Siegel forms of arbitrary number of variables, holomorphic and nonholomorphic; all the congruence subgroups in this method are treated simultaneously.

To expose this technique and these results we need a modification of the classical concept of automorphic form.

Let k be a global field (that is either a field of algebraic numbers or of algebraic functions over finite fields), A its adele ring, ∞ the set of all archimedean valuations of k, $\mathscr{P}$ the set consisting of all nonarchimedean valuations of k and symbol ∞ if char $k = 0$, and the set of all valuations of k otherwise; all the valuations are supposed to be normalized. For any algebraic group X defined over k we denote by X_k, X_A, and X_p, $p \in \mathscr{P}$, respectively, the group of all geometric points of X with values in k, in A, in the completion of k_p of the field k under p if $p \neq \infty$, and in the product of all archimedian completions if $p = \infty$; X_k is often identified with the subgroup of principal adeles in X_A. Let G be a reductive group defined over k, C its center. A complex-valued continuous function f on the factor space $G_k \backslash G_A$ is called a cusp form on G iff:

(1) there exists a unitary character τ of the group C_A such that $f(cg) = \tau(c)f(g)$ $\forall c \in C_A, \forall g \in G_A$;

(2) $\int_{C_A G_k \backslash G_A} |f(g)|^2 \, dg < \infty$;

(3) there exists a maximal compact subgroup $M = M(f)$ in G_A such that the space generated by the right M-translations of the function f is finite-dimensional;

(4) if N is a horocycle subgroup (= the unipotent radical of a parabolic subgroup) of G, then

$$\int_{N_k \backslash N_A} f(ng) \, dn = 0 \qquad \forall g \in G_A.$$

The connection of this definition with the classical one is the following. Any classical automorphic form f_0 on the symmetric space $M_\infty C_\infty \backslash G_\infty$, where M_∞ is a maximal compact subgroup of the semisimple Lie group $C_\infty \backslash G_\infty$, may be considered as a function on the group G_∞ which is invariant under the right Γ-translations for some discrete subgroup $\Gamma \subset G_\infty$. If Γ is an arithmetical subgroup of G_∞, we may assume that its elements are rational over k. Let M denote the closure of Γ in the subgroup $\prod_{p \in \mathscr{P}; p \neq \infty} G_p \subset G_A$; then the natural correspondence $G_k \backslash G_A / M \simeq G_\infty / \Gamma$ allows one to lift f_0 to a function f on the factor space $G_k \backslash G_A$. Now if f_0 was a classical cusp form, then f is a cusp form on G in the sense of the definition given above.

Let K be a semisimple 2-dimensional algebra over k, B a skew-symmetric bilinear form over K, $G = GS_p(4, k)$ the group of similitudes of the k-form $\mathrm{tr}_{K/k} B$, G' its subgroup consisting of all K-linear transformations from G. If K is a field, let W denote a maximal unipotent subgroup in G' and H a maximal split torus in the normalizer of W in G. If $K = k \oplus k$, then

$$G' \simeq \{(g_1, g_2) : g_i \in \mathrm{GL}(2, k), \det g_1 = \det g_2\};$$

choose a maximal unipotent subgroup U in $\mathrm{GL}(2, k)$ and a maximal torus T in its normalizer and set $H = \{(g_1, t) \in G' : t \in T\}$ and $W = \{(e, u) : u \in U\}$, where e is the identity matrix. In both cases let ρ be a nontrivial positive real character of the group H_A which is trivial on the subgroups H_k and C_A, C be the center of G; let ϕ be an automorphic form on H (in the first case it is just a unitary character of the factor group $C_A H_k \backslash H_A$); let, at last, f be a cusp form on the group G. For any complex number s define formally:

$$D_f(s) = \int_{W_k H_k \backslash W_A H_A} f(wh) \phi(h) (\rho(h))^s \, d(w \cdot h). \tag{5}$$

THEOREM 1. *The integral* (5) *is absolutely convergent in a vertical half-plane and defines D_f as a function analytic in this half-plane and bounded in finite vertical strips*;

this function admits a meromorphic continuation satisfying a certain functional equation.

Suppose that the right translations of the form f under G_A generate an irreducible representation π of the group G_A. Moreover, suppose that under the natural decomposition of π into an infinite tensor product of irreducible representations π_p of the groups G_p, $p \in \mathscr{P}$ (cf., for instance, [14, Chapter 3, §3]) the form f is decomposed into a vector of the shape $\xi = \bigotimes_{p \in \mathscr{P}} \xi_p$ where ξ_p is a vector from the space of the representation π_p. (These assumptions are not restrictive, for any cusp form is a finite sum of those satisfying the assumptions.)

THEOREM 2. $D_f(s) = \prod_{p \in \mathscr{P}} D_p(s)$, *where D_p are analytic functions; the infinite product is absolutely convergent in a vertical half-plane.*

When k is the field of rational numbers, K is a field, ϕ is equal to 1 identically, and f is a cusp form obtained from a classical holomorphic cusp form in the way described above, then the functions D_f coincide with the zeta-functions defined by Andrianov **[9]** and these theorems are, essentially, reformulations of his results. When k is a totally real number field, our construction provides zeta-functions associated to Hilbert-Siegel cusp forms. Analogues of the polynomials Q_p (cf. formula (3)) are polynomials of degree 4 if K is a field, and of degree 8 otherwise.

CONJECTURE. *The Hasse-Weil zeta-function of any algebraic variety uniformized by the functions automorphic under a congruence subgroup of $S_p(4, Q)$, where Q is the field of rational numbers, is a product of some powers of the functions D_f and L-functions of the groups* GL(2) *and* GL(1) (*cf.* **[8]**).

Theorem 1 is proved by the method derived from an idea of R. A. Rankin **[10]** and A. Selberg **[11]**. It allows one to reduce the integral (5) to the integral

$$\int_{C_A G'_k \backslash G'_A} f(g) E(g, \phi, s)\, dg, \tag{6}$$

where E is an Eisenstein series on G'. According to Langlands **[12]**, such series as a function of s is analytic in a half-plane and admits a meromorphic continuation satisfying a certain functional equation; since the integral (6) is absolutely convergent for all s, this proves the theorem.

When K is a field, one of the main steps in the proof of Theorem 2 is the following uniqueness theorem. Let Z denote the maximal algebraic subgroup in G which normalizes W and has no k-rational characters; let α be a unitary character of Z_A which is trivial on Z_k and W_A, but nontrivial on the maximal unipotent subgroup of Z_A.

THEOREM 3. *For any irreducible representation π of the group G_p, $p \in \mathscr{P}$, $p \neq \infty$, the space of all linear Z_p-eigenfunctionals with the character $\alpha | Z_p$ on the space of π is of dimension* 1 *at most.*

This theorem is an analogue of Gel'fand-Kajdan's uniqueness theorem **[13]** which is used to prove Theorem 2 when $K = k \oplus k$. Presumably, both uniqueness theorems are true[1] for all archimedean valuations p, too; their proof would allow

[1]This has recently been proven by one of the authors.

one to substitute the set $\mathscr{P}$ with the set of all normalized valuations of the field k.

The adjoint of the symplectic group Sp(4) is isomorphic to the orthogonal group of Dynkin type B_2. The same technique allows one to construct two families of zeta-functions associated to cusp forms on any orthogonal group of Dynkin type B_n and to prove meromorphic continuations and Euler decompositions for them [**19**]; the functions D_f represent these families for $n = 2$.

REFERENCES

1. E. Hecke, *Über modulfunktionen und Dirichletscher Reihen mit Eulerscher Produkten twinklung*. I, II, Math. Ann. **114** (1937), 1–28, 316–351.

2. C. L. Siegel, *Einführung in die Theorie der Modulfunktionen n-ten Grades*, Math. Ann. **116** (1939), 617–657. MR **1**, 203.

3. ———, *Einheiten quadratischer Formen*, Abh. Math. Sem. Univ. Hamburg **13** (1940), 209–239; Gesammelte Abhandlungen, Springer no. 33, vol. 2. MR **2**, 148.

4. ———, *Symplectic geometry*, Amer. J. Math. **65** (1943), 1–86. MR **4**, 242.

5. M. Maass, *Die Primzahlen in der Theorie der Siegelschen Modulfunktionen*, Math. Ann. **124** (1951), 87–122. MR **13**, 823.

6. ———, *Siegel's modular forms and Dirichlet series*, Lecture Notes in Math., vol. 216, Springer-Verlag, Berlin and New York, 1971. MR **48** #8938.

7. A. Weil, *Über die Bestimmung Dirichletscher Reihen durch Funktionalgleichungen*, Math. Ann. **168** (1967), 149–156. MR **34** #7473.

8. H. Jacquet and R. P. Langlands, *Automorphic forms on* GL (2), Lecture Notes in Math., vol. 114, Springer-Verlag, Berlin and New York, 1970.

9. A. N. Andrianov, *Dirichlet series with Euler product in the theory of Siegel modular forms of genus* 2, Trudy Mat. Inst. Steklov. **112** (1971), 73–94 = Proc. Steklov Inst. Math. **112** (1971), 70–93.

10. R. A. Rankin, *Contributions to the theory of Ramanujan's function* $\tau(n)$ *and similar arithmetical functions*, Proc. Cambridge Philos. Soc. **35** (1959), 357–372.

11. A. Selberg, *Bemerkungen über eine Dirichletsche Reihe, die mit der Theorie der Modulformen nahe verbunden ist*, Arch. Math. Naturvid **43** (1940), 47–50. MR **2**, 88.

12. R. P. Langlands, *On functional equations satisfied by Eisenstein series*, 1970 (preprint).

13. I. M. Gel'fand and D. I. Kajdan, *Representations of the group* GL(n, K) *where* K *is a local field*, Funkcional. Anal. i Priložen. **6** (1972), no. 4, 73–74 = Functional Anal. Appl. **6** (1972), 315–317 (1973). MR **48** #11405.

14. I. M. Gel'fand, M. I. Graev and I. I. Pjateckiĭ-Šapiro, *Generalized functions*. Vol. 6: *Theory of representations and automorphic forms*, "Nauka", Moscow, 1966; English transl., Saunders, Philadelphia, Pa., 1969. MR **36** #3725; **38** #2093.

15. I. I. Pjateckiĭ-Šapiro, *Euler subgroups* 1971 (preprint).

16. M. E. Novodvorskiĭ and I. I. Pjateckiĭ-Šapiro, *Generalized Bessel models for a symplectic group of rank* 2, Mat. Sb. **90** (**132**) (1973), 246–256 = Math. USSR Sb. **19** (1973), 243–256. MR **49** #3045.

17. ———, *On zeta-functions of infinite-dimensional representations*, Mat. Sb. **92** (**134**) (1973), 507–517 = Math. USSR Sb. **21** (1973), 499–510.

18. M. E. Novodvorskiĭ, *Fonctions J pour* GSp(4), C. R. Acad. Sci. Paris Sér. A-B **280** (1975), 191–192.

19. ———, *Fonction J pour des groupes orthogonaux*, C. R. Acad. Sci. Paris Sér. A-B **280** (1975), 1421–1422.

MOSCOW, USSR

PURDUE UNIVERSITY

Proceedings of Symposia in Pure Mathematics
Volume 30, 1977

TANGENTIAL CAUCHY-RIEMANN EQUATIONS ASSOCIATED WITH A SIEGEL DOMAIN

HUGO ROSSI AND MICHELE VERGNE

1. Illustrative examples. We shall start by giving some typical examples coming from group representations which motivate our study (the general case will be given in §2).

(a) The upper half-plane:

$$P^+ = \{x + iy: \ x \in \boldsymbol{R}, \ y > 0\}.$$

The group G of holomorphic transformations of P^+ is given by

$$\left\{g = \begin{pmatrix} a & b \\ c & d \end{pmatrix}, \text{ with } a, b, c, d \in \boldsymbol{R}, \ ad - bc = 1\right\};$$

$g \cdot z = (az + b)(cz + d)^{-1}$. This action leads to an action on the boundary $\boldsymbol{R}$ of P^+ and to a natural representation of G on $L^2(\boldsymbol{R})$ given by

$$(T(g)f)(x) = (cx + d)^{-1} f((ax + b)(cx + d)^{-1}),$$

where $g^{-1} = \left(\begin{smallmatrix} a & b \\ c & d \end{smallmatrix}\right)$. This representation is "exceptional" in the sense that it is reducible, whereas nearby representations (slight perturbations of the multiplier $(cx + d)^{-1}$) will be irreducible.

It is easy to see why this representation is reducible: Since the action, as well as the multiplier, comes from holomorphic transformations on the half-plane P^+, the boundary values of functions holomorphic in P^+ will form an invariant subspace, so we will have $L^2(\boldsymbol{R}) = H^2 \oplus \bar{H}^2$, where H^2 is the classical Hardy space.

(b) The cone Ω of positive definite $(n \times n)$ symmetric matrices is an open convex cone in the vector space V all of symmetric $(n \times n)$ matrices. Consider the Siegel upper half-plane: the tube $D(\Omega) = V + i\Omega$ constructed over this cone: $D = D(\Omega)$ $= \{$complex symmetric $(n \times n)$ matrices Z, such that $\operatorname{Im} Z$ is positive definite$\}$.

AMS (MOS) subject classifications (1970). Primary 32M15, 22E25.

This is a bounded homogeneous domain. The group of transformations of this domain is

$$\mathrm{Sp}(n, \boldsymbol{R}) = \left\{\begin{pmatrix} A & B \\ C & D \end{pmatrix}\right\},$$

where A, B, C, D are $n \times n$ matrices satisfying certain relations, and the action is again given by $g \cdot Z = (AZ + B)(CZ + D)^{-1}$. Similarly this action leads to an action on the boundary $\partial D = \{Z; \operatorname{Im} Z \in \partial\Omega\}$. The boundary $\partial\Omega$ of the cone is naturally stratified as $\partial\Omega = \bigcup_{i=0}^{n-1} \mathcal{O}_i$, where $\mathcal{O}_i$ consists of the positive semidefinite matrices of rank i (thus $\mathcal{O}_0 = \{0\}$, and is the vertex of the cone). So the boundary of the domain is naturally stratified as $\partial D = \bigcup_{i=0}^{n-1} \Sigma_i$, where Σ_i consists of all complex symmetric matrices Z such that $\operatorname{Im} Z$ is positive semidefinite and of rank i. (If $i = 0$, we have just $\Sigma_0 = V$ and it is the Šilov boundary.)

The action of G on D leads to an action on ∂D, and each of the Σ_i is invariant (up to a null set) under this action. To each action of G on Σ_i is associated a representation τ_i of G, which is a member of a unitary principal series. We prove that each τ_i is also reducible, by constructing in a similar manner a proper subspace consisting of boundary values on Σ_i of holomorphic functions on $D(\Omega)$; see **[2]**.

(c) The most general example of Siegel domains $D(\Omega, Q)$ can be illustrated by this example: Ω is the cone of $(n \times n)$ positive definite hermitian matrices contained in V, the vector space of all hermitian matrices. Again we have $\partial\Omega = \bigcup_{i=0}^{n-1} \mathcal{O}_i$, where $\mathcal{O}_i$ is the set of all hermitian positive semidefinite matrices of rank i. Let now W_m be the complex vector space of all linear maps $U : \boldsymbol{C}^n \to \boldsymbol{C}^m$. For U and V in W_m, let $Q(U, V) = V^* \circ U$, which is a map from $\boldsymbol{C}^n$ to $\boldsymbol{C}^n$. $D(\Omega, Q)$ is the domain in $V^C \times W_m$ defined by:

$$D(\Omega : Q) = \{(Z, U); Z = X + iY,\ X, Y \in V,\ U \in W_m, \text{ with } Y - Q(U, U) \in \Omega, \text{ i.e., } \operatorname{Im} Z - U^* \circ U \gg 0\}.$$

Again ∂D is stratified by the Σ_i where

$$\Sigma_i = \{(Z, U); \operatorname{Im} Z - Q(U, U) \in \mathcal{O}_i\}$$

and Σ_0 is the Šilov boundary: $\operatorname{Im} Z = Q(U, U)$. To each of the Σ_i is associated a reducible member of a principal unitary series for the group $SU(n + m, n)$, again with a proper subspace defined in terms of boundary values of holomorphic functions or induced Cauchy-Riemann equations.

2. CR functions on Siegel manifolds. Consider $\boldsymbol{C}^n \times \boldsymbol{C}^m$, with variables $(z, u) = (x + iy, u)$. Let E be a domain in $\boldsymbol{C}^m$, N a differentiable function from E to $\boldsymbol{R}^n$, and V a submanifold in $\boldsymbol{R}^n$. We define the *Siegel manifold* $\Sigma(V, N, E) \subset \boldsymbol{C}^n \times \boldsymbol{C}^m$ as $\Sigma = \Sigma(V, N, E) = \{(x + iy, u); x, y \in \boldsymbol{R}^n, u \in E \text{ and } y - N(u) \in V\}$. In the previous examples, V was either $\{0\}$ (corresponding to the Šilov boundary) or some part of a boundary of a convex cone. In this paragraph we shall not assume any particular properties.

For $p \in \Sigma$, let $T_p(\Sigma)^C$ be the complex tangent space to Σ at p. Let $T^{0,1}$ be the space of antiholomorphic vectors $(\partial/\partial\bar{z}_i, \partial/\partial\bar{u}_\alpha)$ and let $A_p(\Sigma) = T_p(\Sigma)^C \cap T^{0,1}$, i.e., $T_p(\Sigma)$ is the space of complex tangent vectors to Σ, which are at the same time antiholomorphic.

If f is in $C^1(\Sigma)$, then we say that f is a *CR function* if $\xi \cdot f = 0$ for every vector field ξ on Σ, such that $\xi_p \in A_p(\Sigma)$. These equations $\xi \cdot f = 0$ are the *tangential Cauchy-Riemann equations*. Of course, if F is a holomorphic function defined in a neighborhood of Σ, the restriction f of F to Σ will be a CR function. Conversely, we want to describe the CR functions on Σ as best as possible by holomorphic functions. These will be boundary values of functions defined in a bigger submanifold $\tilde{\Sigma}$, which are actually holomorphic for some variables.

We prove this by Fourier transformation in the x variables, thereby generalizing the classical Paley-Wiener theorem. The idea formally is this: Parametrize the surface Σ as $\boldsymbol{R}^n \times V \times E$ as follows:

$$\Sigma = \{(z, u) = (x + i(v + N(u)), u); x \in \boldsymbol{R}^n, v \in V, u \in E\}.$$

For ϕ a function defined on D, consider

$$\hat{\phi}(\xi, v, u) = \int e^{-2\pi i\langle \xi, z\rangle} \phi(z, u)\, dx$$

for $\xi \in (\boldsymbol{R}^n)^*$, $v \in V$, $u \in E$. Formally, the CR equations get transformed into the system

$$\partial\hat{\phi}/\partial v_i = 0, \qquad \partial\hat{\phi}/\partial \bar{u}_\alpha = 0,$$

i.e., ϕ is CR if and only if $\hat{\phi}$ is independent of v, and holomorphic in u. In the following section we give precise results.

3. The Paley-Wiener theorem for CR functions. Let $d\mu(v)$ be a C^∞ volume form on V, and let $\nu(u)du$ be a measure on E, where du is the Euclidean measure, and $\nu(u)$ is a positive C^∞ function. Then $dm = dx d\mu(v)\nu(u)du$ is the measure we shall consider on Σ. Let

$$L^2(\Sigma) = \left\{f; \|f\|^2 = \int |f(x + i(v + N(u)), u)|^2\, dx\, d\mu(v)\nu(u)\, du < +\infty\right\}$$

and $H(\Sigma) = \{f$ in $L^2(\Sigma)$ satisfying, in the distribution sense, the tangential Cauchy-Riemann equations$\}$. For $f \in L^2(\Sigma)$ we introduce the partial Fourier-Laplace transform:

$$\hat{f}(\xi, v, u) = \int e^{-2\pi i\langle \xi, z\rangle} f(z, u)\, dx$$

where $z = (x + i(v + N(u)))$, and $\xi \in (\boldsymbol{R}^n)^*$.

By the Plancherel theorem, this transformation is an isomorphism onto the space of functions $f(\xi, v, u)$ measurable on $(\boldsymbol{R}^n)^* \times V \times E$ such that

$$\begin{aligned}\|f\|^2 &= \int e^{-4\pi\langle \xi, v\rangle} e^{-4\pi\langle \xi, N(u)\rangle} |\hat{f}(\xi, v, u)|^2\, d\xi\, d\mu(v)\nu(u)\, du \\ &= \|\hat{f}\|^2 < +\infty.\end{aligned}$$

THEOREM. *If $f \in H(\Sigma)$, then $\hat{f}(\xi, v, u) = \phi(\xi, u)$ almost everywhere, where $\phi(\xi, u)$ is a measurable function on $(\boldsymbol{R}^n)^* \times E$ holomorphic in u.*

Since $\hat{f}$ is independent of v, we can integrate out the v variable after using Plancherel's theorem:

$$(*) \qquad \|f\|^2 = \int_{(R^n)^*\times E} I_\mu(\xi) e^{-4\pi\langle\xi,N(u)\rangle} |\phi(\xi,u)|^2 \, d\xi \nu(u)\, du < +\infty$$

where $I_\mu(\xi) = \int_V e^{-4\pi\langle\xi,v\rangle} d\mu(v)$. Thus

THEOREM (PALEY-WIENER THEOREM FOR CR FUNCTIONS). *The map $f \to \hat{f}(\xi, v, u) = \phi(\xi, u)$ is a unitary isomorphism of $H(\Sigma)$ onto $\hat{H} = \{$classes of measurble functions $\phi(\xi, u)$ on $(R^n)^* \times E$ holomorphic in u such that the integral $(*)$ is finite$\}$.*

4. Extension theorems for CR functions. The Fourier inversion formula

$$\begin{aligned} f(x+iy, u) &= \int e^{2\pi i\langle\xi,x+iy\rangle} \phi(\xi,u)\, d\xi \\ &= \int e^{2\pi i\langle\xi,x\rangle} e^{-2\pi\langle\xi,y\rangle} \phi(\xi,u)\, d\xi \end{aligned}$$

gives the tool to extend CR functions off Σ. The integration above is, of course, over the support S (in ξ) of the function ϕ in $\hat{H}$. Clearly, the smaller the support S is, the larger the set of values of y for which this integral makes sense. In particular, we will be able to define the extension F on $\Sigma + i\varepsilon$ by

$$F(x+i(y+\varepsilon), u) = \int_S e^{2\pi i\langle\xi,x\rangle} e^{-2\pi\langle\xi,y\rangle} e^{-2\pi\langle\xi,\varepsilon\rangle} \phi(\xi,u)\, d\xi,$$

each time ε is such that $\langle\varepsilon, \xi\rangle \geqq 0$ for all $\xi \in S$. Now, the support S (in R^{n*}) is limited by two necessary requirements:

(a) We must have

$$\int_{(R^n)^*\times E} I_\mu(\xi) e^{-4\pi\langle\xi,N(u)\rangle} |\phi(\xi,u)|^2 \, d\xi \nu(u)\, du < +\infty.$$

(b) As $\phi(\xi, u)$ must be holomorphic in u, if

$$(**) \qquad H(\xi) = \left\{ f \in \mathcal{O}(E); \int_E e^{-4\pi\langle\xi,N(u)\rangle} |f(u)|^2 \nu(u)\, du < \infty \right\}$$

then S will have to be contained in the set of ξ where $H(\xi) \neq \{0\}$.

Now, we restrict our general situation, so that we may analyze completely the implications of (a) and (b), and still apply the results to the group-theoretically interesting cases. We assume $\Sigma(V, N, E)$ is of the following form. E will be C^m; $N(u)$ will be a hermitian form $Q(u, u)$ with values in R^n; V will be invariant under the dilations $v \to tv$ ($t \in R^+$) and the measure $d\mu(v)$ will be homogeneous with respect to these homotheties, i.e., $d\mu(tv) = t^\lambda d\mu(v)$; and $dm = dx\, d\mu(v)\, du$. We introduce

$$\hat{V} = \{\xi \in R^{n*}; \langle\xi, v\rangle \geqq 0 \text{ if } v \in V\}.$$

$\hat{V}$ is a convex closed cone in R^{n*}. We denote the interior of $\hat{V}$ by V°. We need the following simple lemmas.

LEMMA 1. *$I_\mu(\xi)$ is infinite off $\hat{V}$.*

LEMMA 2. *Let $H(\xi)$ be as defined in $(**)$ above (with $E = C^m$). $H(\xi) \neq \{0\}$ if and only if $\langle\xi, Q(u, u)\rangle$ is positive definite.*

Finally let

$$U(Q) = \{\xi \in \mathbf{R}^{n*}; \langle \xi, Q(u, u)\rangle > 0, u \in \mathbf{C}^m - \{0\}\}.$$

$U(Q)$ is an open convex cone in $\mathbf{R}^{n*}$. We get, from the preceding theorem, the following corollary:

COROLLARY. (1) *$H(\Sigma) \neq \{0\}$, if and only if the following two conditions are realized:*

(a) $V^\circ \cap U(Q) \neq \emptyset$.

(b) *There is a ξ_0 in V° with $I_\mu(\xi_0) < +\infty$.*

(2) *In this case, $H(\Sigma)$ is isomorphic to*

$$\hat{H} = \{\phi(\xi, u) \text{ defined in } V^\circ \cap U(Q) \times \mathbf{C}^m, \text{ holomorphic in } u, \text{ such that}$$
$$\|\phi\|^2 = \int_{(V^\circ \cap U(Q)) \times \mathbf{C}^m} I_\mu(\xi)|\phi(\xi, u)|^2 e^{-4\pi\langle \xi, Q(u,u)\rangle}\, d\xi\, du < \infty\}.$$

Now let $[V^\circ \cap U(Q)]^\wedge = \{\varepsilon \in \mathbf{R}^n; \langle \xi, \varepsilon\rangle > 0 \text{ for all } \xi \in V^\circ \cap U(Q)\}$. It is easily seen that $[V^\circ \cap U(Q)]^\wedge = C(V, Q)$, the closed convex hull generated by V and all the values $Q(u, u)$, with u in $\mathbf{C}^m$. Let $L(V, Q)$ be the vector subspace of $\mathbf{R}^n$ spanned by $C(V, Q)$ (i.e., by V and the $Q(u, u)$). $C(V, Q)$ has a nonempty interior in $L(V, Q)$ which we denote by C°. Then, for

$$\tilde{\Sigma} = \{(x + iy, u); y - Q(u, u) \in C^\circ\}$$

our principal theorem is that any CR function in $L^2(\Sigma)$ extends to $\tilde{\Sigma}$. To describe these extensions, we need to consider coordinates in $\mathbf{R}^n = \mathbf{R}^{n_1} \times \mathbf{R}^{n_2}$ so that $L(V, Q) = (0, \mathbf{R}^{n_2})$. Then

$$\tilde{\Sigma} = \{(x + iy, u); y - Q(u, u) \in C^\circ, \text{ i.e, } Y_1 = 0, y_2 - Q(u, u) \in C^\circ\} = \mathbf{R}^{n_1} \times \tilde{D},$$

with

$$\tilde{D} = \{(z_2, u); y_2 - Q(u, u) \in C^\circ\}.$$

As C° is open in $\mathbf{R}^{n_2}$, D is a domain in $\mathbf{C}^{n_2} \times \mathbf{C}^m$. If $\varepsilon \in C^\circ$, then $\Sigma + i\varepsilon \subset \tilde{\Sigma}$.

THEOREM. *Let $\tilde{\Sigma}$ be defined as above. Let $H(\tilde{\Sigma})$ be the space of measurable functions on $\tilde{\Sigma}$, $F(x_1, z_2, u)$ which are holomorphic in (z_2, u) in $\tilde{D}$, and such that*

$$\sup_{\varepsilon \in C^\circ} \int_\Sigma |F(\sigma + i\varepsilon)|^2\, dm = \|F\|^2 < +\infty.$$

Then if $F \in H(\tilde{\Sigma})$, and $\varepsilon \in C^\circ$, $F_\varepsilon(\sigma) = F(\sigma + i\varepsilon)$ belongs to $H(\Sigma)$, and converges to a function $f(\sigma)$ in $H(\Sigma)$ as ε tends to 0.

The correspondence $F \to f$ is a unitary isomorphism between $H(\tilde{\Sigma})$ and $H(\Sigma)$.

5. Some remarks. From the point of view of local complex analysis, any function f locally integrable on Σ and satisfying weakly the CR equations should extend to $\tilde{\Sigma}$, but we are unable to prove that. For results related to this problem, see the article by Hill **[1]**.

For the case of irreducible hermitian symmetric domains, we can explicitly compute the surfaces Σ_i. This gives the stratification of the boundary and the appropriate extension properties. In particular (referring to the notation of §1):

(a) $\mathcal{O}_i = \{0\}$ corresponds to the Šilov boundary and the results are due to Vagi **[4]**. If D is of tube type then there are no tangential Cauchy-Riemann equations and so no extension properties. If D is not of tube type, then any CR function in $L^2(\Sigma_0)$ extends to a holomorphic function in D.

(b) If $\mathcal{O}_i \neq \{0\}$, then any CR function on $L^2(\Sigma_i)$ extends to a holomorphic function in $D(\Omega, Q)$ (i.e., this is the corresponding $\tilde{\Sigma}_i$), whether or not $D(\Omega, Q)$ is of tube type.

Bibliography

1. C. Denson Hill and Michael Kazlow, *Function theory on tube manifolds*, Proc. Sympos. Pure Math., vol. 30, part 1, Amer. Math. Soc., Providence, R. I., 1977, pp. 153–156.

2. H. Rossi and M. Vergne, *Analytic continuation of the holomorphic discrete series*, Acta Math. (to appear).

3. ———, *Équations de Cauchy-Riemann tangentielles associées à un domaine de Siegel*, Ann. École Norm. Sup. 4e Série, t.9, 1976, pp. 31–80.

4. S. Vági, *On the boundary values of holomorphic functions*, Rev. Un. Mat. Argentina **25** (1970/71), 123–136. MR **47** #7343.

University of Utah

C.N.R.S. Paris

Proceedings of Symposia in Pure Mathematics
Volume 30, 1977

ON SYMMETRIC AND QUASI-SYMMETRIC SIEGEL DOMAINS

ICHIRO SATAKE

The notion of "Siegel domains", introduced by Pjateckiĭ-Šapiro **[8]**, has turned out to be very useful in several complex variables. It was shown above all that every homogeneous bounded domain can be expressed as a Siegel domain (of the second kind) uniquely determined up to a linear isomorphism **[13]**, **[3]** and that, for a symmetric bounded domain $\mathscr{D}$, such an expression is given explicitly in terms of a maximal parabolic subgroup of $\mathrm{Hol}(\mathscr{D})$ **[5a]**. In this report, we give a survey of more recent developments following the line of **[3]**, **[6]**, **[9c, d]** and **[10b]**.

1. Siegel domains (**[8]**, **[3]**, **[6]**). Let $U = \boldsymbol{R}^n$ (resp. $V = \boldsymbol{C}^m$) be a real (resp. complex) vector space identified with the space of column vectors. Let Ω be a (nonempty) open convex cone in U with vertex at the origin and not containing any straight line. Further, let F be an "Ω-hermitian form", i.e., a sesquilinear map $F : V \times V \to U_C = U \otimes_{\boldsymbol{R}} \boldsymbol{C}$ (which we assume to be $\boldsymbol{C}$-linear in the second variable) such that $F(v, v) \in \bar{\Omega} - \{0\}$ for all $v \in V$, $v \neq 0$. A Siegel domain $\mathscr{D} = \mathscr{D}(U, V, \Omega, F)$ is then defined by

$$\mathscr{D} = \{(u, v) \in U_C \times V \mid \mathrm{Im}\, u - F(v, v) \in \Omega\}. \tag{1}$$

$\mathrm{Hol}(\mathscr{D})$ (resp. $\mathrm{Aff}(\mathscr{D})$) will denote the Lie group of all holomorphic (resp. affine) automorphisms of $\mathscr{D}$. The Lie algebra $\mathfrak{g} = \mathfrak{g}(\mathscr{D})$ of $\mathrm{Hol}(\mathscr{D})$ can be interpreted as the Lie algebra of all complete holomorphic vector fields on $\mathscr{D}$. We also denote by $\mathrm{Aut}(\Omega)$ the linear automorphism group of Ω and by $\mathfrak{g}(\Omega)$ its Lie algebra viewed as a subalgebra of $\mathfrak{gl}(U)$.

It is known **[3]**, **[6]** that $\mathfrak{g}$ has a natural gradation:

$$\mathfrak{g} = \mathfrak{g}_{-1} + \mathfrak{g}_{-1/2} + \mathfrak{g}_0 + \mathfrak{g}_{1/2} + \mathfrak{g}_1, \tag{2}$$

AMS (MOS) subject classifications (1970). Primary 32A07, 32M15, 53C35; Secondary 22E45, 17C35.

such that the nonpositive part $\sum_{\nu\le 0}\mathfrak{g}_\nu$ is the Lie algebra of $\mathrm{Aff}(\mathscr{D})$. To be more precise, we denote the complex variables on U_C and V by $z = (z^k)$ and $w = (w^\alpha)$, respectively, and set $\partial_z = (\partial/\partial z^k)$, $\partial_w = (\partial/\partial w^\alpha)$. For any polynomial function $p(z, w)$ (resp. $q(z, w)$) on $U_C \times V$ with values in U_C (resp. V), we define the corresponding vector field on $U_C \times V$ by

$$p(z, w)\cdot\partial_z = \sum_{k=1}^{n} p(z, w)^k \frac{\partial}{\partial z^k}$$

(resp. $q(z, w)\cdot\partial_w = \sum_{\alpha=1}^{m} q(z, w)^\alpha\, \partial/\partial w^\alpha$). Then one has

$$\mathfrak{g}_{-1} = \{u\cdot\partial_z \mid u \in U\}, \tag{3}$$

$$\mathfrak{g}_{-1/2} = \{2iF(v, w)\cdot\partial_z + v\cdot\partial_w \mid v \in V\}, \tag{4}$$

$$\mathfrak{g}_0 = \{(Az)\cdot\partial_z + (Bw)\cdot\partial_w \mid A \in \mathfrak{g}(\Omega),\ B \in \mathfrak{gl}(V),\ B \text{ associated to } A\}, \tag{5}$$

where "B is associated to A" means that the relation

$$AF(v, v') = F(Bv, v') + F(v, Bv') \tag{6}$$

holds for all $v, v' \in V$. For simplicity, we identify $\mathfrak{g}_{-1}$, $\mathfrak{g}_{-1/2}$, and $\mathfrak{g}_0$ with U, V, and a subalgebra of $\mathfrak{g}(\Omega) \oplus \mathfrak{gl}(V)$, respectively, by the correspondences:

$$-u \leftrightarrow u\cdot\partial_z, \qquad -v \leftrightarrow 2iF(v, w)\cdot\partial_z + v\cdot\partial_w,$$
$$-(A, B) \leftrightarrow (Az)\cdot\partial_z + (Bw)\cdot\partial_w.$$

Then one has

$$[(A, B), u] = Au, \qquad [(A, B), v] = Bv, \qquad [v, v'] = -4\,\mathrm{Im}\, F(v, v'). \tag{7}$$

For $A \in \mathfrak{gl}(U)$ (resp. $B \in \mathfrak{gl}(V)$), the adjoint ${}^tA$ (resp. $B^* = {}^t\bar{B}$) are defined with respect to the standard inner product $\langle\ \rangle$ on $U = \boldsymbol{R}^n$ (resp. the standard hermitian inner product f on $V = \boldsymbol{C}^m$). $\langle\ \rangle$ is extended naturally to a symmetric $\boldsymbol{C}$-bilinear form on $U_C \times U_C$. For a $\boldsymbol{C}$-linear map $\Phi : U_C \to V$, the adjoint Φ^* is defined by

$$f(\Phi(u), v) = \langle \bar{u}, \Phi^*(v)\rangle \qquad (u \in U_C,\ v \in V). \tag{8}$$

Furthermore, for $u \in U_C$, we define $R_u \in \mathfrak{gl}(V)$ by

$$2f(v, R_u v') = \langle u, F(v, v')\rangle \qquad (v, v' \in V). \tag{9}$$

Then clearly one has $R_u^* = R_{\bar{u}}$. In this notation, using a result of Tanaka (J. Math. Soc. Japan **22** (1970); reorganized and extended by Murakami [**6**] and Nakajima [**7**]), we obtain the following theorems (cf. [**9c**]; for another approach, see [**2**]).

THEOREM 1. *$\mathfrak{g}_{1/2}$ consists of vector fields of the form*

$$Y_\Phi = 2iF(\Phi(\bar{z}), w)\cdot\partial_z + (\Phi(z) + 4iR_{\Phi^*(w)}w)\cdot\partial_w, \tag{10}$$

where Φ is a $\boldsymbol{C}$-linear map $U_C \to V$ satisfying the following conditions:

$$\Phi_v = [u \mapsto \mathrm{Im}\, F(v, \Phi(u))] \in \mathfrak{g}(\Omega), \tag{11}$$

$$2F(v, R_{\Phi^*(v')}v') = F(\Phi(F(v', v)), v') \qquad (u \in U;\ v, v' \in V). \tag{12}$$

We identify Y_Φ with $-\Phi$. Then one has

(13) $$[\Phi, u] = \Phi(u), \qquad [\Phi, v] = 4(\Phi_v, \Psi_v),$$

where $\Psi_v(v') = (i/2)\,\Phi(F(v, v')) + iR_{\Phi^*(v)}v' + iR_{\Phi^*(v')}v$. Note that the condition (12) is equivalent to saying that Ψ_v is associated to Φ_v for all $v \in V$.

THEOREM 2. *$\mathfrak{g}_1$ consists of vector fields of the form*

(14) $$Z_{a,b} = a(z, z)\cdot\partial_z + b(z, w)\cdot\partial_w,$$

*where a is a **C**-bilinear extension of a symmetric bilinear map $U \times U \to U$ and b is a **C**-bilinear map $U_C \times V \to V$ satisfying the following conditions*:

(15) $$A_u = [u' \mapsto a(u, u')] \in \mathfrak{g}(\Omega),$$

(16) $$B_u = [v \mapsto \tfrac{1}{2}b(u, v)] \in \mathfrak{gl}^0(V), \quad \textit{and associated to } A_u,$$

(17) $$[u \mapsto \operatorname{Im} F(v, b(u, v'))] \in \mathfrak{g}(\Omega),$$

(18) $$F(v, b(F(v', v''), v'')) = F(b(F(v'', v), v'), v'') \qquad (u, u' \in U,\ v, v', v'' \in V),$$

where we set $\mathfrak{gl}^0(V) = \{B \in \mathfrak{gl}(V) \mid \operatorname{tr} B \in \boldsymbol{R}\}$.

We identify $Z_{a,b}$ with $-(a, b)$. Then one has

(19) $$[(a, b), u] = 2(A_u, B_u), \qquad [(a, b), v] = \Phi \quad (\text{where } \Phi(u) = b(u, v)).$$

2. Symmetric case [9c]. The cone Ω is called *self-dual* if it is homogeneous (i.e., $\operatorname{Aut}(\Omega)$ is transitive) and, for a suitable choice of the basis of U, one has

(20) $$\Omega = \{u \in U \mid \langle u, u'\rangle > 0 \text{ for all } u' \in \bar{\Omega} - \{0\}\},$$

or equivalently, ${}^t\operatorname{Aut}(\Omega) = \operatorname{Aut}(\Omega)$. A self-dual cone Ω becomes a symmetric space (with a flat part) and there exists a point $e \in \Omega$ such that the Cartan involution of $\mathfrak{g}(\Omega)$ at e is given by $A \mapsto -{}^tA$. It follows that, for every $u \in U$, there exists a unique element T_u in $\mathfrak{g}(\Omega)$ such that

(21) $${}^tT_u = T_u, \qquad T_u e = u;$$

in particular, $T_e = 1_U$ (the identity transformation of U). It is known **[1]**, **[12]** that the space U, endowed with a product defined by $u \circ u' = T_u u'$, becomes a formally real Jordan algebra with unit element e. (Actually, the correspondence $\Omega \leftrightarrow U$ is an equivalence between these two categories **[9d]**.) When Ω is self-dual, we always assume that (20) is satisfied for the given inner product and that f is defined by

(22) $$f(v, v') = \langle e, F(v, v')\rangle,$$

or equivalently, $R_e = \frac{1}{2}1_V$. Then clearly one has $R_u \in \mathscr{P}_m(C)$ (= the cone of positive-definite hermitian matrices of degree m) for all $u \in \Omega$.

THEOREM 3 **[9c]**.[1] *A Siegel domain $\mathscr{D} = \mathscr{D}(U, V, \Omega, F)$ is symmetric if and only if the following three conditions are satisfied*:

(i) *Ω is self-dual.*

(ii) *For every $u \in U$, R_u is associated to T_u.*

(iii) *For any $v, v', v'' \in V$, one has*

[1]Theorem 3 was also obtained independently by Dorfmeister **[14]**.

$$F(v, R_{F(v',v'')}v'') = F(R_{F(v'',v)}v', v'').$$

REMARK 1 **[9d]**. Under the condition (i), the condition (ii) (=(ii$_1$)) is equivalent to any of the following conditions:

(ii$_2$) The map $u \mapsto 2R_u$ is a (unitary) Jordan algebra homomorphism of U into $\mathscr{H}_m(C)$ (= the Jordan algebra of hermitian matrices of degree m, endowed with the product $H \circ H' = \frac{1}{2}(HH' + H'H)$).

(ii$_3$) There exists a Lie algebra homomorphism $\beta: \mathfrak{g}(\Omega) \to \mathfrak{gl}(V)$ such that $\beta(A)$ is associated to A and $\beta({}^tA) = \beta(A)^*$ for all $A \in \mathfrak{g}(\Omega)$. ($\beta$ is then uniquely determined by the relation $\beta(T_u) = R_u$ $(u \in U)$.)

(ii$_4$) The natural projection $\mathfrak{g}_0 \to \mathfrak{g}(\Omega)$ is surjective.

REMARK 2. Under the conditions (i), (ii), the condition (iii) (= (iii$_2$)) is equivalent to any of the following conditions:

(iii$_1$) $$R_{F(v,v')}R_u v' = R_{F(R_u v,v')}v',$$

(iii$'_1$) $$R_u R_{F(v,v')}v' = R_{F(v,R_u v')}v',$$

(iii$_3$) $$\langle F(R_u v, v''), F(v', v'')\rangle = \langle F(v, v''), F(v', R_u v'')\rangle$$
$$(u \in U,\ v, v', v'' \in V).$$

When $\mathscr{D}$ is symmetric, the Cartan involution θ of $\mathfrak{g} = \mathfrak{g}(\mathscr{D})$ at $(ie, 0)$ reverses the gradation (2): $\theta\mathfrak{g}_\nu = \mathfrak{g}_{-\nu}$, and on $\mathfrak{g}_0$ one has $\theta: (A, B) \mapsto (-{}^tA, -B^*)$. We put $\theta u = (a^u, b^u)$, $\theta v = \Phi^v$. Then we have the following explicit formulas:

$$a^u(u', u'') = (u \circ u') \circ u'' + (u \circ u'') \circ u' - u \circ (u' \circ u''),$$
$$b^u(u', v) = 4 R_{u'}R_u v,$$

(23) $$\Phi^v(u) = 2iR_u v, \qquad \Phi^{v*}(v') = -iF(v, v').$$

It follows from (13), (19) that one has

(24) $$[\theta u, u'] = 2(A_{u,u'}, B_{u,u'}),$$

(25) $$[\theta v, v'] = 4(\Phi_{v,v'}, \Psi_{v,v'}),$$

where

(26) $$A_{u,u'} = T_{u \circ u'} - [T_u, T_{u'}],$$

(27) $$B_{u,u'} = 2R_{u'}R_u,$$

(28) $$\Phi_{v,v'}(u) = 2\,\mathrm{Re}\,F(v', R_u v),$$

(29) $$\Psi_{v,v'}(v'') = R_{F(v,v')}v'' + R_{F(v,v'')}v' - R_{F(v',v'')}v.$$

We note that under the condition (i), if we define $A_{u,u'}$, $B_{u,u'}$ by (26), (27), then the condition (ii) is equivalent to saying that $B_{u,u'}$ is associated to $A_{u,u'}$ for all u, $u' \in U$. Also, under the conditions (i), (ii), if we define $\Phi_{v,v'}$, $\Psi_{v,v'}$ by (28), (29), then $\Phi_{v,v'} \in \mathfrak{g}(\Omega)$ and the condition (iii) is equivalent to saying that $\Psi_{v,v'}$ is associated to $\Phi_{v,v'}$ for all v, $v' \in V$.[2]

3. Quasi-symmetric Siegel domains [9d], [10b]. We call a Siegel domain $\mathscr{D}$ *quasi-*

[2]To understand the meaning of these conditions, it is useful to observe the relation between symmetric Siegel domains and Jordan triple systems. In this respect, cf. Loos **[15a, b]**.

symmetric if it satisfies the conditions (i), (ii). It is clear that a quasi-symmetric Siegel domain $\mathscr{D}$ is affinely homogeneous (i.e., $\mathrm{Aff}(\mathscr{D})$ is transitive). By Remark 1 in §2, we see that a quasi-symmetric Siegel domain $\mathscr{D}$ is determined by a pair (Ω, β) formed of a self-dual cone Ω and a Lie algebra homomorphism $\beta: \mathfrak{g}(\Omega) \to \mathfrak{gl}(V) = \mathfrak{gl}(m, \boldsymbol{C})$ with the properties mentioned in (ii_3). For brevity, we call such a homomorphism a *special representation* of $\mathfrak{g}(\Omega)$; also we write $\mathscr{D} = \mathscr{D}(\Omega, \beta)$ when $\mathscr{D}$ is determined by a pair (Ω, β). The quasi-symmetric Siegel domain $\mathscr{D}(\Omega, \mathrm{triv})$ is the tube domain $U + i\Omega$, which is necessarily symmetric, for the condition (iii) is trivially satisfied when $V = \{0\}$.

Let $\mathscr{D}' = \mathscr{D}(U', V', \Omega', F') = \mathscr{D}(\Omega', \beta')$ be another quasi-symmetric Siegel domain, where we denote the objects relative to $\mathscr{D}'$ by the corresponding symbols with primes. Then we have

THEOREM 4 **[9d]**. *Two quasi-symmetric Siegel domains $\mathscr{D} = \mathscr{D}(\Omega, \beta)$ and $\mathscr{D}' = \mathscr{D}(\Omega', \beta')$ are holomorphically equivalent if and only if there exist linear isomorphisms $\varphi: U \to U'$, $\psi: V \to V'$ such that $\varphi(\Omega) = \Omega'$, $\varphi(e) = e'$, and*

$$\psi \cdot \beta(A) \cdot \psi^{-1} = \beta'(\varphi \cdot A \cdot \varphi^{-1}) \quad \textit{for all } A \in \mathfrak{g}(\Omega). \tag{30}$$

In particular, when $U = U'$, $\Omega = \Omega'$, $e = e'$, two representations β and β' of $\mathfrak{g}(\Omega)$ are called *automorphically equivalent* (at e) if the condition (30) is satisfied for some $\varphi \in \mathrm{Aut}(\Omega)$ with $\varphi(e) = e$ (or equivalently, ${}^t\varphi = \varphi^{-1}$). By Theorem 4, the classification of quasi-symmetric Siegel domains (up to holomorphic equivalence) is reduced to that of self-dual cones Ω (up to linear isomorphism) and that of special representations β of $\mathfrak{g}(\Omega)$ (up to automorphic equivalence). The classification of self-dual cones Ω (resp. β) is given in **[1]**, **[12c]** (resp. **[9a]**), and that of quasi-symmetric domains $\mathscr{D}$ is given in **[10b]**, **[9d]**. Summarizing the main results, we first note that self-dual cones and quasi-symmetric domains are "completely reducible", i.e., they are decomposed (uniquely) into the direct product of irreducible (i.e., indecomposable) ones, and the irreducible decomposition of Ω gives rise to that of $\mathscr{D}(\Omega, \beta)$. Special representations are also completely reducible. Thus it is sufficient to classify irreducible β's for irreducible Ω's. The isomorphism class of an irreducible self-dual cone Ω is determined by the "invariant" (ν, m), where m is the real rank of $\mathfrak{g}(\Omega)$ and ν is the (common) multiplicity of real roots of $\mathfrak{g}(\Omega)$ if $m \geqq 2$ and $\nu = 1$ if $m = 1$ **[12c]**, **[10b]**. Also, let N be the number of (ordinary) equivalence classes of (nontrivial) irreducible special representations β of $\mathfrak{g}(\Omega)$. Then one has $N \leqq 2$. All possibilities for (ν, m) are shown in the following chart, where the marks $\times, \circ, \odot$ correspond to irreducible self-dual cones with $N = 0, 1, 2$, respectively. In the case $N = 2$, two inequivalent special representations are automorphically equivalent.

THEOREM 5. *For an irreducible quasi-symmetric Siegel domain $\mathscr{D}$, which is not symmetric, one has* $\mathrm{Hol}(\mathscr{D}) = \mathrm{Aff}(\mathscr{D})$.

THEOREM 6. *Let $\mathscr{D}$ be a symmetric bounded domain. Then, for any boundary point t, the fiber $\mathscr{D}_t$ over t in the expression of $\mathscr{D}$ as a Siegel domain of the third kind is quasi-symmetric.*

Theorem 5 is obtained by combining results in **[11]** and **[2]**. Theorem 6 is given in **[9c]**, cf. also **[5b]**, **[9b]**. It will be interesting to find an analytic or differential geo-

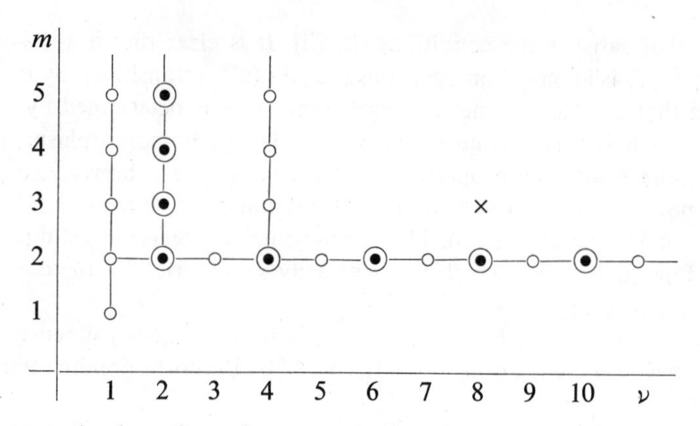

metric characterization of quasi-symmetric domains. It is also desirable to extend Theorem 4 to the case of "morphisms" (e.g., equivariant holomorphic maps) of quasi-symmetric domains.

EXAMPLE. For $\Omega = \mathscr{P}_m(\boldsymbol{C})$ $(m \geqq 2)$, $\mathfrak{g}(\Omega)$ can be identified with $\mathfrak{gl}^0(m, \boldsymbol{C}) = \{B \in \mathfrak{gl}(m, \boldsymbol{C}) \mid \operatorname{tr} B \in \boldsymbol{R}\}$ by the action $(B, H) \mapsto BH + HB^*$ $(B \in \mathfrak{gl}^0(m, \boldsymbol{C}),\ H \in \mathscr{H}_m(\boldsymbol{C}))$. In this case, the invariant is $(2, m)$ and one has $N = 2$. All special representations of $\mathfrak{gl}^0(m, \boldsymbol{C})$ are given by

$$\beta_{r,s} = \overbrace{\mathrm{id} \oplus \cdots \oplus \mathrm{id}}^{r} \oplus \overbrace{\overline{\mathrm{id}} \oplus \cdots \oplus \overline{\mathrm{id}}}^{s}, \tag{31}$$

and two special representations $\beta_{r,s}$ and $\beta_{r',s'}$ are automorphically equivalent if and only if $r = r'$, $s = s'$ or $r = s'$, $s = r'$. We denote by $I_{m;r,s}$ the type (class) of the domain $\mathscr{D}(\mathscr{P}_m(\boldsymbol{C}), \beta_{r,s})$ $(r \geqq s)$. Then $I_{m;r,s}$ is symmetric if and only if $s = 0$, and $I_{m;r,0}$ coincides with $I_{m+r,m}$ in Cartan's notation. (In the case $m = 1$, we have $\mathrm{id} = \overline{\mathrm{id}}$; the invariant is $(1, 1)$ and $N = 1$. Hence in (31) we assume $s = 0$. Then the above results remain true.)

Let $\mathscr{D}$ be a symmetric bounded domain of type $I_{p,q}$ $(p \geqq q)$ and t a boundary point on a boundary component of type $I_{p-k,q-k}$ $(1 \leqq k \leqq q)$. Then the fiber $\mathscr{D}_t$ over t is a quasi-symmetric domain of type $I_{k;p-k,q-k}$ for $k \geqq 2$ and $I_{1;p+q-2,0}$ for $k = 1$. Thus $\mathscr{D}_t$ is symmetric if and only if $k = 1$ or q.

REFERENCES

1. H. Braun and M. Koecher, *Jordan-Algebren*, Die Grundlehren der math. Wissenschaften, Band 128, Springer-Verlag, Berlin and New York, 1966. MR **34** #4310.

2. W. Kaup, *Einige Bemerkungen über polynomiale Vektorfelder, Jordanalgebren und die Automorphismen von Siegelschen Gebieten*, Math. Ann. **204** (1973), 131–144.

3. W. Kaup, Y. Matsushima and T. Ochiai, *On the automorphisms and equivalences of generalized Siegel domains*, Amer. J. Math. **92** (1970), 475–498. MR **42** #2029.

4. M. Koecher, *Jordan algebras and their applications*, Lecture Notes, Univ. of Minnesota, 1962.

5. A. Korányi and J. Wolf, a) *Realization of hermitian symmetric spaces as generalized half-planes*, Ann. of Math. (2) **81** (1965), 265–288. MR **30** #4980.

b) *Generalized Cayley transformations of bounded symmetric domains*, Amer. J. Math. **87** (1965), 899–939. MR **33** #229.

6. S. Murakami, *On automorphisms of Siegel domains*, Lecture Notes in Math., vol. 286, Springer-Verlag, Berlin and New York, 1972.

7. K. Nakajima, *Some studies on Siegel domains*, J. Math. Soc. Japan **27** (1975), 54–75.

8. I. I. Pjateckiĭ-Šapiro, *Geometry of classical domains and theory of automorphic functions*, Fizmatgiz, Moscow, 1961; English transl., Math. and its Applications, vol. 8, Gordon and Breach, New York, 1969. MR **25** #231; **40** #5908.

9. I. Satake, a) *Linear imbeddings of self-dual homogeneous cones*, Nagoya Math. J. **46** (1972), 121–145; corrections, ibid. **60** (1976), 219. MR **46** #9188.

b) *Realization of symmetric domains as Siegel domains of the third kind*, Lecture Notes, Univ. of California, Berkeley, Calif., 1972.

c) *Infinitesimal automorphisms of symmetric Siegel domains* (unpublished note), 1974.

d) *On classification of quasi-symmetric domains*, Nagoya Math. J. **62** (1976).

10. M. Takeuchi, a) *Homogeneous Siegel domains*, Publ. Study Group Geometry, no. 7, Tokyo, 1973.

b) *On symmetric Siegel domains*, Nagoya Math. J. **59** (1975), 9–44.

11. T. Tsuji, *Siegel domains over self-dual cones and their automorphisms*, Nagoya Math. J. **55** (1974), 33–80.

12. È. B. Vinberg, a) *Homogeneous cones*, Dokl. Akad. Nauk SSSR **133** (1960), 9–12 = Soviet Math. Dokl. **1** (1960), 787–790. MR **25** #5077.

b) *The theory of convex homogeneous cones*, Trudy Moskov Mat. Obšč. **12** (1963), 303–358 = Trans. Moscow Math. Soc. **1963**, 340–403. MR **28** #1637.

c) *The structure of the group of automorphisms of a homogeneous convex cone*, Trudy Moskov Mat. Obšč.**13**(1965), 56–83 = Trans. Moscow Math. Soc. **13** (1965), 63–93.

13. È. B. Vinberg, S. G. Gindikin and I. I. Pjateckiĭ-Šapiro, *Classification and canonical realization of complex bounded homogeneous domains*, Trudy Moskov Mat. Obšč. **12** (1963), 359–388 = Trans. Moscow Math. Soc. **1963**, 404–437. MR **28** #1638.

14. J. Dorfmeister, *Infinitesimal automorphisms of homogeneous Siegel domains* (to appear).

15. O. Loos, a) *Jordan triple systems, R-spaces, and bounded symmetric domains*, Bull. Amer. Math. Soc. **77** (1971), 558–561.

b) *Bounded symmetric domains and Jordan triple systems*. I (to appear).

UNIVERSITY OF CALIFORNIA, BERKELEY

Proceedings of Symposia in Pure Mathematics
Volume 30, 1977

COMPLEX HOMOGENEOUS BUNDLES AND FINITE-DIMENSIONAL REPRESENTATION THEORY

FLOYD L. WILLIAMS

1. This article, which is expository in nature, consists of some informal and somewhat nostalgic remarks on some known interrelations between complex manifold theory and the representation theory of compact Lie groups. These relationships exist naturally since the latter theory reduces essentially to the study of finite-dimensional holomorphic representations of the corresponding complexified Lie groups. We shall not consider the important and interesting parallels that exist for noncompact groups as well in connection with the theory of *discrete series* representations.

The classical Cartan-Weyl "highest weight" theory parametrizes the irreducible holomorphic modules via Lie algebra data. However a concrete realization, or "model", of such a module is given by the Borel-Weil theorem. This famous theorem asserts, in effect, that all finite-dimensional, holomorphic, irreducible modules for a simply connected complex semisimple Lie group G are realized by the modules of holomorphic sections of appropriate homogeneous G bundles **[2]**, **[3]**, **[16]**, **[17]**, **[18]**.

If one interprets these sections, as one naturally does, as the zero-dimensional cohomology of the base space, defined by the obvious coefficient sheaf of germs, then one must surely inquire about the nature of the higher cohomology groups; these also have an appropriate module structure. The answer, due to Bott, is provided by the "generalized" Borel-Weil theorem **[3]**, **[6]**, **[8]**, **[14]**, **[15]**.

Now by Dolbeault's theorem the groups in question are isomorphic to the $\bar{\partial}$ cohomology groups of bundled-valued forms of pure type $(0, q)$. In turn, remarkably enough, the latter are isomorphic to the *relative* Lie algebra cohomology groups defined by the Lie algebra of left-invariant, antiholomorphic vector fields

AMS (MOS) sbuject classifications (1970). Primary 32L10, 22E45, 18H25.

on G. We shall begin by recalling, in greater detail and generality, this result which is also due to Bott [3].

2. Let G be a complex Lie group and let g be the (real) Lie algebra of G; i.e., g is the Lie algebra of left-invariant vector fields on G. Then g admits a complex structure J which commutes with the adjoint representation:

$$J[X, Y] = [X, JY] \tag{2.1}$$

for all X, Y in g. We therefore consider g as a complex Lie algebra. Let $g^{1,0}$, $g^{0,1}$ denote the splitting of g^C (the complexification of g) into $\pm\sqrt{-1}$ eigenspaces of J. From the integrability of the complex structure J one concludes that $g^{1,0}$, $g^{0,1}$ are subalgebras of g^C. Moreover (2.1) implies that

$$[g^{1,0}, g^{0,1}] = 0; \tag{2.2}$$

hence $g^{1,0}$, $g^{0,1}$ in fact are ideals.

Now let $B \subset G$ be a closed, connected complex Lie subgroup and suppose a finite-dimensional holomorphic representation λ of B is given, say on a complex vector space W. Since $(G, G/B, B)$ is a principal (holomorphic) fibre bundle with structure group B we can form the *associated* bundle

$$E_\lambda = G \times_B W \to G/B. \tag{2.3}$$

E_λ has a holomorphic structure and G acts on E_λ and G/B equivariantly. Let $H^{0,q}(G/B, E_\lambda)$ denote the corresponding $\bar\partial$ cohomology groups, $q \geqq 0$; i.e., if $\Lambda^{0,q}(G/B, E_\lambda)$ is the space of E_λ-valued forms of pure type $(0, q)$ on G/B, then

$$H^{0,q}(G/B, E_\lambda) = \frac{\ker(\Lambda^{0,q}(G/B, E_\lambda) \xrightarrow{\bar\partial} \Lambda^{0,q+1}(G/B, E_\lambda))}{\bar\partial \Lambda^{0,q-1}(G/B, E_\lambda)}. \tag{2.4}$$

Let $\mathscr{S}E_\lambda$ denote the sheaf of germs of local holomorphic sections of E_λ. Using Dolbeault's theorem we have

$$H^q(G/B, \mathscr{S}E_\lambda) \simeq H^{0,q}(G/B, E_\lambda). \tag{2.5}$$

Denote the Lie algebra of B by b. As above $b^C = b^{1,0} \oplus b^{0,1}$ and $[b^{1,0}, b^{0,1}] = 0$. Define

$$C^\infty(G, W)^{1,0} = \{f \in C^\infty(G, W) \mid xf + \dot\lambda(x)f = 0 \text{ for all } x \text{ in } b^{1,0}\} \tag{2.6}$$

where $\dot\lambda$ is the representation of b on W obtained by differentiation of λ. A key point is that, because of (2.2), $C^\infty(G, W)^{1,0}$ is a module, under differentiation, for the subalgebra $g^{0,1}$. Hence one can form the *relative* cohomology groups $H^q(g^{0,1}, b^{0,1}, C^\infty(G, W)^{1,0})$ of $g^{0,1}$ with respect to the subalgebra $b^{0,1}$; a relative cochain is a cochain which dies by Lie and interior differentiation in the $b^{0,1}$ direction; see [3]. Keeping in mind the above notation and also (2.5) we now state

THEOREM 2.7 (BOTT). $H^{0,q}(G/B, E_\lambda) \simeq H^q(g^{0,1}, b^{0,1}, C^\infty(G, W)^{1,0})$.

Note that in the very special case where B is the identity subgroup and λ is trivial Theorem 2.7 reduces to the familiar statement:

$$H^q(G, \mathscr{S}G) \simeq H^q(g^{0,1}, C^\infty(G)) \tag{2.8}$$

where $\mathscr{S}G$ is the sheaf of germs of local holomorphic functions on G.

Theorem 2.7 leads to a rather interesting reciprocity relation, which is entirely analogous, on the zero-dimensional cohomology level, to the classical Frobenius relation for finite groups. For this assume that G contains a connected compact subgroup K which acts transitively on G/B. For example we shall assume G is semi-simple, B is a Borel subgroup of G and that K is a maximal compact subgroup of G. Since K acts transitively G/B is diffeomorphic to $K/(B \cap K)$. In particular G/B is compact and hence $H^*(G/B, \mathscr{S}E_\lambda)$ (the direct sum of the $H^q(G/B, \mathscr{S}E_\lambda)$) is finite dimensional. Now $H^*(G/B, \mathscr{S}E_\lambda)$ has a G-module structure induced by the left action of G on E_λ. If "$^-$" denotes complex conjugation with respect to K, then one has the following reciprocity formula **[3]**, **[4]**.

THEOREM 2.9 (BOTT). *Let V be a finite-dimensional holomorphic G-module and, as above, let W be a finite-dimensional holomorphic B-module. Let λ be the representation of B on W. Then*

$$\text{Hom}_G(V, H^*(G/B, \mathscr{S}E_\lambda)) = H^*(b, b \cap \bar{b}^c, \text{Hom}(V, W)). \tag{2.10}$$

One may in fact derive the Borel-Weil theorem from Theorem 2.9.

There is an alternate way of viewing the relative cohomology in (2.10). Indeed we can consider $b \cap \bar{b}$ as a Cartan subalgebra h of g. Choose a system of nonzero roots Δ of g relative to h, let g_α be the (one-dimensional) root space corresponding to α in Δ, and let

$$n = \sum_{\alpha \in \Delta^+} g_\alpha \tag{2.11}$$

where $\Delta^+ \subset \Delta$ is a choice of positive roots. Then n is an ideal in b and h is a reductive complement of n in b. Keeping (2.10) in mind one has

$$H^*(b, b \cap \bar{b}^c, \text{Hom}(V, W)) = H^*(n, \text{Hom}(V, W))^h \tag{2.12}$$

where the right-hand side is the space of h-invariant elements in the h-module $H^*(n, \text{Hom}(V, W))$. Thus the cohomology of n makes its debut.

The importance of the latter is revealed in the work of Kostant in **[14]**. Kostant determined the h-module structure of $H^*(n, V)$ for any finite-dimensional irreducible g-module V. For the most part his results show that $H^k(n, V)$ is a direct sum of one-dimensional h-spaces, each space is indexed by a Weyl group element σ of length k (recall that the length of σ is the number of elements in Δ^+ which change sign under σ), and on such a space the eigenvalue of H in h is $[\sigma(\lambda + \delta) - \delta](H)$ where $2\delta = \sum_{\alpha \in \Delta^+} \alpha$ and λ is the highest weight of V. Using this result, moreover, Kostant derives a new proof of the generalized Borel-Weil theorem.

Another beautiful application of the cohomology of n is a derivation of Weyl's celebrated character formula; see **[14]**. We shall resist the great temptation to pursue this at this point and mention only that from the cohomology of n viewpoint Weyl's formula is nothing more, nor is it less, than an Euler-Poincaré principle expressing the equality of an alternating sum of traces on the cochain and cohomology levels.

A basic commutative diagram (to be specified) underlies much of what has been said here. One has a nice reductive object h (or a maximal torus in K with complexified Lie algebra h) acting such that certain h invariant cochains correspond to $(0, q)$-forms. Now under this correspondence, it is a pleasant fact of life that the Lie algebra cohomology operator matches, precisely, the $\bar{\partial}$ operator.

References

1. D. N. Ahiezer, *Cohomology of compact complex homogeneous spaces*, Mat. Sb. **84 (126)** (1971), 290–300 = Math. USSR Sb. **13** (1971), 285–296. MR **44** #2945.

2. A. Borel and A. Weil, *Représentations linéaires et espaces homogènes Kahlerians des groupes de Lie compact*, Séminaire Bourbaki, May 1954 (Exposé by J.-P. Serre).

3. R. Bott, *Homogeneous vector bundles*, Ann. of Math. (2) **66** (1957), 203–248. MR **19**, 681.

4. ———, *Induced representations*, Seminars on Analytic Functions, vol. 2, Institute for Advanced Study, Princeton, N. J., 1957.

5. C. Chevalley and S. Eilenberg, *Cohomology theory of Lie groups and Lie algebras*, Trans. Amer. Math. Soc. **63** (1948), 85–124. MR **9**, 567.

6. M. Demazure, *Une démonstration algébrique d'un théorème de Bott*, Invent. Math. **5** (1968), 349–356. MR **37** #4831.

7. P. Dolbeault, *Forms différentielles et cohomologie sur une variété analytique complex*. I, Ann. of Math. (2) **64** (1956), 83–130. MR **18**, 670.

8. P. Griffiths and W. Schmid, *Locally homogeneous complex manifolds*, Acta. Math. **123** (1969), 253–302. MR **41** #4587.

9. P. Griffiths, *Some geometric and analytic properties of homogeneous complex manifolds*. I. *Sheaves and cohomology*, Acta Math. **110** (1963), 115–155. MR **26** #6993.

10. R. Gunning and H. Rossi, *Analytic functions of several complex variables*, Prentice-Hall, Englewood Cliffs, N. J., 1967. MR **31** #4927.

11. F. Hirzebruch, *Topological methods in algebraic geometry*, Die Grundlehren der math. Wissenschaften, Band 131, Springer-Verlag, New York, 1966. MR **34** #2573.

12. J. Humphreys, *Introduction to Lie algebras and representation theory*, Graduate Texts in Math. vol. 9, Springer-Verlag, New York and Berlin, 1973. MR **48** #2197.

13. G. Hochschild and J.-P. Serre, *Cohomology of Lie algebras*, Ann. of Math. (2) **57** (1953), 591–603. MR **14**, 943.

14. B. Kostant, *Lie algebra cohomology and the generalized Borel-Weil theorem*, Ann. of Math. (2) **74** (1961), 329–387. MR **26** #265

15. S. Ramanan, *Holomorphic vector bundles on homogeneous spaces*, Topology **5** (1966), 159–177. MR **32** #8357.

16. N. Wallach, *Harmonic analysis on homogeneous spaces*, Dekker, New York, 1973.

17. ———, *Induced representations of Lie algebras and a theorem of Borel-Weil*, Trans. Amer. Math. Soc. **136** (1969), 181–187. MR **38** #2258.

18. G. Warner, *Harmonic analysis on semi-simple Lie groups*, Vol. 1, Springer-Verlag, New York, 1972.

19. R. Wells, Jr., *Differential analysis on complex manifolds*, Prentice-Hall, Englewood Cliffs, N. J., 1973.

20. D. P. Želobenko, *Compact Lie groups and their representations*, "Nauka", Moscow, 1970; English transl., Transl. Math. Monographs, vol. 40, Amer. Math. Soc., Providence, R. I., 1973.

University of Massachusetts, Amherst

AUTHOR INDEX

I or II indicates in which part of this volume the pages occur.

Roman numbers refer to pages on which a complete reference to a work by the author is given.

Boldface numbers indicate the first page of the articles in each part of this volume.